DATE DUE

MY 13 '94			
DE 9 '94			
JE 1 '95			
JY 7 '97			
JY 22 '97			
OC 5 '98			
DE 18 '98			
OC '00			
NO 16 '01			
JE 10 '02			
FE 14 '03			

ARCO

Civil Service
ARITHMETIC
and
VOCABULARY

Barbara Erdsneker
Margaret Haller
Eve P. Steinberg

Prentice Hall
New York • London • Toronto • Sydney • Tokyo • Singapore

Eleventh Edition

Prentice Hall General Reference
15 Columbus Circle
New York, NY 10023

An Arco Book

Library of Congress Cataloging-in-Publication Data

Erdsneker, Barbara.
  Civil service arithmetic and vocabulary / Barbara Erdsneker,
Margaret Haller, Eve P. Steinberg. -- 11th ed.
     p.   cm.
  At head of title: Arco
  ISBN: 0-671-86897-7
  1. Arithmetic--Examinations, questions, etc. 2. Vocabulary tests.
3. Civil service--United States--Examinations, questions, etc.
I. Haller, Magaret A. II. Steinberg, Eve P. III. Title.
QA139.E7   1993   93-25332
513'.076--dc20   CIP

Manufactured in the United States of America

1   2   3   4   5   6   7   8   9   10

CONTENTS

WHAT THIS BOOK WILL DO FOR YOU

ARCO Publishing, Inc. has followed testing trends and methods ever since the firm was founded in 1937. We specialize in books that prepare people for tests. Based on this experience, we have prepared the best possible book to help *you* score high.

To write this book we carefully analyzed every detail surrounding the forthcoming examinations . . .

- official and unofficial announcements concerning the examinations
- all the previous examinations, many not available to the public
- related examinations
- technical literature that explains and forecasts the examinations

Can You Prepare Yourself for Your Test?

You want to pass your test. That's why you bought this book. Used correctly, your "self-tutor" will show you what to expect and will give you a speedy brush-up on the subjects tested in your exam. Even if your study time is very limited, you should:

- become familiar with the type of questions you will be asked
- improve your general examination-taking skill
- expand your mathematical ability
- increase your knowledge of words and their meanings

The purpose of this book is to provide you with review and practice material in arithmetic and vocabulary — two areas common to most federal, state and local civil service examinations. The arithmetic section will help you build up your mathematics skills through explanations of fundamental arithmetical principles and practice questions that apply these principles. The vocabulary section offers extensive practice in answering synonym, antonym, and sentence completion questions. A word list and etymology chart provide a foundation for improving your knowledge of words.

Getting Started

It is important to determine how much time you have available to prepare for your exam, and then to make up a realistic schedule of how you will spend this study time. Set aside some time every day for your study, and follow your schedule closely. Use the following study schedules to plan your time and keep a record of your actual study time.

STUDY SCHEDULE — ARITHMETIC

	Sun.	Mon.	Tue.	Wed.	Thu.	Fri.	Sat.
Time Planned							
Time Used							

STUDY SCHEDULE — VOCABULARY

	Sun.	Mon.	Tue.	Wed.	Thu.	Fri.	Sat.
Time Planned							
Time Used							

In the row marked *Time Planned*, write the amount of time you expect to study each day. In the row marked *Time Used*, fill in the amount of time you actually spent working. Make copies of these charts to fill in for each week of study time.

PREPARING FOR THE EXAM

Here are a few suggestions to help you use your study time effectively.

1. *Study alone.*   You will concentrate better when you work by yourself. Keep a list of questions you cannot answer and points you are unsure of to talk over with a friend who is preparing for the same exam. Plan to exchange ideas at a joint review session just before the test.

2. *Eliminate distractions.*   Disturbances caused by family and neighbor activities (telephone calls, chit-chat, TV programs, etc.) work to your disadvantage. Study in a quiet, private room.

3. *Don't try to learn too much in one study period.*   If your mind starts to wander, take a short break and then return to your work.

4. *Review what you have learned.*   Once you have studied something thoroughly, be sure to review it the next day so that the information will be firmly fixed in your mind.

5. *Answer all the questions in this book.*   Don't be satisfied merely with the correct answer to each question. Do additional research on the other choices that are given. You will broaden your background and be more adequately prepared for the actual exam. It's quite possible that a question on the exam that you are going to take may require you to be familiar with the other choices.

6. *Tailor your study to the subject matter. Skim or scan.*   Don't study everything in the same manner. Obviously, certain areas are more important than others.

7. *Organize yourself.*   Make sure that your notes are in good order — valuable time is unnecessarily consumed when you can't find quickly what you are looking for.

8. *Keep physically fit.*   You cannot retain information well when you are uncomfortable, headachy, or tense. Physical health promotes mental efficiency.

How to Take an Exam

1. *Get to the examination room about ten minutes ahead of time.*   You'll get a better start when you are accustomed to the room. If the room is too cold, or too warm, or not well ventilated, call these conditions to the attention of the person in charge.

2. *Make sure that you read the instructions carefully.*   In many cases, test-takers lose credits because they misread some important point in the given directions — example: the *incorrect* choice instead of the *correct* choice.

3. *Skip hard questions and go back later.*   It is a good idea to make a mark on the question sheet next to all questions you cannot answer easily, and to go back to those questions later. First answer the questions you are sure about. Do not panic if you cannot answer a question. Go on and answer the questions you know. Usually the easier questions are presented at the beginning of the exam and the questions become gradually more difficult.

If you do skip ahead on the exam, be sure to skip ahead also on your answer sheet. A good technique is periodically to check the number of the question on the answer sheet with the number of the question on the test. You should do this every time you decide to skip a question. If you fail to skip the corresponding answer blank for that question, all of your following answers will be wrong.

Each student is stronger in some areas than in others. No one is expected to know all the answers. Do not waste time agonizing over a difficult question because it may keep you from getting to other questions that you can answer correctly.

4. *Mark the answer sheet clearly.* When you take the examination, you will mark your answers to the multiple-choice questions on a separate answer sheet that will be given to you at the test center. If you have not worked with an answer sheet before, it is in your best interest to become familiar with the procedures involved. Remember, knowing the correct answer is not enough! If you do not mark the sheet correctly, so that it can be machine-scored, you will not get credit for your answers!

On pages 153 and 154, you will find sample answer sheets for the arithmetic review exams and the sentence completion practice tests. For the bulk of this book, you will circle the letter of the answer you choose for each practice question. However, use the sample answer sheets provided to get practice in using answer sheets correctly as you answer arithmetic review and sentence completion questions. Tear the answer sheet page right out of this book. Choose the best answer to each question and mark the letter of your answer choice in the appropriate space on the answer sheet.

5. *Read each question carefully.* The exam questions are not designed to trick you through misleading or ambiguous alternative choices. It's up to you to read each question carefully so you know what is being asked.

6. *Don't answer too fast.* The multiple-choice questions that you will meet are not superficial exercises. They are designed to test not only your memory but also your understanding and insight. Do not place too much emphasis on speed. The time element is a factor, but it is not all-important. Accuracy should not be sacrificed for speed.

7. *Materials and conduct at the test center.* You need to bring with you to the test center your Admission Form, your social security number, and several No. 2 pencils. Arrive on time as you may not be admitted after testing has begun. Instructions for taking the tests will be read to you by the test supervisor and time will be called when the test is over. If you have questions, you may ask them of the supervisor. Do not give or receive assistance while taking the exams. If you do, you will be asked to turn in all test materials and told to leave the room. You will not be permitted to return and your tests will not be scored.

THE LANGUAGE OF MATHEMATICS

In order to solve a mathematical problem, it is essential to know the mathematical meaning of the words used. There are many expressions having the same meaning in mathematics. These expressions may indicate a relationship between quantities, or an operation (addition, subtraction, multiplication, division) to be performed. This chapter will help you to recognize some of the mathematical synonyms commonly found in word problems.

Equality

The following expressions all indicate that two quantities are equal (=):

> is equal to
> is the same as
> the result is
> yields
> gives

Addition

The following expressions all indicate that the numbers A and B are to be added:

A + B	**2 + 3**
the sum of A and B	the sum of 2 and 3
the total of A and B	the total of 2 and 3
A added to B	2 added to 3
A increased by B	2 increased by 3
A more than B	2 more than 3
A greater than B	2 greater than 3

Subtraction

The following expressions all indicate that the number B is to be subtracted from the number A:

A − B	**10 − 3**
A minus B	10 minus 3
A less B	10 less 3
the difference between A and B	the difference between 10 and 3
from A subtract B	from 10 subtract 3
A take away B	10 take away 3

1

A decreased by B
A diminished by B
B is subtracted from A
B less than A

10 decreased by 3
10 diminished by 3
3 is subtracted from 10
3 less than 10

Multiplication

If the numbers A and B are to be multiplied (A × B), the following expressions may be used.

A × B
A multiplied by B
the product of A and B

2 × 3
2 multiplied by 3
the product of 2 and 3

The parts of a multiplication problem are indicated in the example below:

$$\begin{array}{r} 15 \\ \times\ 10 \\ \hline 150 \end{array}$$ (multiplicand)
(multiplier)
(product)

Division

Division of the numbers A and B (in the order A ÷ B) may be indicated in the following ways.

A ÷ B
A divided by B
the quotient of A and B

14 ÷ 2
14 divided by 2
the quotient of 14 and 2

The parts of a division problem are indicated in the example below:

(divisor) $7\overline{)36}$ (quotient) $5\frac{1}{7}$ (dividend)
$\underline{35}$
1 (remainder)

Factors and Divisors

The relationship A × B = C, for any whole numbers A, B, and C, may be expressed as:

A × B = C
A and B are factors of C
A and B are divisors of C
C is divisible by A and by B
C is a multiple of A and of B

2 × 3 = 6
2 and 3 are factors of 6
2 and 3 are divisors of 6
6 is divisible by 2 and by 3
6 is a multiple of 2 and of 3

FRACTIONS

Fractions and Mixed Numbers

1. A **fraction** is part of a unit.

   a. A fraction has a **numerator** and a **denominator**.

   Example: In the fraction $\frac{3}{4}$, 3 is the numerator and 4 is the denominator.

   b. In any fraction, the numerator is being divided by the denominator.

   Example: The fraction $\frac{2}{7}$ indicates that 2 is being divided by 7.

   c. In a fraction problem, the whole quantity is 1, which may be expressed by a fraction in which the numerator and denominator are the same number.

   Example: If the problem involves $\frac{1}{8}$ of a quantity, then the whole quantity is $\frac{8}{8}$, or 1.

2. A **mixed number** is an integer together with a fraction, such as $2\frac{3}{5}$, $7\frac{3}{8}$, etc. The integer is the integral part, and the fraction is the fractional part.

3. An **improper fraction** is one in which the numerator is equal to or greater than the denominator, such as $\frac{19}{6}$, $\frac{25}{4}$, or $\frac{10}{10}$.

4. To change a mixed number to an improper fraction:

   a. Multiply the denominator of the fraction by the integer.

   b. Add the numerator to this product.

   c. Place this sum over the denominator of the fraction.

   Illustration: Change $3\frac{4}{7}$ to an improper fraction.

   SOLUTION:
   $$7 \times 3 = 21$$
   $$21 + 4 = 25$$
   $$3\tfrac{4}{7} = \tfrac{25}{7}$$

   Answer: $\frac{25}{7}$

5. To change an improper fraction to a mixed number:

   a. Divide the numerator by the denominator. The quotient, disregarding the remainder, is the integral part of the mixed number.

   b. Place the remainder, if any, over the denominator. This is the fractional part of the mixed number.

   Illustration: Change $\frac{36}{13}$ to a mixed number.

   SOLUTION:
   $$
   \begin{array}{r}
   2 \\
   13\overline{)36} \\
   \underline{26} \\
   10 \quad \text{remainder}
   \end{array}
   $$
   $$\tfrac{36}{13} = 2\tfrac{10}{13}$$

   Answer: $2\frac{10}{13}$

6. The numerator and denominator of a fraction may be changed by multiplying both by the same number, without affecting the value of the fraction.

Example: The value of the fraction $\frac{2}{5}$ will not be altered if the numerator and the denominator are multiplied by 2, to result in $\frac{4}{10}$.

7. The numerator and the denominator of a fraction may be changed by dividing both by the same number, without affecting the value of the fraction. This process is called **reducing the fraction**. A fraction that has been reduced as much as possible is said to be in **lowest terms**.

Example: The value of the fraction $\frac{3}{12}$ will not be altered if the numerator and denominator are divided by 3, to result in $\frac{1}{4}$.

Example: If $\frac{6}{30}$ is reduced to lowest terms (by dividing both numerator and denominator by 6), the result is $\frac{1}{5}$.

8. As a final answer to a problem:

   a. Improper fractions should be changed to mixed numbers.

   b. Fractions should be reduced as far as possible.

Addition of Fractions

9. Fractions cannot be added unless the denominators are all the same.

   a. If the denominators are the same, add all the numerators and place this sum over the common denominator. In the case of mixed numbers, follow the above rule for the fractions and then add the integers.

   Example: The sum of $2\frac{3}{8} + 3\frac{1}{8} + \frac{3}{8} = 5\frac{7}{8}$.

   b. If the denominators are not the same, the fractions, in order to be added, must be converted to ones having the same denominator. To do this, it is first necessary to find the lowest common denominator.

10. The **lowest common denominator** (henceforth called the L.C.D.) is the lowest number that can be divided evenly by all the given denominators. If no two of the given denominators can be divided by the same number, then the L.C.D. is the product of all the denominators.

    Example: The L.C.D. of $\frac{1}{2}$, $\frac{1}{3}$, and $\frac{1}{5}$ is $2 \times 3 \times 5 = 30$.

11. To find the L.C.D. when two or more of the given denominators can be divided by the same number:

    a. Write down the denominators, leaving plenty of space between the numbers.

    b. Select the smallest number (other than 1) by which one or more of the denominators can be divided evenly.

    c. Divide the denominators by this number, copying down those that cannot be divided evenly. Place this number to one side.

    d. Repeat this process, placing each divisor to one side until there are no longer any denominators that can be divided evenly by any selected number.

e. Multiply all the divisors to find the L.C.D.

Illustration: Find the L.C.D. of $\frac{1}{5}$, $\frac{1}{7}$, $\frac{1}{10}$, and $\frac{1}{14}$.

SOLUTION:

$$2 \overline{)\ 5 \qquad 7 \qquad 10 \qquad 14}$$
$$5 \overline{)\ 5 \qquad 7 \qquad 5 \qquad 7}$$
$$7 \overline{)\ 1 \qquad 7 \qquad 1 \qquad 7}$$
$$\quad\ 1 \qquad 1 \qquad 1 \qquad 1$$
$$7 \times 5 \times 2 = 70$$

Answer: The L.C.D. is 70.

12. To add fractions having different denominators:

a. Find the L.C.D. of the denominators.

b. Change each fraction to an equivalent fraction having the L.C.D. as its denominator.

c. When all of the fractions have the same denominator, they may be added, as in the example following item 9a.

Illustration: Add $\frac{1}{4}$, $\frac{3}{10}$, and $\frac{2}{5}$.

SOLUTION: Find the L.C.D.:

$$2 \overline{)\ 4 \qquad 10 \qquad 5}$$
$$2 \overline{)\ 2 \qquad 5 \qquad 5}$$
$$5 \overline{)\ 1 \qquad 5 \qquad 5}$$
$$\quad\ 1 \qquad 1 \qquad 1$$
$$\text{L.C.D.} = 2 \times 2 \times 5 = 20$$

$$\frac{1}{4} = \frac{5}{20}$$
$$\frac{3}{10} = \frac{6}{20}$$
$$+ \frac{2}{5} = + \frac{8}{20}$$
$$\overline{\qquad\quad \frac{19}{20}}$$

Answer: $\frac{19}{20}$

13. To add mixed numbers in which the fractions have different denominators, add the fractions by following the rules in item 12 above, then add the integers.

Illustration: Add $2\frac{5}{7}$, $5\frac{1}{2}$, and 8.

SOLUTION: L.C.D. = 14

$$2\frac{5}{7} = 2\frac{10}{14}$$
$$5\frac{1}{2} = 5\frac{7}{14}$$
$$+ 8 = + 8$$
$$\overline{\qquad 15\frac{17}{14} = 16\frac{3}{14}}$$

Answer: $16\frac{3}{14}$

Subtraction of Fractions

14. a. Unlike addition, which may involve adding more than two numbers at the same time, subtraction involves only two numbers.

    b. In subtraction, as in addition, the denominators must be the same.

15. To subtract fractions:

    a. Find the L.C.D.

    b. Change both fractions so that each has the L.C.D. as the denominator.

    c. Subtract the numerator of the second fraction from the numerator of the first, and place this difference over the L.C.D.

    d. Reduce, if possible.

    Illustration: Find the difference of $\frac{5}{8}$ and $\frac{1}{4}$.

    SOLUTION: L.C.D. = 8

    $$\begin{array}{r} \frac{5}{8} = \frac{5}{8} \\ -\frac{1}{4} = -\frac{2}{8} \\ \hline \frac{3}{8} \end{array}$$

    Answer: $\frac{3}{8}$

16. To subtract mixed numbers:

    a. It may be necessary to "borrow," so that the fractional part of the first term is larger than the fractional part of the second term.

    b. Subtract the fractional parts of the mixed numbers and reduce.

    c. Subtract the integers.

    Illustration: Subtract $16\frac{4}{5}$ from $29\frac{1}{3}$.

    SOLUTION: L.C.D. = 15

    $$\begin{array}{r} 29\frac{1}{3} = 29\frac{5}{15} \\ -16\frac{4}{5} = -16\frac{12}{15} \end{array}$$

    Note that $\frac{5}{15}$ is less than $\frac{12}{15}$. Borrow 1 from 29, and change to $\frac{15}{15}$.

    $$\begin{array}{r} 29\frac{5}{15} = 28\frac{20}{15} \\ -16\frac{12}{15} = -16\frac{12}{15} \\ \hline 12\frac{8}{15} \end{array}$$

    Answer: $12\frac{8}{15}$

Multiplication of Fractions

17. a. To be multiplied, fractions need not have the same denominators.

    b. A whole number has the denominator 1 understood.

18. To multiply fractions:

    a. Change the mixed numbers, if any, to improper fractions.

    b. Multiply all the numerators, and place this product over the product of the denominators.

    c. Reduce, if possible.

    Illustration: Multiply $\frac{2}{3} \times 2\frac{4}{7} \times \frac{5}{9}$.

    SOLUTION:
    $$2\frac{4}{7} = \frac{18}{7}$$
    $$\frac{2}{3} \times \frac{18}{7} \times \frac{5}{9} = \frac{180}{189}$$
    $$= \frac{20}{21}$$

    Answer: $\frac{20}{21}$

19. a. **Cancellation** is a device to facilitate multiplication. To cancel means to divide a numerator and a denominator by the same number in a multiplication problem.

    Example: In the problem $\frac{4}{7} \times \frac{5}{6}$, the numerator 4 and the denominator 6 may be divided by 2.

    $$\overset{2}{\cancel{4}} \times \frac{5}{\underset{3}{\cancel{6}}} = \frac{10}{21}$$

    b. The word "of" is often used to mean "multiply."

    Example: $\frac{1}{2}$ of $\frac{1}{2} = \frac{1}{2} \times \frac{1}{2} = \frac{1}{4}$

20. To multiply a whole number by a mixed number:

    a. Multiply the whole number by the fractional part of the mixed number.

    b. Multiply the whole number by the integral part of the mixed number.

    c. Add both products.

    Illustration: Multiply $23\frac{3}{4}$ by 95.

    SOLUTION:
    $$\frac{95}{1} \times \frac{3}{4} = \frac{285}{4}$$
    $$= 71\frac{1}{4}$$
    $$95 \times 23 = 2185$$
    $$2185 + 71\frac{1}{4} = 2256\frac{1}{4}$$

    Answer: $2256\frac{1}{4}$

Division of Fractions

21. The **reciprocal** of a fraction is that fraction inverted.

    a. When a fraction is inverted, the numerator becomes the denominator and the denominator becomes the numerator.

    Example: The reciprocal of $\frac{3}{8}$ is $\frac{8}{3}$.

    Example: The reciprocal of $\frac{1}{3}$ is $\frac{3}{1}$, or simply 3.

    b. Since every whole number has the denominator 1 understood, the reciprocal of a whole number is a fraction having 1 as the numerator and the number itself as the denominator.

    Example: The reciprocal of 5 (expressed fractionally as $\frac{5}{1}$) is $\frac{1}{5}$.

22. To divide fractions:

    a. Change all the mixed numbers, if any, to improper fractions.

    b. Invert the second fraction and multiply.

    c. Reduce, if possible.

    Illustration: Divide $\frac{2}{3}$ by $2\frac{1}{4}$.

    SOLUTION:
    $$2\frac{1}{4} = \frac{9}{4}$$
    $$\frac{2}{3} \div \frac{9}{4} = \frac{2}{3} \times \frac{4}{9}$$
    $$= \frac{8}{27}$$

    Answer: $\frac{8}{27}$

23. A **complex fraction** is one that has a fraction as the numerator, or as the denominator, or as both.

    Example: $\frac{\frac{2}{3}}{5}$ is a complex fraction.

24. To clear (simplify) a complex fraction:

    a. Divide the numerator by the denominator.

    b. Reduce, if possible.

    Illustration: Clear $\frac{\frac{3}{7}}{\frac{5}{14}}$.

    SOLUTION: $\frac{3}{7} \div \frac{5}{14} = \frac{3}{7} \times \frac{14}{5} = \frac{42}{35}$
    $$= \frac{6}{5}$$
    $$= 1\frac{1}{5}$$

    Answer: $1\frac{1}{5}$

Comparing Fractions

25. If two fractions have the same denominator, the one having the larger numerator is the greater fraction.

    Example: $\frac{3}{7}$ is greater than $\frac{2}{7}$.

26. If two fractions have the same numerator, the one having the larger denominator is the smaller fraction.

    Example: $\frac{5}{12}$ is smaller than $\frac{5}{11}$.

27. To compare two fractions having different numerators and different denominators:

    a. Change the fractions to equivalent fractions having their L.C.D. as their new denominator.

b. Compare, as in the example following item 25.

Illustration: Compare $\frac{4}{7}$ and $\frac{5}{8}$.

SOLUTION: L.C.D. $= 7 \times 8 = 56$

$$\frac{4}{7} = \frac{32}{56}$$
$$\frac{5}{8} = \frac{35}{56}$$

Answer: Since $\frac{35}{56}$ is larger than $\frac{32}{56}$, $\frac{5}{8}$ is larger than $\frac{4}{7}$.

Fraction Problems

28. Most fraction problems can be arranged in the form: "What fraction of a number is another number?" This form contains three important parts:

- The fractional part
- The number following "of"
- The number following "is"

a. If the fraction and the "of" number are given, multiply them to find the "is" number.

Illustration: What is $\frac{3}{4}$ of 20?

SOLUTION: Write the question as "$\frac{3}{4}$ of 20 is what number?" Then multiply the fraction $\frac{3}{4}$ by the "of" number, 20:

$$\frac{3}{\underset{1}{\cancel{4}}} \times \overset{5}{\cancel{20}} = 15$$

Answer: 15

b. If the fractional part and the "is" number are given, divide the "is" number by the fraction to find the "of" number.

Illustration: $\frac{4}{5}$ of what number is 40?

SOLUTION: To find the "of" number, divide 40 by $\frac{4}{5}$:

$$40 \div \frac{4}{5} = \frac{\overset{10}{\cancel{40}}}{1} \times \frac{5}{\underset{1}{\cancel{4}}}$$
$$= 50$$

Answer: 50

c. To find the fractional part when the other two numbers are known, divide the "is" number by the "of" number.

Illustration: What part of 12 is 9?

SOLUTION: $9 \div 12 = \frac{9}{12}$
$$= \frac{3}{4}$$

Answer: $\frac{3}{4}$

Practice Problems Involving Fractions

1. Reduce to lowest terms: $\frac{60}{108}$.
   (A) $\frac{1}{48}$ (C) $\frac{5}{9}$
   (B) $\frac{1}{3}$ (D) $\frac{10}{18}$

2. Change $\frac{27}{7}$ to a mixed number.
   (A) $2\frac{1}{7}$ (C) $6\frac{1}{3}$
   (B) $3\frac{6}{7}$ (D) $7\frac{1}{2}$

3. Change $4\frac{2}{3}$ to an improper fraction.
   (A) $\frac{10}{3}$ (C) $\frac{14}{3}$
   (B) $\frac{11}{3}$ (D) $\frac{42}{3}$

4. Find the L.C.D. of $\frac{1}{6}$, $\frac{1}{10}$, $\frac{1}{15}$, and $\frac{1}{21}$.
   (A) 160   (C) 630
   (B) 330   (D) 1260

5. Add $16\frac{3}{8}$, $4\frac{1}{5}$, $12\frac{3}{4}$, and $23\frac{5}{6}$.
   (A) $57\frac{91}{120}$ (C) 58
   (B) $57\frac{1}{4}$ (D) 59

6. Subtract $27\frac{5}{14}$ from $43\frac{1}{6}$.
   (A) 15   (C) $15\frac{8}{21}$
   (B) 16   (D) $15\frac{17}{21}$

7. Multiply $17\frac{3}{8}$ by 128.
   (A) 2200   (C) 2356
   (B) 2305   (D) 2256

8. Divide $1\frac{2}{3}$ by $1\frac{1}{9}$.
   (A) $\frac{2}{3}$ (C) $1\frac{23}{27}$
   (B) $1\frac{1}{2}$ (D) 6

9. What is the value of $12\frac{1}{6} - 2\frac{3}{8} - 7\frac{2}{3} + 19\frac{3}{4}$?
   (A) 21   (C) $21\frac{1}{8}$
   (B) $21\frac{7}{8}$ (D) 22

10. Simplify the complex fraction $\frac{\frac{4}{9}}{\frac{2}{5}}$
    (A) $\frac{1}{2}$ (C) $\frac{2}{5}$
    (B) $\frac{9}{10}$ (D) $1\frac{1}{9}$

11. Which fraction is largest?
    (A) $\frac{9}{16}$ (C) $\frac{5}{8}$
    (B) $\frac{7}{10}$ (D) $\frac{4}{5}$

12. One brass rod measures $3\frac{5}{16}$ inches long and another brass rod measures $2\frac{3}{4}$ inches long. Together their length is
    (A) $6\frac{9}{16}$ in.   (C) $6\frac{1}{16}$ in.
    (B) $5\frac{1}{8}$ in.   (D) $5\frac{1}{16}$ in.

13. The number of half-pound packages of tea that can be weighed out of a box that holds $10\frac{1}{2}$ lb. of tea is
    (A) 5   (C) $20\frac{1}{2}$
    (B) $10\frac{1}{2}$ (D) 21

14. If each bag of tokens weighs $5\frac{3}{4}$ pounds, how many pounds do 3 bags weigh?
    (A) $7\frac{1}{4}$ (C) $16\frac{1}{2}$
    (B) $15\frac{3}{4}$ (D) $17\frac{1}{4}$

15. During one week, a man traveled $3\frac{1}{2}$, $1\frac{1}{4}$, $1\frac{1}{6}$, and $2\frac{3}{8}$ miles. The next week he traveled $\frac{1}{4}$, $\frac{3}{8}$, $\frac{9}{16}$, $3\frac{1}{16}$, $2\frac{5}{8}$, and $3\frac{3}{16}$ miles. How many more miles did he travel the second week than the first week?
    (A) $1\frac{37}{48}$ (C) $1\frac{3}{4}$
    (B) $1\frac{1}{2}$ (D) 1

16. A certain type of board is sold only in lengths of multiples of 2 feet. The shortest board sold is 6 feet and the longest is 24 feet. A builder needs a large quantity of this type of board in $5\frac{1}{2}$-foot lengths. For minimum waste the lengths to be ordered should be
    (A) 6 ft   (C) 22 ft
    (B) 12 ft   (D) 24 ft

17. A man spent $\frac{15}{16}$ of his entire fortune in buying a car for $7500. How much money did he possess?
    (A) $6000   (C) $7000
    (B) $6500   (D) $8000

18. The population of a town was 54,000 in the last census. It has increased $\frac{2}{3}$ since then. Its present population is
    (A) 18,000   (C) 72,000
    (B) 36,000   (D) 90,000

19. If one third of the liquid contents of a can evaporates on the first day and three fourths of the remainder evaporates on the second day, the fractional part of the original contents remaining at the close of the second day is

(A) $\frac{5}{12}$ (C) $\frac{1}{6}$

(B) $\frac{7}{12}$ (D) $\frac{1}{2}$

20. A car is run until the gas tank is $\frac{1}{8}$ full. The tank is then filled to capacity by putting in 14 gallons. The capacity of the gas tank of the car is

(A) 14 gal   (C) 16 gal

(B) 15 gal   (D) 17 gal

Fraction Problems — Correct Answers

1. **(C)**	6. **(D)**	11. **(D)**	16. **(C)**
2. **(B)**	7. **(D)**	12. **(C)**	17. **(D)**
3. **(C)**	8. **(B)**	13. **(D)**	18. **(D)**
4. **(C)**	9. **(B)**	14. **(D)**	19. **(C)**
5. **(A)**	10. **(D)**	15. **(A)**	20. **(C)**

Problem Solutions — Fractions

1. Divide the numerator and denominator by 12:

$$\frac{60 \div 12}{108 \div 12} = \frac{5}{9}$$

One alternate method (there are several) is to divide the numerator and denominator by 6 and then by 2:

$$\frac{60 \div 6}{108 \div 6} = \frac{10}{18}$$

$$\frac{10 \div 2}{18 \div 2} = \frac{5}{9}$$

Answer: **(C)** $\frac{5}{9}$

2. Divide the numerator (27) by the denominator (7):

$$\begin{array}{r} 3 \\ 7\overline{)27} \\ \underline{21} \\ 6 \end{array} \text{ remainder}$$

$$\frac{27}{7} = 3\frac{6}{7}$$

Answer: **(B)** $3\frac{6}{7}$

3.
$$4 \times 3 = 12$$
$$12 + 2 = 14$$
$$4\frac{2}{3} = \frac{14}{3}$$

Answer: **(C)** $\frac{14}{3}$

4.

2) 6   10   18   21   (2 is a divisor of 6, 10, and 18)

3) 3   5   9   21   (3 is a divisor of 3, 9, and 21)

3) 1   5   3   7   (3 is a divisor of 3)

5) 1   5   1   7   (5 is a divisor of 5)

7) 1   1   1   7   (7 is a divisor of 7)

   1   1   1   1

L.C.D. = 2 × 3 × 3 × 5 × 7 = 630

Answer: **(C)** 630

5. L.C.D. = 120

$$16\frac{3}{8} = 16\frac{45}{120}$$
$$4\frac{4}{5} = 4\frac{96}{120}$$
$$12\frac{3}{4} = 12\frac{90}{120}$$
$$+ 23\frac{5}{6} = + 23\frac{100}{120}$$
$$55\frac{331}{120} = 57\frac{91}{120}$$

Answer: **(A)** $57\frac{91}{120}$

6. L.C.D. = 42

$$43\frac{1}{6} = 43\frac{7}{42} = 42\frac{49}{42}$$
$$- 27\frac{5}{14} = - 27\frac{15}{42} = - 27\frac{15}{42}$$
$$15\frac{34}{42} = 15\frac{17}{21}$$

Answer: **(D)** $15\frac{17}{21}$

7.
$$17\frac{5}{8} = \frac{141}{8}$$
$$\frac{141}{\cancel{8}} \times \frac{\cancel{128}^{16}}{1} = 2256$$

Answer: **(D)** 2256

8.
$$1\frac{2}{3} \div 1\frac{1}{9} = \frac{5}{3} \div \frac{10}{9}$$
$$= \frac{\cancel{5}^{1}}{\cancel{3}_{1}} \times \frac{\cancel{9}^{3}}{\cancel{10}_{2}}$$
$$= \frac{3}{2}$$
$$= 1\frac{1}{2}$$

Answer: **(B)** $1\frac{1}{2}$

9. L.C.D. = 24

$$12\frac{1}{6} = 12\frac{4}{24} = 11\frac{28}{24}$$
$$- 2\frac{3}{8} = - 2\frac{9}{24} = - 2\frac{9}{24}$$
$$9\frac{19}{24} = 9\frac{19}{24}$$
$$- 7\frac{2}{3} = - 7\frac{16}{24}$$
$$2\frac{3}{24} = 2\frac{3}{24}$$
$$+ 19\frac{3}{4} = + 19\frac{18}{24}$$
$$21\frac{21}{24}$$

$$21\frac{21}{24} = 21\frac{7}{8}$$

Answer: **(B)** $21\frac{7}{8}$

10. To simplify a complex fraction, divide the numerator by the denominator:

$$\frac{4}{9} \div \frac{2}{5} = \frac{\cancel{4}^{2}}{9} \times \frac{5}{\cancel{2}_{1}}$$
$$= \frac{10}{9}$$
$$= 1\frac{1}{9}$$

Answer: **(D)** $1\frac{1}{9}$

11. Write all of the fractions with the same denominator. L.C.D. = 80

$$\frac{9}{16} = \frac{45}{80}$$
$$\frac{7}{10} = \frac{56}{80}$$
$$\frac{5}{8} = \frac{50}{80}$$
$$\frac{4}{5} = \frac{64}{80}$$

Answer: **(D)** $\frac{4}{5}$

12.
$$3\frac{5}{16} = 3\frac{5}{16}$$
$$+ 2\frac{3}{4} = + 2\frac{12}{16}$$
$$5\frac{17}{16}$$
$$= 6\frac{1}{16}$$

Answer: **(C)** $6\frac{1}{16}$ in.

13.
$$10\frac{1}{2} \div \frac{1}{2} = \frac{21}{2} \div \frac{1}{2}$$
$$= \frac{21}{\cancel{2}} \times \frac{\cancel{2}^{1}}{1}$$
$$= 21$$

Answer: **(D)** 21

14.
$$5\frac{3}{4} \times 3 = \frac{23}{4} \times \frac{3}{1}$$
$$= \frac{69}{4}$$
$$= 17\frac{1}{4}$$

Answer: **(D)** $17\frac{1}{4}$

15. First week:
L.C.D. = 24

$$3\frac{1}{2} = 3\frac{12}{24} \text{ miles}$$
$$1\frac{1}{4} = 1\frac{6}{24}$$
$$1\frac{1}{6} = 1\frac{4}{24}$$
$$+ 2\frac{3}{8} = + 2\frac{9}{24}$$
$$7\frac{31}{24} = 8\frac{7}{24} \text{ miles}$$

Second week:
L.C.D. = 16

$$\frac{1}{4} = \frac{4}{16} \text{ miles}$$
$$\frac{3}{8} = \frac{6}{16}$$
$$\frac{9}{16} = \frac{9}{16}$$
$$3\frac{1}{16} = 3\frac{1}{16}$$
$$2\frac{5}{8} = 2\frac{10}{16}$$
$$+ 3\frac{3}{16} = + 3\frac{3}{16}$$
$$8\frac{33}{16} = 10\frac{1}{16} \text{ miles}$$

L.C.D. = 48

$$10\frac{1}{16} = 9\frac{51}{48} \text{ miles second week}$$
$$- 8\frac{7}{24} = - 8\frac{14}{48} \text{ miles first week}$$
$$1\frac{37}{48} \text{ miles more traveled}$$

Answer: **(A)** $1\frac{37}{48}$

16. Consider each choice:

    Each 6-ft board yields one $5\frac{1}{2}$-ft board with $\frac{1}{2}$ ft waste.

    Each 12-ft board yields two $5\frac{1}{2}$-ft boards with 1 ft waste. $(2 \times 5\frac{1}{2} = 11; 12 - 11 = 1$ ft waste)

    Each 24-ft board yields four $5\frac{1}{2}$-ft boards with 2 ft waste. $(4 \times 5\frac{1}{2} = 22; 24 - 22 = 2$ ft waste)

    Each 22 ft board may be divided into four $5\frac{1}{2}$-ft boards with no waste. $(4 \times 5\frac{1}{2} = 22$ exactly)

    Answer: **(C)** 22 ft

17. $\frac{15}{16}$ of fortune is \$7500.

    Therefore, his fortune $= 7500 \div \frac{15}{16}$

    $$= \frac{\overset{500}{\cancel{7500}}}{1} \times \frac{16}{\cancel{15}}$$

    $$= 8000$$

    Answer: **(D)** \$8000

18. $\frac{2}{3}$ of 54,000 = increase

    $$\text{Increase} = \frac{2}{\cancel{3}} \times \overset{18,000}{\cancel{54,000}}$$

    $$= 36,000$$

    $$\text{Present population} = 54,000 + 36,000$$

    $$= 90,000$$

    Answer: **(D)** 90,000

19. First day: $\frac{1}{3}$ evaporates

    $\frac{2}{3}$ remains

    Second day: $\frac{3}{4}$ of $\frac{2}{3}$ evaporates

    $\frac{1}{4}$ of $\frac{2}{3}$ remains

    The amount remaining is

    $$\frac{1}{\cancel{4}} \times \overset{1}{\cancel{\frac{2}{3}}} = \frac{1}{6} \text{ of original contents}$$

    Answer: **(C)** $\frac{1}{6}$

20. $\frac{7}{8}$ of capacity = 14 gal

    therefore, capacity $= 14 \div \frac{7}{8}$

    $$= \frac{\overset{2}{\cancel{14}}}{1} \times \frac{8}{\cancel{7}}$$

    $$= 16 \text{ gal}$$

    Answer: **(C)** 16 gal

DECIMALS

1. A **decimal**, which is a number with a decimal point (.), is actually a fraction, the denominator of which is understood to be 10 or some power of 10.

   a. The number of digits, or places, after a decimal point determines which power of 10 the denominator is. If there is one digit, the denominator is understood to be 10; if there are two digits, the denominator is understood to be 100, etc.

   Example: $.3 = \frac{3}{10}$, $.57 = \frac{57}{100}$, $.643 = \frac{643}{1000}$

   b. The addition of zeros after a decimal point does not change the value of the decimal. The zeros may be removed without changing the value of the decimal.

   Example: $.7 = .70 = .700$ and vice versa, $.700 = .70 = .7$

   c. Since a decimal point is understood to exist after any whole number, the addition of any number of zeros after such a decimal point does not change the value of the number.

   Example: $2 = 2.0 = 2.00 = 2.000$

Addition of Decimals

2. Decimals are added in the same way that whole numbers are added, with the provision that the decimal points must be kept in a vertical line, one under the other. This determines the place of the decimal point in the answer.

   Illustration: Add 2.31, .037, 4, and 5.0017

   SOLUTION:
   ```
        2.3100
         .0370
        4.0000
     +  5.0017
       11.3487
   ```

   Answer: 11.3487

Subtraction of Decimals

3. Decimals are subtracted in the same way that whole numbers are subtracted, with the provision that, as in addition, the decimal points must be kept in a vertical line, one under the other. This determines the place of the decimal point in the answer.

   Illustration: Subtract 4.0037 from 15.3

   SOLUTION:
   ```
       15.3000
     -  4.0037
       11.2963
   ```

   Answer: 11.2963

14

Multiplication of Decimals

4. Decimals are multiplied in the same way that whole numbers are multiplied.

   a. The number of decimal places in the product equals the sum of the decimal places in the multiplicand and in the multiplier.

   b. If there are fewer places in the product than this sum, then a sufficient number of zeros must be added in front of the product to equal the number of places required, and a decimal point is written in front of the zeros.

Illustration: Multiply 2.372 by .012

SOLUTION:

$$\begin{array}{r} 2.372 \quad \text{(3 decimal places)} \\ \times \quad .012 \quad \text{(3 decimal places)} \\ \hline 4744 \\ 2372 \\ \hline .028464 \quad \text{(6 decimal places)} \end{array}$$

Answer: .028464

5. A decimal can be multiplied by a power of 10 by moving the decimal point to the *right* as many places as indicated by the power. If multiplied by 10, the decimal point is moved one place to the right; if multiplied by 100, the decimal point is moved two places to the right; etc.

   Example:
   $.235 \times 10 = 2.35$
   $.235 \times 100 = 23.5$
   $.235 \times 1000 = 235$

Division of Decimals

6. There are four types of division involving decimals:
   • When the dividend only is a decimal.
   • When the divisor only is a decimal.
   • When both are decimals.
   • When neither dividend nor divisor is a decimal.

   a. When the dividend only is a decimal, the division is the same as that of whole numbers, except that a decimal point must be placed in the quotient exactly above that in the dividend.

Illustration: Divide 12.864 by 32

SOLUTION:

$$\begin{array}{r} .402 \\ 32 \overline{) 12.864} \\ 12\ 8 \\ \hline 64 \\ 64 \\ \hline \end{array}$$

Answer: .402

b. When the divisor only is a decimal, the decimal point in the divisor is omitted and as many zeros are placed to the right of the dividend as there were decimal places in the divisor.

Illustration: Divide 211327 by 6.817

$$
\begin{array}{r}
31000 \\
SOLUTION:\quad 6.817\overline{\smash{)}211327} = 6817\overline{\smash{)}211327000} \\
\text{(3 decimal places)}\qquad \underline{20451} \qquad \text{(3 zeros added)}\\
6817\\
\underline{6817}
\end{array}
$$

Answer: 31000

c. When both divisor and dividend are decimals, the decimal point in the divisor is omitted and the decimal point in the dividend must be moved to the right as many decimal places as there were in the divisor. If there are not enough places in the dividend, zeros must be added to make up the difference.

Illustration: Divide 2.62 by .131

$$
\begin{array}{r}
20\\
SOLUTION:\quad .131\overline{\smash{)}2.62} = 131\overline{\smash{)}2620}\\
\underline{262}
\end{array}
$$

Answer: 20

d. In instances when neither the divisor nor the dividend is a decimal, a problem may still involve decimals. This occurs in two cases: when the dividend is a smaller number than the divisor; and when it is required to work out a division to a certain number of decimal places. In either case, write in a decimal point after the dividend, add as many zeros as necessary, and place a decimal point in the quotient above that in the dividend.

Illustration: Divide 7 by 50.

$$
\begin{array}{r}
.14\\
SOLUTION:\quad 50\overline{\smash{)}7.00}\\
\underline{5\,0}\\
2\,00\\
\underline{2\,00}
\end{array}
$$

Answer: .14

Illustration: How much is 155 divided by 40, carried out to 3 decimal places?

$$
\begin{array}{r}
3.875\\
SOLUTION:\quad 40\overline{\smash{)}155.000}\\
\underline{120}\\
35\,0\\
\underline{32\,0}\\
3\,00\\
\underline{2\,80}\\
200
\end{array}
$$

Answer: 3.875

7. A decimal can be divided by a power of 10 by moving the decimal to the *left* as many places as indicated by the power. If divided by 10, the decimal point is moved one place to the left; if divided by 100, the decimal point is moved two places to the left; etc. If there are not enough places, add zeros in front of the number to make up the difference and add a decimal point.

Example: .4 divided by 10 = .04
.4 divided by 100 = .004

Rounding Decimals

8. To round a number to a given decimal place:

    a. Locate the given place.

    b. If the digit to the right is less than 5, omit all digits following the given place.

    c. If the digit to the right is 5 or more, raise the given place by 1 and omit all digits following the given place.

    Examples: 4.27   = 4.3 to the nearest tenth
    .71345 = .713 to the nearest thousandth

9. In problems involving money, answers are usually rounded to the nearest cent.

Conversion of Fractions to Decimals

10. A fraction can be changed to a decimal by dividing the numerator by the denominator and working out the division to as many decimal places as required.

Illustration: Change $\frac{5}{11}$ to a decimal of 2 places.

$$SOLUTION: \quad \frac{5}{11} = 11 \overline{\smash{)}\,5.00} \quad .45\tfrac{5}{11}$$

$$\begin{array}{r} .45\tfrac{5}{11} \\ 11\overline{\smash{)}\,5.00} \\ \underline{4.44} \\ 60 \\ \underline{55} \\ 5 \end{array}$$

Answer: $.45\tfrac{5}{11}$

11. To clear fractions containing a decimal in either the numerator or the denominator, or in both, divide the numerator by the denominator.

Illustration: What is the value of $\frac{2.34}{.6}$?

SOLUTION: $\frac{2.34}{.6}$ = $.6\overline{)2.34}$ = $6\overline{)23.4}$

$$\begin{array}{r} 3.9 \\ 6\overline{)23.4} \\ \underline{18} \\ 5\,4 \\ \underline{5\,4} \end{array}$$

Answer: 3.9

Conversion of Decimals to Fractions

12. Since a decimal point indicates a number having a denominator that is a power of 10, a decimal can be expressed as a fraction, the numerator of which is the number itself and the denominator of which is the power indicated by the number of decimal places in the decimal.

Example: $.3 = \frac{3}{10}$, $.47 = \frac{47}{100}$

13. When the decimal is a mixed number, divide by the power of 10 indicated by its number of decimal places. The fraction does not count as a decimal place.

Illustration: Change $.25\frac{1}{3}$ to a fraction.

SOLUTION: $.25\frac{1}{3} = 25\frac{1}{3} \div 100$
$= \frac{76}{3} \times \frac{1}{100}$
$= \frac{76}{300} = \frac{19}{75}$

Answer: $\frac{19}{75}$

14. When to change decimals to fractions:

a. When dealing with whole numbers, do not change the decimal.

Example: In the problem $12 \times .14$, it is better to keep the decimal:
$$12 \times .14 = 1.68$$

b. When dealing with fractions, change the decimal to a fraction.

Example: In the problem $\frac{3}{5} \times .17$, it is best to change the decimal to a fraction:
$$\frac{3}{5} \times .17 = \frac{3}{5} \times \frac{17}{100} = \frac{51}{500}$$

15. Because decimal equivalents of fractions are often used, it is helpful to be familiar with the most common conversions.

$\frac{1}{2} = .5$	$\frac{1}{3} = .3333$
$\frac{1}{4} = .25$	$\frac{2}{3} = .6667$
$\frac{3}{4} = .75$	$\frac{1}{6} = .1667$
$\frac{1}{5} = .2$	$\frac{1}{7} = .1429$
$\frac{1}{8} = .125$	$\frac{1}{9} = .1111$
$\frac{1}{16} = .0625$	$\frac{1}{12} = .0833$

Note that the left column contains exact values. The values in the right column have been rounded to the nearest ten-thousandth.

Practice Problems Involving Decimals

1. Add 37.03, 11.5627, 3.4005, 3423, and 1.141.
   (A) 3476.1342
   (B) 3500
   (C) 3524.4322
   (D) 3424.1342

2. Subtract 4.64324 from 7.
   (A) 3.35676
   (B) 2.35676
   (C) 2.45676
   (D) 2.36676

3. Multiply 27.34 by 16.943.
   (A) 463.22162
   (B) 453.52162
   (C) 462.52162
   (D) 462.53162

4. How much is 19.6 divided by 3.2, carried out to 3 decimal places?
   (A) 6.125
   (B) 6.124
   (C) 6.123
   (D) 5.123

5. What is $\frac{5}{11}$ in decimal form (to the nearest hundredth)?
   (A) .44
   (B) .55
   (C) .40
   (D) .45

6. What is $.64\frac{2}{3}$ in fraction form?
   (A) $\frac{97}{120}$
   (B) $\frac{97}{150}$
   (C) $\frac{97}{130}$
   (D) $\frac{98}{130}$

7. What is the difference between $\frac{3}{5}$ and $\frac{9}{8}$ expressed decimally?
   (A) .525
   (B) .425
   (C) .520
   (D) .500

8. A boy saved up $4.56 the first month, $3.82 the second month, and $5.06 the third month. How much did he save altogether?
   (A) $12.56
   (B) $13.28
   (C) $13.44
   (D) $14.02

9. The diameter of a certain rod is required to be 1.51 ± .015 inches. The rod would not be acceptable if the diameter measured
   (A) 1.490 in
   (B) 1.500 in
   (C) 1.510 in
   (D) 1.525 in

10. After an employer figures out an employee's salary of $190.57, he deducts $3.05 for social security and $5.68 for pension. What is the amount of the check after these deductions?
    (A) $181.84
    (B) $181.92
    (C) $181.93
    (D) $181.99

11. If the outer diameter of a metal pipe is 2.84 inches and the inner diameter is 1.94 inches, the thickness of the metal is
    (A) .45 in
    (B) .90 in
    (C) 1.94 in
    (D) 2.39 in

12. A boy earns $20.56 on Monday, $32.90 on Tuesday, $20.78 on Wednesday. He spends half of all that he earned during the three days. How much has he left?
    (A) $29.19
    (B) $31.23
    (C) $34.27
    (D) $37.12

13. The total cost of $3\frac{1}{2}$ pounds of meat at $1.69 a pound and 20 lemons at $.60 a dozen will be
    (A) $6.00
    (B) $6.40
    (C) $6.52
    (D) $6.92

14. A reel of cable weighs 1279 lb. If the empty reel weighs 285 lb and the cable weighs 7.1 lb per foot, the number of feet of cable on the reel is
    (A) 220
    (B) 180
    (C) 140
    (D) 100

15. 345 fasteners at $4.15 per hundred will cost
    (A) $ .1432
    (B) $1.4320
    (C) $ 14.32
    (D) $143.20

Decimal Problems — Correct Answers

1. (A)		6. (B)		11. (B)	
2. (B)		7. (A)		12. (D)	
3. (A)		8. (C)		13. (D)	
4. (A)		9. (A)		14. (C)	
5. (D)		10. (A)		15. (C)	

Problem Solutions — Decimals

1. Line up all the decimal points one under the other. Then add:

```
     37.03
     11.5627
      3.4005
   3423.0000
 +    1.141
   3476.1342
```

Answer: **(A)** 3476.1342

2. Add a decimal point and five zeros to the 7. Then subtract:

```
   7.00000
 - 4.64324
   2.35676
```

Answer: **(B)** 2.35676

3. Since there are two decimal places in the multiplicand and three decimal places in the multiplier, there will be 2 + 3 = 5 decimal places in the product.

```
      27.34
    × 16.943
       8202
      1 0936
     24 606
    164 04
    273 4
    463.22162
```

Answer: **(A)** 463.22162

4. Omit the decimal point in the divisor by moving it one place to the right. Move the decimal point in the dividend one place to the right and add three zeros in order to carry your answer out to three decimal places, as instructed in the problem.

```
           6.125
  3.2. ) 19.6.000
          19 2
           4 0
           3 2
            80
            64
           160
           160
```

Answer: **(A)** 6.125

5. To convert a fraction to a decimal, divide the numerator by the denominator:

```
          .454
  11 ) 5.000
       4 4
        60
        55
        50
        44
         6
```

Answer: **(D)** .45 to the nearest hundredth

6. To convert a decimal to a fraction, divide by the power of 10 indicated by the number of decimal places. (The fraction does not count as a decimal place.)

$$64\tfrac{2}{3} \div 100 = \tfrac{194}{3} \div \tfrac{100}{1}$$
$$= \tfrac{194}{3} \times \tfrac{1}{100}$$
$$= \tfrac{194}{300}$$
$$= \tfrac{97}{150}$$

Answer: **(B)** $\tfrac{97}{150}$

7. Convert each fraction to a decimal and subtract to find the difference:

$\tfrac{9}{8} = 1.125 \qquad \tfrac{3}{5} = .60$

$$\begin{array}{r} 1.125 \\ -\ .60 \\ \hline .525 \end{array}$$

Answer: **(A)** .525

8. Add the savings for each month:

$$\begin{array}{r} \$4.56 \\ 3.82 \\ +\ 5.06 \\ \hline \$13.44 \end{array}$$

Answer: **(C)** $13.44

9.
$$\begin{array}{r} 1.51 \\ +\ .015 \\ \hline 1.525 \end{array} \qquad \begin{array}{r} 1.510 \\ -\ .015 \\ \hline 1.495 \end{array}$$

The rod may have a diameter of from 1.495 inches to 1.525 inches inclusive.

Answer: **(A)** 1.490 in.

10. Add to find total deductions:

$$\begin{array}{r} \$3.05 \\ +\ 5.68 \\ \hline \$8.73 \end{array}$$

Subtract total deductions from salary to find amount of check:

$$\begin{array}{r} \$190.57 \\ -\ 8.73 \\ \hline \$181.84 \end{array}$$

Answer: **(A)** $181.84

11. Outer diameter minus inner diameter equals thickness of metal:

$$\begin{array}{r} 2.84 \\ -\ 1.94 \\ \hline .90 \end{array}$$

Answer: **(B)** .90 in

12. Add daily earnings to find total earnings:

$$\begin{array}{r} \$20.56 \\ 32.90 \\ +\ 20.78 \\ \hline \$74.24 \end{array}$$

Divide total earnings by 2 to find out what he has left:

$$2\,)\overline{\,\$74.24\,} \quad = \$37.12$$

Answer: **(D)** $37.12

13. Find cost of $3\tfrac{1}{2}$ pounds of meat:

$$\begin{array}{r} \$1.69 \\ \times\ 3.5 \\ \hline 845 \\ 5\ 07 \\ \hline \$5.915 \end{array} = \$5.92 \text{ to the nearest cent}$$

Find cost of 20 lemons:
$.60 \div 12 = \$.05$ (for 1 lemon)
$.05 \times 20 = \$1.00$ (for 20 lemons)

Add cost of meat and cost of lemons:

$$\begin{array}{r} \$5.92 \\ +\ 1.00 \\ \hline \$6.92 \end{array}$$

Answer: **(D)** $6.92

14. Subtract weight of empty reel from total weight to find weight of cable:

$$\begin{array}{r} 1279 \text{ lb} \\ -\ 285 \text{ lb} \\ \hline 994 \text{ lb} \end{array}$$

Each foot of cable weighs 7.1 lb. Therefore, to find the number of feet of cable on the reel, divide 994 by 7.1:

```
                14 0.
       7.1. ) 994.0.
               71
              284
              284
                0 0
```

Answer: **(C)** 140

15. Each fastener costs:

$$\$4.15 \div 100 = \$.0415$$

345 fasteners cost:

```
            345
         × .0415
           1725
            345
          13 80
         14.3175
```

Answer: **(C)** $14.32

PERCENTS

1. The **percent symbol** (%) means "parts of a hundred." Some problems involve expressing a fraction or a decimal as a percent. In other problems, it is necessary to express a percent as a fraction or a decimal in order to perform the calculations.

2. To change a whole number or a decimal to a percent:

   a. Multiply the number by 100.

   b. Affix a % sign.

   Illustration:　Change 3 to a percent.

   SOLUTION:　$3 \times 100 = 300$
   $$3 = 300\%$$

   Answer:　300%

   Illustration:　Change .67 to a percent.

   SOLUTION:　$.67 \times 100 = 67$
   $$.67 = 67\%$$

   Answer:　67%

3. To change a fraction or a mixed number to a percent:

   a. Multiply the fraction or mixed number by 100.

   b. Reduce, if possible.

   c. Affix a % sign.

   Illustration:　Change $\frac{1}{7}$ to a percent.

   SOLUTION:　$\frac{1}{7} \times 100 = \frac{100}{7}$
   $$= 14\frac{2}{7}$$
   $$\frac{1}{7} = 14\frac{2}{7}\%$$

   Answer:　$14\frac{2}{7}\%$

   Illustration:　Change $4\frac{2}{3}$ to a percent.

   SOLUTION:　$4\frac{2}{3} \times 100 = \frac{14}{3} \times 100 = \frac{1400}{3}$
   $$= 466\frac{2}{3}$$
   $$4\frac{2}{3} = 466\frac{2}{3}\%$$

   Answer:　$466\frac{2}{3}\%$

23

4. To remove a % sign attached to a decimal, divide the decimal by 100. If necessary, the resulting decimal may then be changed to a fraction.

Illustration: Change .5% to a decimal and to a fraction.

SOLUTION: .5% = .5 ÷ 100 = .005

.005 = $\frac{5}{1000}$ = $\frac{1}{200}$

Answer: .5% = .005

.5% = $\frac{1}{200}$

5. To remove a % sign attached to a fraction or mixed number, divide the fraction or mixed number by 100, and reduce, if possible. If necessary, the resulting fraction may then be changed to a decimal.

Illustration: Change $\frac{3}{4}$% to a fraction and to a decimal.

SOLUTION: $\frac{3}{4}$% = $\frac{3}{4}$ ÷ 100 = $\frac{3}{4}$ × $\frac{1}{100}$

= $\frac{3}{400}$

$$\frac{3}{400} = 400 \overline{)3.0000}^{.0075}$$

Answer: $\frac{3}{4}$% = $\frac{3}{400}$

$\frac{3}{4}$% = .0075

6. To remove a % sign attached to a decimal that includes a fraction, divide the decimal by 100. If necessary, the resulting number may then be changed to a fraction.

Illustration: Change .5$\frac{1}{3}$% to a fraction.

SOLUTION: .5$\frac{1}{3}$% = .005$\frac{1}{3}$

= $\dfrac{5\frac{1}{3}}{1000}$

= 5$\frac{1}{3}$ ÷ 1000

= $\frac{16}{3}$ × $\frac{1}{1000}$

= $\frac{16}{3000}$

= $\frac{2}{375}$

Answer: .5$\frac{1}{3}$% = $\frac{2}{375}$

7. Some fraction-percent equivalents are used so frequently that it is helpful to be familiar with them.

$\frac{1}{25}$ = 4%	$\frac{1}{5}$ = 20%
$\frac{1}{20}$ = 5%	$\frac{1}{4}$ = 25%
$\frac{1}{12}$ = 8$\frac{1}{3}$%	$\frac{1}{3}$ = 33$\frac{1}{3}$%
$\frac{1}{10}$ = 10%	$\frac{1}{2}$ = 50%
$\frac{1}{8}$ = 12$\frac{1}{2}$%	$\frac{2}{3}$ = 66$\frac{2}{3}$%
$\frac{1}{6}$ = 16$\frac{2}{3}$%	$\frac{3}{4}$ = 75%

Solving Percent Problems

8. Most percent problems involve three quantities:
    • The rate, R, which is followed by a % sign.
    • The base, B, which follows the word "of."
    • The amount or percentage, P, which usually follows the word "is."

    a. If the rate (R) and the base (B) are known, then the percentage (P) = R × B.

    Illustration: Find 15% of 50.

    SOLUTION: Rate = 15%
    Base = 50
    $$P = R \times B$$
    $$P = 15\% \times 50$$
    $$= .15 \times 50$$
    $$= 7.5$$

    Answer: 15% of 50 is 7.5.

    b. If the rate (R) and the percentage (P) are known, then the base (B) = $\frac{P}{R}$.

    Illustration: 7% of what number is 35?

    SOLUTION: Rate = 7%
    Percentage = 35
    $$B = \frac{P}{R}$$
    $$B = \frac{35}{7\%}$$
    $$= 35 \div .07$$
    $$= 500$$

    Answer: 7% of 500 is 35.

    c. If the percentage (P) and the base (B) are known, the rate (R) = $\frac{P}{B}$.

    Illustration: There are 96 men in a group of 150 people. What percent of the group are men?

    SOLUTION: Base = 150
    Percentage (amount) = 96
    Rate = $\frac{96}{150}$
    $$= .64$$
    $$= 64\%$$
    Answer: 64% of the group are men.

    Illustration: In a tank holding 20 gallons of solution, 1 gallon is alcohol. What is the strength of the solution in percent?

    SOLUTION: Percentage (amount) = 1 gallon
    Base = 20 gallons
    Rate = $\frac{1}{20}$
    $$= .05$$
    $$= 5\%$$
    Answer: The solution is 5% alcohol.

9. In a percent problem, the whole is 100%.

Example: If a problem involves 10% of a quantity, the rest of the quantity is 90%.

Example: If a quantity has been increased by 5%, the new amount is 105% of the original quantity.

Example: If a quantity has been decreased by 15%, the new amount is 85% of the original quantity.

Practice Problems Involving Percents

1. 10% written as a decimal is
(A) 1.0   (C) 0.001
(B) 0.01   (D) 0.1

2. What is 5.37% in fraction form?
(A) $\frac{537}{10,000}$ (C) $\frac{537}{1000}$
(B) $5\frac{37}{10,000}$ (D) $5\frac{37}{100}$

3. What percent of $\frac{5}{8}$ is $\frac{3}{4}$?
(A) 75%   (C) 80%
(B) 60%   (D) 90%

4. What percent is 14 of 24?
(A) $62\frac{1}{4}$%   (C) $41\frac{2}{3}$%
(B) $58\frac{1}{3}$%   (D) $33\frac{3}{5}$%

5. 200% of 800 equals
(A) 2500   (C) 1600
(B) 16   (D) 4

6. If John must have a mark of 80% to pass a test of 35 items, the number of items he may miss and still pass the test is
(A) 7   (C) 11
(B) 8   (D) 28

7. The regular price of a TV set that sold for $118.80 at a 20% reduction sale is
(A) $148.50   (C) $138.84
(B) $142.60   (D) $ 95.04

8. A circle graph of a budget shows the expenditure of 26.2% for housing, 28.4% for food, 12% for clothing, 12.7% for taxes, and the balance for miscellaneous items. The percent for miscellaneous items is
(A) 31.5   (C) 20.7
(B) 79.3   (D) 68.5

9. Two dozen shuttlecocks and four badminton rackets are to be purchased for a playground. The shuttlecocks are priced at $.35 each and the rackets at $2.75 each. The playground receives a discount of 30% from these prices. The total cost of this equipment is
(A) $ 7.29   (C) $13.58
(B) $11.43   (D) $18.60

10. A piece of wood weighing 10 ounces is found to have a weight of 8 ounces after drying. The moisture content was
(A) 25%   (C) 20%
(B) $33\frac{1}{3}$%   (D) 40%

11. A bag contains 800 coins. Of these, 10 percent are dimes, 30 percent are nickels, and the rest are quarters. The amount of money in the bag is
(A) less than $150
(B) between $150 and $300
(C) between $301 and $450
(D) more than $450

12. Six quarts of a 20% solution of alcohol in water are mixed with 4 quarts of a 60% solution of alcohol in water. The alcoholic strength of the mixture is
    (A) 80%   (C) 36%
    (B) 40%   (D) 72%

13. A man insures 80% of his property and pays a $2\frac{1}{2}$% premium amounting to $348. What is the total value of his property?
    (A) $17,000   (C) $18,400
    (B) $18,000   (D) $17,400

14. A clerk divided his 35-hour work week as follows: $\frac{1}{5}$ of his time was spent in sorting mail; $\frac{1}{2}$ of his time in filing letters; and $\frac{1}{7}$ of his time in reception work. The rest of his time was devoted to messenger work. The percent of time spent on messenger work by the clerk during the week was most nearly
    (A) 6%   (C) 14%
    (B) 10%   (D) 16%

15. In a school in which 40% of the enrolled students are boys, 80% of the boys are present on a certain day. If 1152 boys are present, the total school enrollment is
    (A) 1440   (C) 3600
    (B) 2880   (D) 5400

Percent Problems — Correct Answers

1.	(D)	6.	(A)	11.	(A)
2.	(A)	7.	(A)	12.	(C)
3.	(D)	8.	(C)	13.	(D)
4.	(B)	9.	(C)	14.	(D)
5.	(C)	10.	(C)	15.	(C)

Problem Solutions — Percents

1. $10\% = .10 = .1$

   Answer: **(D)** 0.1

2. $5.37\% = .0537 = \dfrac{537}{10,000}$

   Answer: **(A)** $\dfrac{537}{10,000}$

3. Base (number following "of") = $\frac{5}{6}$
   Percentage (number following "is") = $\frac{3}{4}$,
   $$\text{Rate} = \frac{\text{Percentage}}{\text{Base}}$$
   $$= \text{Percentage} \div \text{Base}$$

   $$\text{Rate} = \tfrac{3}{4} \div \tfrac{5}{6}$$
   $$= \tfrac{3}{4} \times \tfrac{6}{5}$$
   $$= \tfrac{9}{10}$$
   $$\tfrac{9}{10} = .9 = 90\%$$

   Answer: **(D)** 90%

4. Base (number following "of") = 24
   Percentage (number following "is") = 14

   Rate = Percentage ÷ Base
   Rate = 14 ÷ 24
   $$= .58\tfrac{1}{3}$$
   $$= 58\tfrac{1}{3}\%$$

   Answer: **(B)** $58\tfrac{1}{3}\%$

5. 200% of 800 = 2.00 × 800
              = 1600

    Answer: (C) 1600

6. He must answer 80% of 35 correctly. Therefore, he may miss 20% of 35.

    20% of 35 = .20 × 35
              = 7

    Answer: (A) 7

7. Since $118.80 represents a 20% reduction, $118.80 = 80% of the regular price.

    Regular price = $\dfrac{\$118.80}{80\%}$

            = $118.80 ÷ .80
            = $148.50

    Answer: (A) $148.50

8. All the items in a circle graph total 100%. Add the figures given for housing, food, clothing, and taxes:

         26.2%
         28.4%
         12 %
      + 12.7%
         79.3%

    Subtract this total from 100% to find the percent for miscellaneous items:

        100.0%
      − 79.3%
         20.7%

    Answer: (C) 20.7%

9. Price of shuttlecocks = 24 × $.35 = $ 8.40
    Price of rackets   = 4 × $2.75 = $11.00
    Total price   = $19.40
    Discount is 30%, and 100% − 30% = 70%

    Actual cost = 70% of 19.40
               = .70 × 19.40
               = 13.58

    Answer: (C) $13.58

10. Subtract weight of wood after drying from original weight of wood to find amount of moisture in wood:

         10
      − 8
       2 ounces of moisture in wood

Moisture content = $\dfrac{2 \text{ ounces}}{10 \text{ ounces}}$ = .2 = 20%

Answer: (C) 20%

11. Find the number of each kind of coin:

      10% of 800 = .10 × 800 = 80 dimes
      30% of 800 = .30 × 800 = 240 nickels
      60% of 800 = .60 × 800 = 480 quarters

    Find the value of the coins:

        80 dimes =   80 × .10 = $ 8.00
     240 nickels = 240 × .05 =   12.00
    480 quarters = 480 × .25 = 120.00
                   Total $140.00

    Answer: (A) less than $150

12. First solution contains 20% of 6 quarts of alcohol.

       Alcohol content = .20 × 6
                   = 1.2 quarts

    Second solution contains 60% of 4 quarts of alcohol.

       Alcohol content = .60 × 4
                   = 2.4 quarts

    Mixture contains: 1.2 + 2.4 = 3.6 quarts
                               alcohol
             6 + 4 = 10 quarts
                               liquid

    Alcoholic strength of mixture = $\dfrac{3.6}{10}$ = 36%

    Answer: (C) 36%

13. 2½% of insured value = $348

        Insured value = $\dfrac{348}{2\frac{1}{2}\%}$

                  = 348 ÷ .025
                  = $13,920

    $13,920 is 80% of total value

        Total value = $\dfrac{\$13,920}{80\%}$

               = $13,920 ÷ .80
               = $17,400

    Answer: (D) $17,400

14. $\frac{1}{5}$ = 20% sorting mail
      $\frac{1}{2}$ = 50% filing
    + $\frac{1}{7}$ = 14.3% reception
          84.3% accounted for

$$\begin{array}{r} 100\% \\ -\ 84.3\% \\ \hline 15.7\% \end{array}$$

Answer: **(D)** most nearly 16%

15. 80% of the boys = 1152

$$\text{Number of boys} = \frac{1152}{80\%}$$
$$= 1152 \div .80$$
$$= 1440$$

40% of students = 1440

$$\text{Total number of students} = \frac{1440}{40\%}$$
$$= 1440 \div .40$$
$$= 3600$$

Answer: **(C)** 3600

SHORTCUTS IN MULTIPLICATION AND DIVISION

There are several shortcuts for simplifying multiplication and division. Following the description of each shortcut, practice problems are provided.

Dropping Final Zeros

1. a. A zero in a whole number is considered a "final zero" if it appears in the units column or if all columns to its right are filled with zeros. A final zero may be omitted in certain kinds of problems.

    b. In decimal numbers a zero appearing in the extreme right column may be dropped with no effect on the solution of a problem.

2. In multiplying whole numbers, the final zero(s) may be dropped during computation and simply transferred to the answer.

Examples:

$$
\begin{array}{r}
2310 \\
\times\ 150 \\
\hline
1155 \\
231 \\
\hline
346500
\end{array}
\qquad
\begin{array}{r}
129 \\
\times\ 210 \\
\hline
129 \\
258 \\
\hline
27090
\end{array}
\qquad
\begin{array}{r}
1760 \\
\times\ 205 \\
\hline
880 \\
352 \\
\hline
360800
\end{array}
$$

Practice Problems

Solve the following multiplication problems, dropping the final zeros during computation.

1.
$$
\begin{array}{r}
230 \\
\times\ 12 \\
\hline
\end{array}
$$

2.
$$
\begin{array}{r}
175 \\
\times\ 130 \\
\hline
\end{array}
$$

3.
$$
\begin{array}{r}
203 \\
\times\ 14 \\
\hline
\end{array}
$$

4.
$$
\begin{array}{r}
621 \\
\times\ 140 \\
\hline
\end{array}
$$

5. 430
    × 360

6. 132
    × 310

7. 350
    × 24

8. 520
    × 410

9. 634
    × 120

10. 431
    × 230

Solutions to Practice Problems

1. 230
    × 12
    ─────
     46
     23
    ─────
    2760

2. 175
    × 130
    ─────
    525
    175
    ─────
    22750

3. 203
    × 14
    ─────
    812
    203
    ─────
    2842
    (no final zeros)

4. 621
    × 140
    ─────
    2484
    621
    ─────
    86940

5. 430
    × 360
    ─────
    258
    129
    ─────
    154800

6. 132
    × 310
    ─────
    132
    396
    ─────
    40920

7. 350
    × 24
    ─────
    140
    70
    ─────
    8400

8. 520
    × 410
    ─────
    52
    208
    ─────
    213200

9. 634
    × 120
    ─────
    1268
    634
    ─────
    76080

10. 431
    × 230
    ─────
    1293
    862
    ─────
    99130

Multiplying Whole Numbers by Decimals

3. In multiplying a whole number by a decimal number, if there are one or more final zeros in the multiplicand, move the decimal point in the multiplier to the right the same number of places as there are final zeros in the multiplicand. Then cross out the final zero(s) in the multiplicand.

Examples:

$$\frac{27500}{\times\ \ \ \ .15} = \frac{275}{\times\ \ 15}$$

$$\frac{1250}{\times\ .345} = \frac{125}{\times\ 3.45}$$

Practice Problems

Rewrite the following problems, dropping the final zeros and moving decimal points the appropriate number of spaces. Then compute the answers.

1. $\begin{array}{r} 2400 \\ \times\ \ \ .02 \\ \hline \end{array}$

6. $\begin{array}{r} 480 \\ \times\ \ \ .4 \\ \hline \end{array}$

2. $\begin{array}{r} 620 \\ \times\ .04 \\ \hline \end{array}$

7. $\begin{array}{r} 400 \\ \times\ .04 \\ \hline \end{array}$

3. $\begin{array}{r} 800 \\ \times\ .005 \\ \hline \end{array}$

8. $\begin{array}{r} 5300 \\ \times\ \ \ \ .5 \\ \hline \end{array}$

4. $\begin{array}{r} 600 \\ \times\ .002 \\ \hline \end{array}$

9. $\begin{array}{r} 930 \\ \times\ \ \ .3 \\ \hline \end{array}$

5. $\begin{array}{r} 340 \\ \times\ .08 \\ \hline \end{array}$

10. $\begin{array}{r} 9000 \\ \times\ .001 \\ \hline \end{array}$

Solutions to Practice Problems

The rewritten problems are shown, along with the answers.

1. $\begin{array}{r} 24 \\ \times\ 2 \\ \hline 48 \end{array}$

2. $\begin{array}{r} 62 \\ \times\ .4 \\ \hline 24.8 \end{array}$

3.
$$
\begin{array}{r}
8 \\
\times\ .5 \\
\hline
4.0
\end{array}
$$

7.
$$
\begin{array}{r}
4 \\
\times\ 4 \\
\hline
16
\end{array}
$$

4.
$$
\begin{array}{r}
6 \\
\times\ .2 \\
\hline
1.2
\end{array}
$$

8.
$$
\begin{array}{r}
53 \\
\times\ 50 \\
\hline
2650
\end{array}
$$

5.
$$
\begin{array}{r}
34 \\
\times\ .8 \\
\hline
27.2
\end{array}
$$

9.
$$
\begin{array}{r}
93 \\
\times\ 3 \\
\hline
279
\end{array}
$$

6.
$$
\begin{array}{r}
48 \\
\times\ 4 \\
\hline
192
\end{array}
$$

10.
$$
\begin{array}{r}
9 \\
\times\ 1 \\
\hline
9
\end{array}
$$

Dividing by Whole Numbers

4. a. When there are final zeros in the divisor but no final zeros in the dividend, move the decimal point in the dividend to the left as many places as there are final zeros in the divisor, then omit the final zeros.

Example: $2700. \overline{)\ 37523.} = 27. \overline{)\ 375.23}$

b. When there are fewer final zeros in the divisor than there are in the dividend, drop the same number of final zeros from the dividend as there are final zeros in the divisor.

Example: $250. \overline{)\ 45300.} = 25. \overline{)\ 4530.}$

c. When there are more final zeros in the divisor than there are in the dividend, move the decimal point in the dividend to the left as many places as there are final zeros in the divisor, then omit the final zeros.

Example: $2300. \overline{)\ 690.} = 23. \overline{)\ 6.9}$

d. When there are no final zeros in the divisor, no zeros can be dropped in the dividend.

Example: $23. \overline{)\ 690.} = 23. \overline{)\ 690.}$

Practice Problems

Rewrite the following problems, dropping the final zeros and moving the decimal points the appropriate number of places. Then compute the quotients.

1. $600. \overline{)\ 72.}$

3. $7600 \overline{)\ 1520.}$

5. $11.0 \overline{)\ 220.}$

2. $310. \overline{)\ 6200.}$

4. $46. \overline{)\ 920.}$

6. $700. \overline{)\ 84.}$

7. $90.\overline{)8100.}$ 10. $41.0\overline{)820.}$ 13. $5500.\overline{)110.}$

8. $8100.\overline{)1620.}$ 11. $800.\overline{)96.}$ 14. $36.\overline{)720.}$

9. $25.\overline{)5250.}$ 12. $650.\overline{)1300.}$ 15. $87.0\overline{)1740.}$

Rewritten Practice Problems

1. $6.\overline{).72}$ 6. $7.\overline{).84}$ 11. $8.\overline{).96}$

2. $31.\overline{)620.}$ 7. $9.\overline{)810.}$ 12. $65.\overline{)130.}$

3. $76.\overline{)15.2}$ 8. $81.\overline{)16.2}$ 13. $55.\overline{)1.1}$

4. $46.\overline{)920.}$ 9. $25.\overline{)5250.}$ 14. $36.\overline{)720.}$

5. $11.\overline{)220.}$ 10. $41.\overline{)820.}$ 15. $87.\overline{)1740.}$

Solutions to Practice Problems

1. $6.\overline{).72}$ = .12   6. $7.\overline{).84}$ = .12   11. $8.\overline{).96}$ = .12

2. $31.\overline{)620.}$ = 20; 62; 00   7. $9.\overline{)810.}$ = 90   12. $65.\overline{)130.}$ = 2; 130; 00

3. $76.\overline{)15.2}$ = .2; 15 2; 0 0   8. $81.\overline{)16.2}$ = .2; 16 2; 0 0   13. $55.\overline{)1.10}$ = .02; 1.10; 00

4. $46.\overline{)920.}$ = 20; 92; 00   9. $25.\overline{)5250.}$ = 210; 50; 25; 25; 00   14. $36.\overline{)720.}$ = 20; 72; 00

5. $11.\overline{)220.}$ = 20; 22; 00   10. $41.\overline{)820.}$ = 20; 82; 00   15. $87.\overline{)1740.}$ = 20; 174; 00

Division by Multiplication

5. Instead of dividing by a particular number, the same answer is obtained by multiplying by the equivalent multiplier.

6. To find the equivalent multiplier of a given divisor, divide 1 by the divisor.

   Example: The equivalent multiplier of $12\frac{1}{2}$ is $1 \div 12\frac{1}{2}$ or .08. The division problem $100 \div 12\frac{1}{2}$ may be more easily solved as the multiplication problem $100 \times .08$. The answer will be the same.

7. Common divisors and their equivalent multipliers are shown below:

Divisor	Equivalent Multiplier
$11\frac{1}{9}$	.09
$12\frac{1}{2}$	.08
$14\frac{2}{7}$	.07
$16\frac{2}{3}$	.06
20	.05
25	.04
$33\frac{1}{3}$	.03
50	.02

8. A divisor may be multiplied or divided by any power of 10, and the only change in its equivalent multiplier will be in the placement of the decimal point, as may be seen in the following table:

Divisor	Equivalent Multiplier
.025	40.
.25	4.
2.5	.4
25.	.04
250.	.004
2500.	.0004

Practice Problems

Rewrite and solve each of the following problems by using equivalent multipliers. Drop the final zeros where appropriate.

1. $100 \div 16\frac{2}{3} =$

2. $200 \div 25 =$

3. $300 \div 33\frac{1}{3} =$

4. $250 \div 50 =$

5. $80 \div 12\frac{1}{2} =$

6. $800 \div 14\frac{2}{7} =$

7. $620 \div 20 =$

8. $500 \div 11\frac{1}{9} =$

9. $420 \div 16\frac{2}{3} =$

10. $1200 \div 33\frac{1}{3} =$

11. $955 \div 50 =$

12. $300 \div 33\frac{1}{3} =$

13. $275 \div 12\frac{1}{2} =$

14. $625 \div 25 =$

15. $244 \div 20 =$

16. $350 \div 16\frac{2}{3} =$

17. $400 \div 33\frac{1}{3} =$

18. $375 \div 25 =$

19. $460 \div 20 =$

20. $250 \div 12\frac{1}{2} =$

Solutions to Practice Problems

The rewritten problems and their solutions appear below:

1. $100 \times .06 = 1 \times 6 = 6$

2. $200 \times .04 = 2 \times 4 = 8$

3. $300 \times .03 = 3 \times 3 = 9$

4. $250 \times .02 = 25 \times .2 = 5$

5. $80 \times .08 = 8 \times .8 = 6.4$

6. $800 \times .07 = 8 \times 7 = 56$

7. $620 \times .05 = 62 \times .5 = 31$

8. $500 \times .09 = 5 \times 9 = 45$

9. $420 \times .06 = 42 \times .6 = 25.2$

10. $1200 \times .03 = 12 \times 3 = 36$

11. $955 \times .02 = 19.1$

12. $300 \times .03 = 3 \times 3 = 9$

13. $275 \times .08 = 22$

14. $625 \times .04 = 25$

15. $244 \times .05 = 12.2$

16. $350 \times .06 = 35 \times .6 = 21$

17. $400 \times .03 = 4 \times 3 = 12$

18. $375 \times .04 = 15$

19. $460 \times .05 = 46 \times .5 = 23$

20. $250 \times .08 = 25 \times .8 = 20$

Multiplication by Division

9. Just as some division problems are made easier by changing them to equivalent multiplication problems, certain multiplication problems are made easier by changing them to equivalent division problems.

10. Instead of arriving at an answer by multiplying by a particular number, the same answer is obtained by dividing by the equivalent divisor.

11. To find the equivalent divisor of a given multiplier, divide 1 by the multiplier.

12. Common multipliers and their equivalent divisors are shown below:

Multiplier	Equivalent Divisor
$11\frac{1}{9}$	.09
$12\frac{1}{2}$	.08
$14\frac{2}{7}$	.07
$16\frac{2}{3}$	.06
20	.05
25	.04
$33\frac{1}{3}$	.03
50	.02

Notice that the multiplier-equivalent divisor pairs are the same as the divisor-equivalent multiplier pairs given earlier.

Practice Problems

Rewrite and solve each of the following problems by using division. Drop the final zeros where appropriate.

1. $77 \times 14\frac{2}{7} =$

2. $81 \times 11\frac{1}{9} =$

3. $475 \times 20 =$

4. $42 \times 50 =$

5. $36 \times 33\frac{1}{3} =$

6. $96 \times 12\frac{1}{2} =$

7. $126 \times 16\frac{2}{3} =$

8. $48 \times 25 =$

9. $33 \times 33\frac{1}{3} =$

10. $84 \times 14\frac{2}{7} =$

11. $99 \times 11\frac{1}{9} =$

12. $126 \times 33\frac{1}{3} =$

13. $168 \times 12\frac{1}{2} =$

14. $654 \times 16\frac{2}{3} =$

15. $154 \times 14\frac{2}{7} =$

16. $5250 \times 50 =$

17. $324 \times 25 =$

18. $625 \times 20 =$

19. $198 \times 11\frac{1}{9} =$

20. $224 \times 14\frac{2}{7} =$

Solutions to Practice Problems

The rewritten problems and their solutions appear below:

1. $.07\overline{)77.} = 7\overline{)7700.}\ ^{1100.}$

2. $.09\overline{)81.} = 9\overline{)8100.}\ ^{900.}$

3. $.05\overline{)475.} = 5\overline{)47500.}\ ^{9500.}$

4. $.02\overline{)42.} = 2\overline{)4200.}\ ^{2100.}$

5. $.03\overline{)36.} = 3\overline{)3600.}\ ^{1200.}$

6. $.08\overline{)96.} = 8\overline{)9600.}\ ^{1200.}$

7. $.06\overline{)126.} = 6\overline{)12600.}\ ^{2100.}$

8. $.04\overline{)48.} = 4\overline{)4800.}\ ^{1200.}$

9. $.03\overline{)33.} = 3\overline{)3300.}\ ^{1100.}$

10. $.07\overline{)84.} = 7\overline{)8400.}\ ^{1200.}$

11. $.09\overline{)99.} = 9\overline{)9900.}\ ^{1100.}$

12. $.03\overline{)126.} = 3\overline{)12600.}\ ^{4200.}$

13. $.08\overline{)168.} = 8\overline{)16800.}\ ^{2100.}$

14. $.06\overline{)654.} = 6\overline{)65400.}\ ^{10900.}$

15. $.07\overline{)154.} = 7\overline{)15400.}\ ^{2200.}$

16. $.02\overline{)5250.} = 2\overline{)525000.}\ ^{262500.}$

17. $.04\overline{)324.} = 4\overline{)32400.}\ ^{8100.}$

18. $.05\overline{)625.} = 5\overline{)62500.}\ ^{12500.}$

19. $.09\overline{)198.} = 9\overline{)19800.}\ ^{2200.}$

20. $.07\overline{)224.} = 7\overline{)22400.}\ ^{3200.}$

POWERS AND ROOTS

1. The numbers that are multiplied to give a product are called the **factors** of the product.

   Example: In $2 \times 3 = 6$, 2 and 3 are factors.

2. If the factors are the same, an **exponent** may be used to indicate the number of times the factor appears.

   Example: In $3 \times 3 = 3^2$, the number 3 appears as a factor twice, as is indicated by the exponent 2.

3. When a product is written in exponential form, the number the exponent refers to is called the **base**. The product itself is called the **power**.

   Example: In 2^5, the number 2 is the base and 5 is the exponent.
   $2^5 = 2 \times 2 \times 2 \times 2 \times 2 = 32$, so 32 is the power.

4. a. If the exponent used is 2, we say that the base has been **squared**, or raised to the second power.

      Example: 6^2 is read "six squared" or "six to the second power."

   b. If the exponent used is 3, we say that the base has been **cubed**, or raised to the third power.

      Example: 5^3 is read "five cubed" or "five to the third power."

   c. If the exponent is 4, we say that the base has been raised to the fourth power. If the exponent is 5, we say the base has been raised to the fifth power, etc.

      Example: 2^8 is read "two to the eighth power."

5. A number that is the product of a number squared is called a **perfect square**.

   Example: 25 is a perfect square because $25 = 5^2$.

6. a. If a number has exactly two equal factors, each factor is called the **square root** of the number.

      Example: $9 = 3 \times 3$; therefore, 3 is the square root of 9.

   b. The symbol $\sqrt{}$ is used to indicate square root.

      Example: $\sqrt{9} = 3$ means that the square root of 9 is 3, or $3 \times 3 = 9$.

7. The square root of the most common perfect squares may be found by using the following table, or by trial and error; that is, by finding the number that, when squared, yields the given perfect square.

Number	Perfect Square	Number	Perfect Square
1	1	10	100
2	4	11	121
3	9	12	144
4	16	13	169
5	25	14	196
6	36	15	225
7	49	20	400
8	64	25	625
9	81	30	900

Example: To find $\sqrt{81}$, note that 81 is the perfect square of 9, or $9^2 = 81$. Therefore, $\sqrt{81} = 9$.

8. To find the square root of a number that is not a perfect square, use the following method:

   a. Locate the decimal point.

   b. Mark off the digits in groups of two in both directions beginning at the decimal point.

   c. Mark the decimal point for the answer just above the decimal point of the number whose square root is to be taken.

   d. Find the largest perfect square contained in the left-hand group of two.

   e. Place its square root in the answer. Subtract the perfect square from the first digit or pair of digits.

   f. Bring down the next pair.

   g. Double the partial answer.

   h. Add a trial digit to the right of the doubled partial answer. Multiply this new number by the trial digit. Place the correct new digit in the answer.

   i. Subtract the product.

   j. Repeat steps f–i as often as necessary.

You will notice that you get one digit in the answer for every group of two you marked off in the original number.

Illustration: Find the square root of 138,384.

SOLUTION:

$$
\begin{array}{r}
3. \\
\sqrt{13'83'84.} \\
3^2 = 9 \\
\hline
4\,83
\end{array}
$$

$$
\begin{array}{r}
3\ \ 7\ \ 2. \\
\sqrt{13'83'84.} \\
3^2 = 9 \\
\hline
4\,83 \\
7 \times 67 = 4\,69 \\
\hline
14\,84 \\
2 \times 742 = 14\,84 \\
\hline
\end{array}
$$

The number must first be marked off in groups of two figures each, beginning at the decimal point, which, in the case of a whole number, is at the right. The number of figures in the root will be the same as the number of groups so obtained.

The largest square less than 13 is 9. $\sqrt{9} = 3$

Place its square root in the answer. Subtract the perfect square from the first digit or pair of digits. Bring down the next pair. To form our trial divisor, annex 0 to this root "3" (making 30) and multiply by 2.

   $483 \div 60 = 8$. Multiplying the trial divisor 68 by 8, we obtain 544, which is too large. We then try multiplying 67 by 7. This is correct. Add the trial digit to the right of the doubled partial answer. Place the new digit in the answer. Subtract the product. Bring down the final group. Annex 0 to the new root 37 and multiply by 2 for the trial divisor:

$$2 \times 370 = 740$$
$$1484 \div 740 = 2$$

Place the 2 in the answer.

Answer: The square root of 138,384 is 372.

Illustration: Find the square root of 3 to the nearest hundredth.

$$
\begin{array}{rl}
 & \quad\ \ 1.\ 7\ 3\ 2 \\
\textit{SOLUTION:} & \sqrt{3.00'00'00} \\
1^2 = & \underline{1} \\
20 & 2\ 00 \\
7 \times 27 = & \underline{1\ 89} \\
340 & \quad 11\ 00 \\
3 \times 343 = & \quad \underline{10\ 29} \\
3460 & \qquad\quad 71\ 00 \\
2 \times 3462 = & \qquad\quad \underline{69\ 24}
\end{array}
$$

Answer: The square root of 3 is 1.73 to the nearest hundredth.

9. To find the square root of a fraction, find the square root of its numerator and of its denominator.

    Example: $\sqrt{\frac{4}{9}} = \dfrac{\sqrt{4}}{\sqrt{9}} = \frac{2}{3}$

10. a. If a number has exactly three equal factors, each factor is called the **cube root** of the number.

    b. The symbol $\sqrt[3]{}$ is used to indicate the cube root.

    Example: $8 = 2 \times 2 \times 2$; therefore, $\sqrt[3]{8} = 2$

Practice Problems Involving Powers and Roots

1. The square of 10 is
   (A) 1
   (C) 5
   (B) 2
   (D) 100

2. The cube of 9 is
   (A) 3
   (C) 81
   (B) 27
   (D) 729

3. The fourth power of 2 is
   (A) 2
   (C) 8
   (B) 4
   (D) 16

4. In exponential form, the product $7 \times 7 \times 7 \times 7 \times 7$ may be written
   (A) 5^7
   (C) 2^7
   (B) 7^5
   (D) 7^2

5. The value of 3^5 is
   (A) 243
   (C) 35
   (B) 125
   (D) 15

6. The square root of 1175, to the nearest whole number, is
   (A) 32
   (C) 34
   (B) 33
   (D) 35

7. Find $\sqrt{503}$ to the nearest tenth.
   (A) 22.4
   (C) 22.6
   (B) 22.5
   (D) 22.7

8. Find $\sqrt{\frac{1}{4}}$.
   (A) 2
   (C) $\frac{1}{8}$
   (B) $\frac{1}{2}$
   (D) $\frac{1}{16}$

9. Find $\sqrt[3]{64}$.
   (A) 3
   (C) 8
   (B) 4
   (D) 32

10. The sum of 2^2 and 2^3 is
    (A) 9
    (C) 12
    (B) 10
    (D) 32

Powers and Roots Problems — Correct Answers

1. **(D)**	5. **(A)**	8. **(B)**
2. **(D)**	6. **(C)**	9. **(B)**
3. **(D)**	7. **(A)**	10. **(C)**
4. **(B)**		

Problem Solutions — Powers and Roots

1. $10^2 = 10 \times 10 = 100$

   Answer: **(D)** 100

2. $9^3 = 9 \times 9 \times 9$
   $= 81 \times 9$
   $= 729$

   Answer: **(D)** 729

3. $2^4 = 2 \times 2 \times 2 \times 2$
    $= 4 \times 2 \times 2$
    $= 8 \times 2$
    $= 16$

    Answer: **(D)** 16

4. $7 \times 7 \times 7 \times 7 \times 7 = 7^5$

    Answer: **(B)** 7^5

5. $3^5 = 3 \times 3 \times 3 \times 3 \times 3$
    $= 243$

    Answer: **(A)** 243

6.

$$\begin{array}{r} 3\ \ 4.\ 2 \\ \sqrt{11'75.00} \end{array} \quad \text{= 34 to the nearest whole number}$$

$3^2 =$	9
	2 75
$4 \times 64 =$	2 56
	19 00
$2 \times 682 =$	13 64
	5 36

Answer: **(C)** 34

7.

$$\begin{array}{r} 2\ \ 2.\ 4\ \ 2 \\ \sqrt{5'03.00'00} \end{array} \quad \text{= 22.4 to the nearest tenth}$$

$2^2 =$	4
	1 03
$2 \times 42 =$	84
	19 00
$4 \times 444 =$	17 76
	1 24 00
$2 \times 4482 =$	89 64
	34 36

Answer: **(A)** 22.4

8. $\sqrt{\frac{1}{4}} = \frac{\sqrt{1}}{\sqrt{4}} = \frac{1}{2}$

    Answer: **(B)** $\frac{1}{2}$

9. Since $4 \times 4 \times 4 = 64$, $\sqrt[3]{64} = 4$

    Answer: **(B)** 4

10. $2^2 + 2^3 = 4 + 8 = 12$

    Answer: **(C)** 12

TABLE OF MEASURES

American Measures

Length

1 foot (ft or ') = 12 inches (in or ")
1 yard (yd) = 36 inches
1 yard = 3 feet
1 rod (rd) = 16½ feet
1 mile (mi) = 5280 feet
1 mile = 1760 yards
1 mile = 320 rods

Liquid Measure

1 cup (c) = 8 fluid ounces (fl oz)
1 pint (pt) = 2 cups
1 pint = 4 gills (gi)
1 quart (qt) = 2 pints
1 gallon (gal) = 4 quarts
1 barrel (bl) = 31½ gallons

Weight

1 pound (lb) = 16 ounces (oz)
1 hundredweight (cwt) = 100 pounds
1 ton (T) = 2000 pounds

Dry Measure

1 quart (qt) = 2 pints (pt)
1 peck (pk) = 8 quarts
1 bushel (bu) = 4 pecks

Area

1 square foot (ft²) = 144 square inches (in²)
1 square yard (yd²) = 9 square feet

Volume

1 cubic foot (ft³ or cu ft) = 1728 cubic inches
1 cubic yard (yd³ or cu yd) = 27 cubic feet
1 gallon = 231 cubic inches

General Measures

Time

1 minute (min) = 60 seconds (sec)
1 hour (hr) = 60 minutes
1 day = 24 hours
1 week = 7 days
1 year = 52 weeks
1 calendar year = 365 days

Angles and Arcs

1 minute (') = 60 seconds (")
1 degree (°) = 60 minutes
1 circle = 360 degrees

Counting

1 dozen (doz) = 12 units
1 gross (gr) = 12 dozen
1 gross = 144 units

Table of American—Metric Conversions (Approximate)

American to Metric	*Metric to American*
1 inch = 2.54 centimeters	1 centimeter = .39 inches
1 yard = .9 meters	1 meter = 1.1 yards
1 mile = 1.6 kilometers	1 kilometer = .6 miles
1 ounce = 28 grams	1 kilogram = 2.2 pounds
1 pound = 454 grams	1 liter = 1.06 liquid quart
1 fluid ounce = 30 milliliters	
1 liquid quart = .95 liters	

Table of Metric Conversions*

1 liter = 1000 cubic centimeters (cm^3)
1 milliliter = 1 cubic centimeter
1 liter of water weighs 1 kilogram
1 milliliter of water weighs 1 gram

---

*These conversions are exact only under specific conditions. If the conditions are not met, the conversions are approximate.

THE METRIC SYSTEM

———— LENGTH ————

Unit	Abbreviation	Number of Meters
myriameter	mym	10,000
kilometer	km	1,000
hectometer	hm	100
dekameter	dam	10
meter	m	1
decimeter	dm	0.1
centimeter	cm	0.01
millimeter	mm	0.001

———— AREA ————

Unit	Abbreviation	Number of Square Meters
square kilometer	sq km *or* km^2	1,000,000
hectare	ha	10,000
are	a	100
centare	ca	1
square centimeter	sq cm *or* cm^2	0.0001

———— VOLUME ————

Unit	Abbreviation	Number of Cubic Meters
dekastere	das	10
stere	s	1
decistere	ds	0.10
cubic centimeter	cu cm *or* cm^3 *or* cc	0.000001

———— CAPACITY ————

Unit	Abbreviation	Number of Liters
kiloliter	kl	1,000
hectoliter	hl	100
dekaliter	dal	10
liter	l	1
deciliter	dl	0.10
centiliter	cl	0.01
milliliter	ml	0.001

———— MASS AND WEIGHT ————

Unit	Abbreviation	Number of Grams
metric ton	MT *or* t	1,000,000
quintal	q	100,000
kilogram	kg	1,000
hectogram	hg	100
dekagram	dag	10
gram	g *or* gm	1
decigram	dg	0.10
centigram	cg	0.01
milligram	mg	0.001

DENOMINATE NUMBERS (MEASUREMENT)

1. A **denominate number** is a number that specifies a given measurement. The unit of measure is called the **denomination**.

   Example: 7 miles, 3 quarts, and 5 grams are denominate numbers.

2. a. The American system of measurement uses such denominations as pints, ounces, pounds, and feet.

   b. The metric system of measurement uses such denominations as grams, liters, and meters.

American System of Measurement

3. To convert from one unit of measure to another, find in the Table of Measures how many units of the smaller denomination equal one unit of the larger denomination. This number is called the **conversion number**.

4. To convert from one unit of measure to a smaller unit, multiply the given number of units by the conversion number.

   Illustration: Convert 7 yards to inches.

   SOLUTION: 1 yard = 36 inches (conversion number)
   7 yards = 7 × 36 inches
   = 252 inches

   Answer: 252 in

   Illustration: Convert 2 hours 12 minutes to minutes.

   SOLUTION: 1 hour = 60 minutes (conversion number)
   2 hr 12 min = 2 hr + 12 min
   2 hr = 2 × 60 min = 120 min
   2 hr 12 min = 120 min + 12 min
   = 132 min

   Answer: 132 min

47

5. To convert from one unit of measure to a larger unit:

   a. Divide the given number of units by the conversion number.

   Illustration: Convert 48 inches to feet.

   SOLUTION: 1 foot = 12 inches (conversion number)
   $$48 \text{ in} \div 12 = 4 \text{ ft}$$

   Answer: 4 ft

   b. If there is a remainder it is expressed in terms of the smaller unit of measure.

   Illustration: Convert 35 ounces to pounds and ounces.

   SOLUTION: 1 pound = 16 ounces (conversion number)

   $$35 \text{ oz} \div 16 = 16 \overline{)\begin{array}{l} 2 \text{ lb} \\ 35 \text{ oz} \end{array}}$$
   $$\underline{32}$$
   $$3 \text{ oz}$$
   $$= 2 \text{ lb } 3 \text{ oz}$$

   Answer: 2 lb 3 oz

6. To add denominate numbers, arrange them in columns by common unit, then add each column. If necessary, simplify the answer, starting with the smallest unit.

   Illustration: Add 1 yd 2 ft 8 in, 2 yd 2 ft 10 in, and 3 yd 1 ft 9 in.

   SOLUTION:
   ```
        1 yd  2 ft   8 in
        2 yd  2 ft  10 in
      + 3 yd  1 ft   9 in
        6 yd  5 ft  27 in
   ```
   = 6 yd 7 ft 3 in   (since 27 in = 2 ft 3 in)
   = 8 yd 1 ft 3 in   (since 7 ft = 2 yd 1 ft)

   Answer: 8 yd 1 ft 3 in

7. To subtract denominate numbers, arrange them in columns by common unit, then subtract each column starting with the smallest unit. If necessary, borrow to increase the number of a particular unit.

   Illustration: Subtract 2 gal 3 qt from 7 gal 1 qt.

   SOLUTION:
   ```
        7 gal 1 qt =    6 gal 5 qt
      − 2 gal 3 qt = −  2 gal 3 qt
                        4 gal 2 qt
   ```
   Note that 1 gal was borrowed from 7 gal.
   1 gal = 4 qt
   Therefore, 7 gal 1 qt = 6 gal 5 qt

   Answer: 4 gal 2 qt

8. To multiply a denominate number by a given number:

   a. If the denominate number contains only one unit, multiply the numbers and write the unit.

   Example: 3 oz × 4 = 12 oz

b. If the denominate number contains more than one unit of measurement, multiply the number of each unit by the given number and simplify the answer, if necessary.

Illustration: Multiply 4 yd 2 ft 8 in by 2.

SOLUTION:
$$
\begin{array}{l}
 \text{4 yd 2 ft\ \ 8 in} \\
\underline{\times \phantom{\text{4 yd 2 ft\ \ }} 2} \\
 \text{8 yd 4 ft 16 in} \\
= \text{8 yd 5 ft\ \ 4 in \ (since 16 in = 1 ft 4 in)} \\
= \text{9 yd 2 ft\ \ 4 in \ (since 5 ft = 1 yd 2 ft)}
\end{array}
$$

Answer: 9 yd 2 ft 4 in

9. To divide a denominate number by a given number, convert all units to the smallest unit, then divide. Simplify the answer, if necessary.

Illustration: Divide 5 lb 12 oz by 4.

SOLUTION:
$$
\begin{array}{l}
 \text{1 lb = 16 oz,\ therefore} \\
\text{5 lb 12 oz} = \text{92 oz} \\
\text{92 oz} \div 4 = \text{23 oz} \\
\phantom{\text{92 oz} \div 4} = \text{1 lb 7 oz}
\end{array}
$$

Answer: 1 lb 7 oz

10. Alternate method of division:

a. Divide the number of the largest unit by the given number.

b. Convert any remainder to the next largest unit.

c. Divide the total number of that unit by the given number.

d. Again convert any remainder to the next unit and divide.

e. Repeat until no units remain.

Illustration: Divide 9 hr 21 min 40 sec by 4.

SOLUTION:
$$
\begin{array}{r}
\text{2 hr}\quad \text{20 min}\quad \text{25 sec} \\
\hline
4\,)\overline{\text{9 hr}\quad \text{21 min}\quad \text{40 sec}} \\
\underline{\text{8 hr}}\phantom{\quad \text{21 min}\quad \text{40 sec}} \\
\text{1 hr} = \underline{\text{60 min}}\phantom{\quad \text{40 sec}} \\
\text{81 min}\phantom{\quad \text{40 sec}} \\
\underline{\text{80 min}}\phantom{\quad \text{40 sec}} \\
\text{1 min} = \underline{\text{60 sec}} \\
\text{100 sec} \\
\underline{\text{100 sec}} \\
\text{0 sec}
\end{array}
$$

Answer: 2 hr 20 min 25 sec

Metric Measurement

11. The basic units of the metric system are the meter (m), which is used for length; the gram (g), which is used for weight; and the liter (*l*), which is used for capacity, or volume.

12. The prefixes that are used with the basic units, and their meanings, are:

Prefix	Abbreviation	Meaning
micro	**m**	one millionth of (.000001)
milli	m	one thousandth of (.001)
centi	c	one hundredth of (.01)
deci	d	one tenth of (.1)
deka	da or dk	ten times (10)
hecto	h	one hundred times (100)
kilo	k	one thousand times (1000)
mega	M	one million times (1,000,000)

13. To convert *to* a basic metric unit from a prefixed metric unit, multiply by the number indicated in the prefix.

    Example: Convert 72 millimeters to meters.

    $$72 \text{ millimeters} = 72 \times .001 \text{ meters}$$
    $$= .072 \text{ meters}$$

    Example: Convert 4 kiloliters to liters.

    $$4 \text{ kiloliters} = 4 \times 1000 \text{ liters}$$
    $$= 4000 \text{ liters}$$

14. To convert *from* a basic unit to a prefixed unit, divide by the number indicated in the prefix.

    Example: Convert 300 liters to hectoliters.

    $$300 \text{ liters} = 300 \div 100 \text{ hectoliters}$$
    $$= 3 \text{ hectoliters}$$

    Example: Convert 4.5 meters to decimeters.

    $$4.5 \text{ meters} = 4.5 \div .1 \text{ decimeters}$$
    $$= 45 \text{ decimeters}$$

15. To convert from any prefixed metric unit to another prefixed unit, first convert to a basic unit, then convert the basic unit to the desired unit.

    Illustration: Convert 420 decigrams to kilograms.

    SOLUTION: $420 \text{ dg} = 420 \times .1 \text{ g} = 42 \text{ g}$
    $42 \text{ g} = 42 \div 1000 \text{ kg} = .042 \text{ kg}$

    Answer: .042 kg

16. To add, subtract, multiply, or divide using metric measurement, first convert all units to the same unit, then perform the desired operation.

    Illustration: Subtract 1200 g from 2.5 kg.

    SOLUTION:
    $$\begin{array}{rcr} 2.5 \text{ kg} = & 2500 \text{ g} \\ -\ 1200 \text{ g} = & -\ 1200 \text{ g} \\ \hline & 1300 \text{ g} \end{array}$$

    Answer: 1300 g or 1.3 kg

17. To convert from a metric measure to an American measure, or the reverse:

    a. In the Table of American–Metric Conversions, find how many units of the desired measure are equal to one unit of the given measure.

    b. Multiply the given number by the number found in the table.

    Illustration: Find the number of pounds in 4 kilograms.

    SOLUTION: From the table, 1 kg = 2.2 lb.
    $$\begin{aligned} 4 \text{ kg} &= 4 \times 2.2 \text{ lb} \\ &= 8.8 \text{ lb} \end{aligned}$$

    Answer: 8.8 lb

    Illustration: Find the number of meters in 5 yards.

    SOLUTION:
    $$\begin{aligned} 1 \text{ yd} &= .9 \text{ m} \\ 5 \text{ yd} &= 5 \times .9 \text{ m} \\ &= 4.5 \text{ m} \end{aligned}$$

    Answer: 4.5 m

Temperature Measurement

18. The temperature measurement currently used in the United States is the degree Fahrenheit (°F). The metric measurement for temperature is the degree Celsius (°C), also called degree Centigrade.

19. Degrees Celsius may be converted to degrees Fahrenheit by the formula:

    $$°F = \tfrac{9}{5}°C + 32°$$

    Illustration: Water boils at 100°C. Convert this to °F.

    SOLUTION:
    $$\begin{aligned} °F &= \tfrac{9}{\cancel{5}} \times \overset{20}{\cancel{100}}° + 32° \\ &= 180° + 32° \\ &= 212° \end{aligned}$$

    Answer: 100°C = 212°F

20. Degrees Fahrenheit may be converted to degrees Celsius by the formula:

$$°C = \tfrac{5}{9}(°F - 32°)$$

In using this formula, perform the subtraction in the parentheses first, then multiply by $\tfrac{5}{9}$.

Illustration: If normal body temperature is 98.6°F, what is it on the Celsius scale?

SOLUTION: $°C = \tfrac{5}{9}(98.6° - 32°)$
$= \tfrac{5}{9} \times 66.6°$
$= \tfrac{333}{9}°$
$= 37°$

Answer: Normal body temperature = 37°C.

Practice Problems Involving Measurement

1. A carpenter needs boards for 4 shelves, each 2'9" long. How many feet of board should he buy?
   (A) 11
   (B) $11\tfrac{1}{6}$
   (C) 13
   (D) $15\tfrac{1}{2}$

2. The number of half-pints in 19 gallons of milk is
   (A) 76
   (B) 152
   (C) 304
   (D) 608

3. The product of 8 ft 7 in multiplied by 8 is
   (A) 69 ft 6 in
   (B) 68.8 ft
   (C) $68\tfrac{2}{3}$ ft
   (D) 68 ft 2 in

4. $\tfrac{1}{3}$ of 7 yards is
   (A) 2 yd
   (B) 4 ft
   (C) $3\tfrac{1}{2}$ yd
   (D) 7 ft

5. Six gross of special drawing pencils were purchased for use in an office. If the pencils were used at the rate of 24 a week, the maximum number of weeks that the six gross of pencils would last is
   (A) 6 weeks
   (B) 12 weeks
   (C) 24 weeks
   (D) 36 weeks

6. If 7 ft 9 in is cut from a piece of wood that is 9 ft 6 in, the piece left is
   (A) 1 ft 9 in
   (B) 1 ft 10 in
   (C) 2 ft 2 in
   (D) 2 ft 5 in

7. Take 3 hours 49 minutes from 5 hours 13 minutes.
   (A) 1 hr 5 min
   (B) 1 hr 10 min
   (C) 1 hr 18 min
   (D) 1 hr 24 min

8. A piece of wood 35 feet 6 inches long was used to make 4 shelves of equal lengths. The length of each shelf was
   (A) 8.9 in
   (B) 8 ft 9 in
   (C) 8 ft $9\tfrac{1}{2}$ in
   (D) 8 ft $10\tfrac{1}{2}$ in

9. The number of yards equal to 126 inches is
   (A) 3.5
   (B) 10.5
   (C) 1260
   (D) 1512

10. If there are 231 cubic inches in one gallon, the number of cubic inches in 3 pints is closest to which one of the following?
    (A) 24
    (B) 29
    (C) 57
    (D) 87

11. The sum of 5 feet $2\tfrac{3}{4}$ inches, 8 feet $\tfrac{1}{2}$ inch, and $12\tfrac{1}{2}$ inches is
    (A) 14 ft $3\tfrac{3}{4}$ in
    (B) 14 ft $5\tfrac{3}{4}$ in
    (C) 14 ft $9\tfrac{1}{4}$ in
    (D) 15 ft $\tfrac{1}{2}$ in

12. Add 5 hr 13 min, 3 hr 49 min, and 14 min. The sum is
    (A) 8 hr 16 min
    (B) 9 hr 16 min
    (C) 9 hr 76 min
    (D) 8 hr 6 min

13. Assuming that 2.54 centimeters = 1 inch, a metal rod that measures 1½ feet would most nearly equal which one of the following?
    (A) 380 cm   (C) 30 cm
    (B) 46 cm   (D) 18 cm

14. A micromillimeter is defined as one millionth of a millimeter. A length of 17 micromillimeters may be represented as
    (A) .00017 mm   (C) .000017 mm
    (B) .0000017 mm   (D) .00000017 mm

15. How many liters are equal to 4200 m*l*?
    (A) .42   (C) 420
    (B) 4.2   (D) 420,000

16. Add 26 dg, .4 kg, 5 g, and 184 cg.
    (A) 215.40 g   (C) 409.44 g
    (B) 319.34 g   (D) 849.00 g

17. Four full bottles of equal size contain a total of 1.28 liters of cleaning solution. How many milliliters are in each bottle?
    (A) 3.20   (C) 320
    (B) 5.12   (D) 512

18. How many liters of water can be held in a 5-gallon jug? (See Conversion Table.)
    (A) 19   (C) 40
    (B) 38   (D) 50

19. To the nearest degree, what is a temperature of 12°C equal to on the Fahrenheit scale?
    (A) 19°   (C) 57°
    (B) 54°   (D) 79°

20. A company requires that the temperature in its offices be kept at 68°F. What is this in °C?
    (A) 10°   (C) 20°
    (B) 15°   (D) 25°

Measurement Problems — Correct Answers

1. (A)	6. (A)	11. (A)	16. (C)
2. (C)	7. (D)	12. (B)	17. (C)
3. (C)	8. (D)	13. (B)	18. (A)
4. (D)	9. (A)	14. (C)	19. (B)
5. (D)	10. (D)	15. (B)	20. (C)

Problem Solutions — Measurement

1.
   2 ft 9 in
   ×   4
   ─────────────
   8 ft 36 in = 11 ft

   Answer: **(A)** 11

2. Find the number of half-pints in 1 gallon:
   1 gal = 4 qts

   4 qts = 4 × 2 pts = 8 pts
   8 pts = 8 × 2 = 16 half-pints

   Multiply to find the number of half-pints in 19 gallons:

   19 gal = 19 × 16 half-pints
          = 304 half-pints

   Answer: **(C)** 304

3. $$\begin{array}{r} 8 \text{ ft} \quad 7 \text{ in} \\ \times \qquad\quad 8 \\ \hline \end{array}$$
    64 ft 56 in = 68 ft 8 in
    (since 56 in = 4 ft 8 in)
    8 in = $\frac{8}{12}$ ft = $\frac{2}{3}$ ft
    68 ft 8 in = $68\frac{2}{3}$ ft

    Answer: **(C)** $68\frac{2}{3}$ ft

4. $\frac{1}{3} \times 7$ yd = $2\frac{1}{3}$ yd
    = 2 yd 1 ft
    = 2 × 3 ft + 1 ft
    = 7 ft

    Answer: **(D)** 7 ft

5. Find the number of units in 6 gross:
    1 gross = 144 units
    6 gross = 6 × 144 units
    = 864 units
    Divide units by rate of use:
    864 ÷ 24 = 36

    Answer: **(D)** 36 weeks

6. $$\begin{array}{rcl} 9 \text{ ft } 6 \text{ in} = & 8 \text{ ft } 18 \text{ in} \\ - 7 \text{ ft } 9 \text{ in} = & - 7 \text{ ft} \quad 9 \text{ in} \\ \hline & 1 \text{ ft} \quad 9 \text{ in} \end{array}$$

    Answer: **(A)** 1 ft 9 in

7. $$\begin{array}{rcl} 5 \text{ hours } 13 \text{ minutes} = & 4 \text{ hours } 73 \text{ minutes} \\ - 3 \text{ hours } 49 \text{ minutes} = & - 3 \text{ hours } 49 \text{ minutes} \\ \hline & 1 \text{ hour} \quad 24 \text{ minutes} \end{array}$$

    Answer: **(D)** 1 hr 24 min

8. $$\begin{array}{r} 8 \text{ feet} \quad 10 \text{ inches} + \frac{2}{4} \text{ inches} = 8 \text{ ft } 10\frac{1}{2} \text{ in} \\ 4 \overline{)\, 35 \text{ feet} \qquad 6 \text{ inches}} \\ \underline{32 \text{ feet}} \\ 3 \text{ feet} = \underline{36 \text{ inches}} \\ 42 \text{ inches} \\ \underline{40 \text{ inches}} \\ 2 \text{ inches} \end{array}$$

    Answer: **(D)** 8 ft $10\frac{1}{2}$ in

9. 1 yd = 36 in
    126 ÷ 36 = 3.5

    Answer: **(A)** 3.5

10. 1 gal = 4 qt = 8 pt
    Therefore, 1 pt = 231 cubic inches ÷ 8
    = 28.875 cubic inches
    3 pts = 3 × 28.875 cubic inches
    = 86.625 cubic inches

    Answer: **(D)** 87

11. $$\begin{array}{r} 5 \text{ feet} \quad 2\frac{3}{4} \text{ inches} \\ 8 \text{ feet} \quad \frac{1}{2} \text{ inches} \\ + \qquad 12\frac{1}{2} \text{ inches} \\ \hline 13 \text{ feet } 15\frac{3}{4} \text{ inches} \\ = 14 \text{ feet} \quad 3\frac{3}{4} \text{ inches} \end{array}$$

    Answer: **(A)** 14 feet $3\frac{3}{4}$ inches

12. $$\begin{array}{r} 5 \text{ hr } 13 \text{ min} \\ 3 \text{ hr } 49 \text{ min} \\ + \qquad 14 \text{ min} \\ \hline 8 \text{ hr } 76 \text{ min} \\ = 9 \text{ hr } 16 \text{ min} \end{array}$$

    Answer: **(B)** 9 hr 16 min

13. 1 foot = 12 inches
    $1\frac{1}{2}$ feet = $1\frac{1}{2}$ × 12 inches = 18 inches
    1 inch = 2.54 cm
    Therefore,
    18 inches = 18 × 2.54 cm
    = 45.72 cm

    Answer: **(B)** 46 cm

14. 1 micromillimeter = .000001 mm
    17 micromillimeters = 17 × .000001 mm
    = .000017 mm

    Answer: **(C)** .000017 mm

15. 4200 ml = 4200 × .001 *l*
    = 4.200 *l*

    Answer: **(B)** 4.2

16. Convert all of the units to grams:
    $$\begin{array}{rcl} 26 \text{ dg} = 26 \times .1 \text{ g} & = & 2.6 \text{ g} \\ .4 \text{ kg} = .4 \times 1000 \text{ g} & = & 400 \quad\text{ g} \\ 5 \text{ g} & = & 5 \quad\text{ g} \\ 184 \text{ cg} = 184 \times .01 \text{ g} & = & \underline{1.84 \text{ g}} \\ & & 409.44 \text{ g} \end{array}$$

    Answer: **(C)** 409.44 g

17.   1.28 liters ÷ 4 = .32 liters
                .32 liters = .32 ÷ .001 m*l*
                         = 320 m*l*

Answer: **(C)**   320

18.   Find the number of liters in 1 gallon:

             1 qt = .95 *l*
             1 gal = 4 qts
             1 gal = 4 × .95 *l* = 3.8 *l*

Multiply to find the number of liters in 5 gallons:

             5 gal = 5 × 3.8 *l* = 19 *l*

Answer: **(A)**   19

19.   °F = $\frac{9}{5}$ × 12° + 32°
                  = $\frac{108°}{5}$ + 32°
                  = 21.6° + 32°
                  = 53.6°

Answer: **(B)**   54°

20.   °C = $\frac{5}{9}$(68° − 32°)
                  = $\frac{5}{9}$ × $\overset{4}{\cancel{36°}}$
                  = 20°

Answer: **(C)**   20°

GEOMETRY

Angles

1. a. An **angle** is the figure formed by two lines meeting at a point.

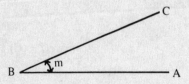

   b. The point B is the **vertex** of the angle and the lines BA and BC are the **sides** of the angle.

2. There are three common ways of naming an angle:

   a. By a small letter or figure written within the angle, as ∠m.

   b. By the capital letter at its vertex, as ∠B.

   c. By three capital letters, the middle letter being the vertex letter, as ∠ABC.

3. a. When two straight lines intersect (cut each other), four angles are formed. If these four angles are equal, each angle is a **right angle** and contains 90°. The symbol ⌐ is used to indicate a right angle.

   Example:

   A

   B⌐_____ C   ∠ABC is a right angle.

   b. An angle less than a right angle is an **acute angle**.

   c. If the two sides of an angle extend in opposite directions forming a straight line, the angle is a **straight angle** and contains 180°.

   d. An angle greater than a right angle (90°) and less than a straight angle (180°) is an **obtuse angle**.

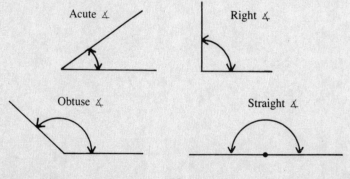

56

4. a. Two angles are **complementary** if their sum is 90°.

    b. To find the complement of an angle, subtract the given number of degrees from 90°.

    Example: The complement of 60° is 90° − 60° = 30°.

5. a. Two angles are **supplementary** if their sum is 180°.

    b. To find the supplement of an angle, subtract the given number of degrees from 180°.

    Example: The supplement of 60° is 180° − 60° = 120°.

Lines

6. a. Two lines are **perpendicular** to each other if they meet to form a right angle. The symbol ⊥ is used to indicate that the lines are perpendicular.

    Example: ∡ABC is a right angle. Therefore, AB ⊥ BC.

    b. Lines that do not meet no matter how far they are extended are called **parallel lines**. The symbol ‖ is used to indicate that two lines are parallel.

    Example:
    A————————B   AB ‖ CD

    C————————D

Triangles

7. A **triangle** is a closed, three-sided figure. The figures below are all triangles.

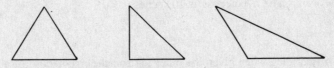

8. a. The sum of the three angles of a triangle is 180°.

    b. To find an angle of a triangle when you are given the other two angles, add the given angles and subtract their sum from 180°.

    Illustration: Two angles of a triangle are 60° and 40°. Find the third angle.

    SOLUTION: 60° + 40° = 100°
                 180° − 100° = 80°

    Answer: The third angle is 80°.

9. a. A triangle that has two equal sides is called an **isosceles triangle**.

    b. In an isosceles triangle, the angles opposite the equal sides are also equal.

10. a. A triangle that has all three sides equal is called an **equilateral triangle**.

    b. Each angle of an equilateral triangle is 60°.

11. a. A triangle that has a right angle is called a **right triangle**.

    b. In a right triangle, the two acute angles are complementary.

    c. In a right triangle, the side opposite the right angle is called the **hypotenuse** and is the longest side. The other two sides are called **legs**.

Example:

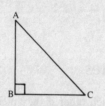

AC is the hypotenuse.
AB and BC are the legs.

12. The **Pythagorean Theorem** states that in a right triangle, the square of the hypotenuse equals the sum of the squares of the legs.

13. To find the hypotenuse of a right triangle when given the legs:

    a. Square each leg.

    b. Add the squares.

    c. Extract the square root of this sum.

Illustration: In a right triangle the legs are 6 inches and 8 inches. Find the hypotenuse.

SOLUTION: $\quad 6^2 = 36 \qquad 8^2 = 64$
$$36 + 64 = 100$$
$$\sqrt{100} = 10$$

Answer: The hypotenuse is 10 inches.

14. To find a leg when given the other leg and the hypotenuse of a right triangle:

    a. Square the hypotenuse and the given leg.

    b. Subtract the square of the leg from the square of the hypotenuse.

    c. Extract the square root of this difference.

Illustration: One leg of a right triangle is 12 feet and the hypotenuse is 20 feet. Find the other leg.

SOLUTION: $\quad 12^2 = 144 \qquad 20^2 = 400$
$$400 - 144 = 256$$
$$\sqrt{256} = 16$$

Answer: The other leg is 16 feet.

Quadrilaterals

15. a. A **quadrilateral** is a closed, four-sided figure in two dimensions. Common quadrilaterals are the **parallelogram**, **rectangle**, and **square**.

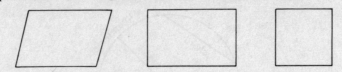

   b. The sum of the four angles of a quadrilateral is 360°.

16. a. A **parallelogram** is a quadrilateral in which both pairs of opposite sides are parallel.

   b. Opposite sides of a parallelogram are also equal.

   c. Opposite angles of a parallelogram are equal.

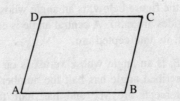

   In parallelogram ABCD,
   AB || CD, AD || BC
   AB = CD, AD = BC
   ∡A = ∡C, ∡B = ∡D

17. A **rectangle** has all of the properties of a parallelogram. In addition, all four of its angles are right angles.

18. A **square** is a rectangle having the additional property that all four of its sides are equal.

Circles

19. A **circle** is a closed plane curve, all points of which are equidistant from a point within called the **center**.

20. a. A **complete circle** contains 360°.

   b. A **semicircle** contains 180°.

21. a. A **chord** is a line segment connecting any two points on the circle.

   b. A **radius** of a circle is a line segment connecting the center with any point on the circle.

   c. A **diameter** is a chord passing through the center of the circle.

   d. A **secant** is a chord extended in either one or both directions.

e. A **tangent** is a line touching a circle at one and only one point.

f. The **circumference** is the curved line bounding the circle.

g. An **arc** of a circle is any part of the circumference.

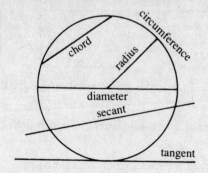

22. a. A **central angle**, as ∡AOB in the figure below, is an angle whose vertex is the center of the circle and whose sides are radii. A central angle is equal to, or has the same number of degrees as, its intercepted arc.

 b. An **inscribed angle**, as ∡MNP, is an angle whose vertex is on the circle and whose sides are chords. An inscribed angle has half the number of degrees as its intercepted arc. ∡MNP intercepts arc MP and has half the degrees of arc MP.

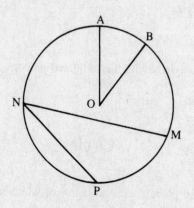

Perimeter

23. The **perimeter** of a two-dimensional figure is the distance around the figure.

Example: The perimeter of the figure above is 9 + 8 + 4 + 3 + 5 = 29.

24. a. The perimeter of a triangle is found by adding all of its sides.

    Example: If the sides of a triangle are 4, 5, and 7, its perimeter is $4 + 5 + 7 = 16$.

    b. If the perimeter and two sides of a triangle are given, the third side is found by adding the two given sides and subtracting this sum from the perimeter.

    Illustration: Two sides of a triangle are 12 and 15, and the perimeter is 37. Find the other side.

    SOLUTION: $12 + 15 = 27$
    $37 - 27 = 10$

    Answer: The third side is 10.

25. The perimeter of a rectangle equals twice the sum of the length and the width. The formula is $P = 2(l + w)$.

    Example: The perimeter of a rectangle whose length is 7 feet and width is 3 feet equals $2 \times 10 = 20$ feet.

26. The perimeter of a square equals one side multiplied by 4. The formula is $P = 4s$.

    Example: The perimeter of a square, one side of which is 5 feet, is 4×5 feet $= 20$ feet.

27. a. The circumference of a circle is equal to the product of the diameter multiplied by π. The formula is $C = \pi d$.

    b. The number π ("pi") is approximately equal to $\frac{22}{7}$, or 3.14 (3.1416 for greater accuracy). A problem will usually state which value to use; otherwise, express the answer in terms of "pi," π.

    Example: The circumference of a circle whose diameter is 4 inches $= 4\pi$ inches; or, if it is stated that $\pi = \frac{22}{7}$, then the circumference is $4 \times \frac{22}{7} = \frac{88}{7} = 12\frac{4}{7}$ inches.

    c. Since the diameter is twice the radius, the circumference equals twice the radius multiplied by π. The formula is $C = 2\pi r$.

    Example: If the radius of a circle is 3 inches, then the circumference $= 6\pi$ inches.

    d. The diameter of a circle equals the circumference divided by π.

    Example: If the circumference of a circle is 11 inches, then, assuming

$$\pi = \tfrac{22}{7},$$
$$\text{diameter} = 11 \div \tfrac{22}{7} \text{ inches}$$
$$= \overset{1}{\cancel{11}} \times \tfrac{7}{\underset{2}{\cancel{22}}} \text{ inches}$$
$$= \tfrac{7}{2} \text{ inches, or } 3\tfrac{1}{2} \text{ inches}$$

Area

28. a. In a figure of two dimensions, the total space within the figure is called the **area**.

    b. Area is expressed in square denominations, such as square inches, square centimeters, and square miles.

    c. In computing area, all dimensions must be expressed in the same denomination.

29. The area of a square is equal to the square of the length of any side. The formula is $A = s^2$.

    Example: The area of a square, one side of which is 6 inches, is $6 \times 6 = 36$ square inches.

30. a. The area of a rectangle equals the product of the length multiplied by the width. The length is any side; the width is the side next to the length. The formula is $A = l \times w$.

    Example: If the length of a rectangle is 6 feet and its width 4 feet, then the area is $6 \times 4 = 24$ square feet.

    b. If given the area of a rectangle and one dimension, divide the area by the given dimension to find the other dimension.

    Example: If the area of a rectangle is 48 square feet and one dimension is 4 feet, then the other dimension is $48 \div 4 = 12$ feet.

31. a. The altitude, or height, of a parallelogram is a line drawn from a vertex perpendicular to the opposite side, or base.

    Example: 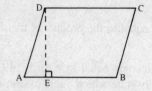 DE is the height.
    AB is the base.

    b. The area of a parallelogram is equal to the product of its base and its height: $A = b \times h$.

    Example: If the base of a parallelogram is 10 centimeters and its height is 5 centimeters, its area is $5 \times 10 = 50$ square centimeters.

    c. If given one of these dimensions and the area, divide the area by the given dimension to find the base or the height of a parallelogram.

    Example: If the area of a parallelogram is 40 square inches and its height is 8 inches, its base is $40 \div 8 = 5$ inches.

32. a. The altitude, or height, of a triangle is a line drawn from a vertex perpendicular to the opposite side, called the base.

    b. The area of a triangle is equal to one-half the product of the base and the height: $A = \frac{1}{2}b \times h$.

    Example: The area of a triangle having a height of 5 inches and a base of 4 inches is $\frac{1}{2} \times 5 \times 4 = \frac{1}{2} \times 20 = 10$ square inches.

    c. In a right triangle, one leg may be considered the height and the other leg the base. Therefore, the area of a right triangle is equal to one-half the product of the legs.

    Example: The legs of a right triangle are 3 and 4. Its area is $\frac{1}{2} \times 3 \times 4 = 6$ square units.

33. a. The area of a circle is equal to the radius squared, multiplied by π: $A = \pi r^2$.

    Example: If the radius of a circle is 6 inches, then the area = 36π square inches.

    b. To find the radius of a circle given the area, divide the area by π and find the square root of the quotient.

    Example: To find the radius of a circle of area 100π:

$$\frac{100\pi}{\pi} = 100$$
$$\sqrt{100} = 10 = \text{radius}.$$

34. Some figures are composed of several geometric shapes. To find the area of such a figure it is necessary to find the area of each of its parts.

    Illustration: Find the area of the figure below:

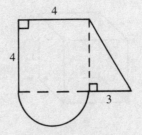

    SOLUTION: The figure is composed of three parts: a square of side 4, a semicircle of diameter 4 (the lower side of the square), and a right triangle with legs 3 and 4 (the right side of the square).

$$\text{Area of square} = 4^2 = 16$$
$$\text{Area of triangle} = \tfrac{1}{2} \times 3 \times 4 = 6$$
$$\text{Area of semicircle is } \tfrac{1}{2} \text{ area of circle} = \tfrac{1}{2}\pi r^2$$
$$\text{Radius} = \tfrac{1}{2} \times 4 = 2$$
$$\text{Area} = \tfrac{1}{2}\pi r^2$$
$$= \tfrac{1}{2} \times \pi \times 2^2$$
$$= 2\pi$$

    Answer: Total area = $16 + 6 + 2\pi = 22 + 2\pi$.

Three-Dimensional Figures

35. a. In a three-dimensional figure, the total space contained within the figure is called the **volume**; it is expressed in **cubic denominations**.

    b. The total outside surface is called the **surface area**; it is expressed in **square denominations**.

    c. In computing volume and surface area, all dimensions must be expressed in the same denomination.

36.   a. A **rectangular solid** is a figure of three dimensions having six rectangular faces

meeting each other at right angles. The three dimensions are length, width and height.

The figure at the left is a rectangular solid; "l" is the length, "w" is the width, and "h" is the height.

b. The volume of a rectangular solid is the product of the length, width, and height; $V = l \times w \times h$.

Example: The volume of a rectangular solid whose length is 6 feet, width 3 feet, and height 4 feet is $6 \times 3 \times 4 = 72$ cubic feet.

37.   a. A **cube** is a rectangular solid whose edges are equal. The figure below is a cube; the length, width, and height are all equal to "e."

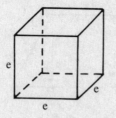

b. The volume of a cube is equal to the edge cubed: $V = e^3$.

Example: The volume of a cube whose height is 6 inches equals $6^3 = 6 \times 6 \times 6 = 216$ cubic inches.

c. The surface area of a cube is equal to the area of any side multiplied by 6.

Example: The surface area of a cube whose length is 5 inches $= 5^2 \times 6 = 25 \times 6 = 150$ square inches.

38.   The volume of a **circular cylinder** is equal to the product of π, the radius squared, and the height.

$$V = \pi r^2 h$$

Example: A circular cylinder has a radius of 7 inches and a height of $\frac{1}{2}$ inch. Using $\pi = \frac{22}{7}$, its volume is

$$\tfrac{22}{7} \times 7 \times 7 \times \tfrac{1}{2} = 77 \text{ cubic inches}$$

39. The volume of a **sphere** is equal to $\frac{4}{3}$ the product of π and the radius cubed.

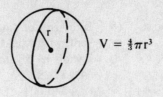

$$V = \tfrac{4}{3}\pi r^3$$

Example: If the radius of a sphere is 3 cm, its volume in terms of π is
$$\tfrac{4}{3} \times \pi \times 3 \text{ cm} \times 3 \text{ cm} \times 3 \text{ cm} = 36\pi \text{ cm}^3$$

40. The volume of a **cone** is given by the formula $V = \frac{1}{3}\pi r^2 h$, where r is the radius and h is the height.

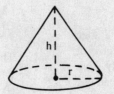

Example: In the cone shown, if h = 9 cm, r = 10 cm, and π = 3.14, then the volume is

$$\tfrac{1}{3} \times 3.14 \times 10 \times 10 \times 9 \text{ cm}^3 = 3.14 \times 300 \text{ cm}^3$$
$$= 942 \text{ cm}^3$$

41. The volume of a **pyramid** is given by the formula $V = \frac{1}{3}Bh$, where B is the area of the base, and h is the height.

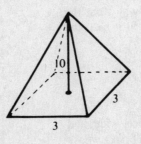

Example: In the pyramid shown, the height is 10″ and the side of the base is 3″. Since the base is a square, B = 3^2 = 9 square inches.
$$V = \tfrac{1}{3} \times 9 \times 10 = 30 \text{ cubic inches}$$

Practice Problems Involving Geometry

1. If the perimeter of a rectangle is 68 yards and the width is 48 feet, the length is
   (A) 10 yd   (C) 20 ft
   (B) 18 yd   (D) 56 ft

2. The total length of fencing needed to enclose a rectangular area 46 feet by 34 feet is
   (A) 26 yd 1 ft   (C) 52 yd 2 ft
   (B) $26\frac{2}{3}$ yd   (D) $53\frac{1}{3}$ yd

3. An umbrella 50″ long can lie on the bottom of a trunk whose length and width are, respectively,
   (A) 36″, 30″   (C) 42″, 36″
   (B) 42″, 24″   (D) 39″, 30″

4. A road runs 1200 ft from A to B, and then makes a right angle going to C, a distance of 500 ft. A new road is being built directly from A to C. How much shorter will the new road be?
   (A) 400 ft   (C) 850 ft
   (B) 609 ft   (D) 1300 ft

5. A certain triangle has sides that are, respectively, 6 inches, 8 inches, and 10 inches long. A rectangle equal in area to that of the triangle has a width of 3 inches. The perimeter of the rectangle, expressed in inches, is
   (A) 11   (C) 22
   (B) 16   (D) 24

6. A ladder 65 feet long is leaning against the wall. Its lower end is 25 feet away from the wall. How much further away will it be if the upper end is moved down 8 feet?
   (A) 60 ft   (C) 14 ft
   (B) 52 ft   (D) 10 ft

7. A rectangular bin 4 feet long, 3 feet wide, and 2 feet high is solidly packed with bricks whose dimensions are 8 inches, 4 inches, and 2 inches. The number of bricks in the bin is
   (A) 54   (C) 1296
   (B) 648   (D) none of these

8. If the cost of digging a trench is $2.12 a cubic yard, what would be the cost of digging a trench 2 yards by 5 yards by 4 yards?
   (A) $21.20   (C) $64.00
   (B) $40.00   (D) $84.80

9. A piece of wire is shaped to enclose a square, whose area is 121 square inches. It is then reshaped to enclose a rectangle whose length is 13 inches. The area of the rectangle, in square inches, is
   (A) 64   (C) 117
   (B) 96   (D) 144

10. The area of a 2-foot-wide walk around a garden that is 30 feet long and 20 feet wide is
    (A) 104 sq ft   (C) 680 sq ft
    (B) 216 sq ft   (D) 704 sq ft

11. The area of a circle is 49π. Find its circumference, in terms of π.
    (A) 14π (C) 49π
    (B) 28π (D) 98π

12. In two hours, the minute hand of a clock rotates through an angle of
    (A) 90°   (C) 360°
    (B) 180°   (D) 720°

13. A box is 12 inches in width, 16 inches in length, and 6 inches in height. How many square inches of paper would be required to cover it on all sides?
    (A) 192   (C) 720
    (B) 360   (D) 1440

14. If the volume of a cube is 64 cubic inches, the sum of its edges is
    (A) 48 in   (C) 16 in
    (B) 32 in   (D) 24 in

15. The diameter of a conical pile of cement is 30 feet and its height is 14 feet. If $\frac{3}{4}$ cubic yard of cement weighs 1 ton, the number of tons of cement in the cone to the nearest ton is

    (Volume of cone $= \frac{1}{3}\pi r^2 h$; use $\pi = \frac{22}{7}$)

    (A) 92   (C) 489
    (B) 163   (D) 652

Geometry Problems — Correct Answers

1. **(B)**	6. **(C)**	11. **(A)**
2. **(D)**	7. **(B)**	12. **(D)**
3. **(C)**	8. **(D)**	13. **(C)**
4. **(A)**	9. **(C)**	14. **(A)**
5. **(C)**	10. **(B)**	15. **(B)**

Problem Solutions — Geometry

1.

Perimeter = 68 yards
Each width = 48 feet = 16 yards
Both widths = 16 yd + 16 yd = 32 yd
Perimeter = sum of all sides
Remaining two sides must total 68 − 32 = 36 yards.
Since the remaining two sides are equal, they are each 36 ÷ 2 = 18 yards.

Answer: **(B)** 18 yd

2. Perimeter = 2(46 + 34) feet
   = 2 × 80 feet
   = 160 feet
   160 feet = 160 ÷ 3 yards = $53\frac{1}{3}$ yards

Answer: **(D)** $53\frac{1}{3}$ yd

3. The umbrella would be the hypotenuse of a right triangle whose legs are the dimensions of the trunk.

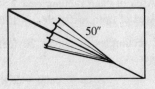

The Pythagorean Theorem states that in a right triangle, the square of the hypotenuse equals the sum of the squares of the legs. Therefore, the sum of the dimensions of the trunk squared must at least equal the length of the umbrella squared, which is 50^2 or 2500.

The only set of dimensions filling this condition is **(C)**:

$$(42)^2 + (36)^2 = 1764 + 1296$$
$$= 3060$$

Answer: **(C)** 42″, 36″

4. The new road is the hypotenuse of a right triangle, whose legs are the old road.

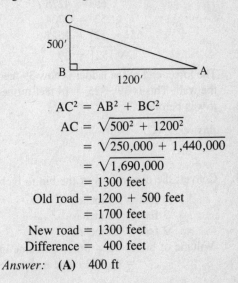

$$AC^2 = AB^2 + BC^2$$
$$AC = \sqrt{500^2 + 1200^2}$$
$$= \sqrt{250,000 + 1,440,000}$$
$$= \sqrt{1,690,000}$$
$$= 1300 \text{ feet}$$

Old road = 1200 + 500 feet
= 1700 feet
New road = 1300 feet
Difference = 400 feet

Answer: **(A)** 400 ft

5. Since $6^2 + 8^2 = 10^2$ (36 + 64 = 100), the triangle is a right triangle. The area of the triangle is $\frac{1}{2} \times 6 \times 8 = 24$ square inches. Therefore, the area of the rectangle is 24 square inches.

If the width of the rectangle is 3 inches, the length is $24 \div 3 = 8$ inches. Then the perimeter of the rectangle is $2(3 + 8) = 2 \times 11 = 22$ inches.

Answer: **(C)** 22

6. The ladder forms a right triangle with the wall and the ground.

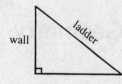

First, find the height that the ladder reaches when the lower end of the ladder is 25 feet from the wall:

$$65^2 = 4225$$
$$25^2 = 625$$
$$65^2 - 25^2 = 3600$$
$$\sqrt{3600} = 60$$

The ladder reaches 60 feet up the wall when its lower end is 25 feet from the wall.

If the upper end is moved down 8 feet, the ladder will reach a height of $60 - 8 = 52$ feet.

The new triangle formed has a hypotenuse of 65 feet and one leg of 52 feet. Find the other leg:

$$65^2 = 4225$$
$$52^2 = 2704$$
$$65^2 - 52^2 = 1521$$
$$\sqrt{1521} = 39$$

The lower end of the ladder is now 39 feet from the wall. This is $39 - 25 = 14$ feet further than it was before.

Answer: **(C)** 14 ft

7. Convert the dimensions of the bin to inches:
$$4 \text{ feet} = 48 \text{ inches}$$
$$3 \text{ feet} = 36 \text{ inches}$$
$$2 \text{ feet} = 24 \text{ inches}$$
Volume of bin = $48 \times 36 \times 24$ cubic inches
= 41,472 cubic inches

Volume of
each brick = $8 \times 4 \times 2$ cubic inches
= 64 cubic inches
$41,472 \div 64 = 648$ bricks

Answer: **(B)** 648

8. The trench contains
$$2 \text{ yd} \times 5 \text{ yd} \times 4 \text{ yd} = 40 \text{ cubic yards}$$
$$40 \times \$2.12 = \$84.80$$

Answer: **(D)** \$84.80

9. Find the dimensions of the square: If the area of the square is 121 square inches, each side is $\sqrt{121} = 11$ inches, and the perimeter is $4 \times 11 = 44$ inches.

Next, find the dimensions of the rectangle: The perimeter of the rectangle is the same as the perimeter of the square, since the same length of wire is used to enclose either figure. Therefore, the perimeter of the rectangle is 44 inches. If the two lengths are each 13 inches, their total is 26 inches, and $44 - 26$ inches, or 18 inches, remain for the two widths. Each width is equal to $18 \div 2 = 9$ inches.

The area of a rectangle with length 13 in and width 9 in is $13 \times 9 = 117$ sq in.

Answer: **(C)** 117

10.

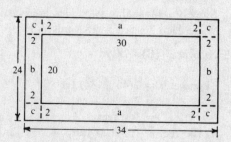

The walk consists of:

a. 2 rectangles of length 30 feet and width 2 feet.

Area of each rectangle = $2 \times 30 = 60$ sq ft
Area of both rectangles = 120 sq ft

b. 2 rectangles of length 20 feet and width 2 feet.

Area of each = $2 \times 20 = 40$ sq ft
Area of both = 80 sq ft

c. 4 squares, each having sides measuring 2 feet.

Area of each square = 2^2 = 4 sq ft
Area of 4 squares = 16 sq ft

Total area of walk = 120 + 80 + 16
= 216 sq ft

Alternate solution:

Area of walk = Area of large rectangle
− area of small rectangle
= 34 × 24 − 30 × 20
= 816 − 600
= 216 sq ft

Answer: **(B)** 216 sq ft

11. If the area of a circle is 49π, its radius is $\sqrt{49}$ = 7. Then, the circumference is equal to $2 \times 7 \times \pi = 14\pi$.

Answer: **(A)** 14π

12. In one hour, the minute hand rotates through 360°. In two hours, it rotates through 2 × 360° = 720°.

Answer: **(D)** 720°

13. Find the area of each surface:

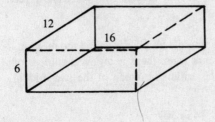

Area of top = 12 × 16 = 192 sq in
Area of bottom = 12 × 16 = 192 sq in
Area of front = 6 × 16 = 96 sq in
Area of back = 6 × 16 = 96 sq in
Area of right side = 6 × 12 = 72 sq in
Area of left side = 6 × 12 = + 72 sq in
Total surface area = 720 sq in

Answer: **(C)** 720

14. For a cube, V = e^3. If the volume is 64 cubic inches, each edge is $\sqrt[3]{64}$ = 4 inches.
A cube has 12 edges. If each edge is 4 inches, the sum of the edges is 4 × 12 = 48 inches.

Answer: **(A)** 48 in

15. If the diameter = 30 feet, the radius = 15 feet.

$$V = \tfrac{1}{2} \times \tfrac{22}{7} \times 15 \times 15 \times 14$$
= 3300 cubic feet

27 cubic feet = 1 cubic yard
3300 ÷ 27 cu ft = $122\tfrac{2}{9}$ cu yd
$122\tfrac{2}{9} \div \tfrac{3}{4} = \tfrac{1100}{9} \times \tfrac{4}{3} = \tfrac{4400}{27}$ = 163 tons to the nearest ton

Answer: **(B)** 163

STATISTICS AND PROBABILITY

Statistics

1. The **averages** used in statistics include the **arithmetic mean**, the **median** and the **mode**.

2. a. The most commonly used average of a group of numbers is the **arithmetic mean**. It is found by adding the numbers given and then dividing this sum by the number of items being averaged.

   Illustration: Find the arithmetic mean of 2, 8, 5, 9, 6, and 12.

   SOLUTION: There are 6 numbers.

   $$\text{Arithmetic mean} = \frac{2 + 8 + 5 + 9 + 6 + 12}{6}$$
   $$= \frac{42}{6}$$
   $$= 7$$

   Answer: The arithmetic mean is 7.

   b. If a problem calls for simply the "average" or the "mean," it is referring to the arithmetic mean.

3. If a group of numbers is arranged in order, the middle number is called the **median**. If there is no single middle number (this occurs when there is an even number of items), the median is found by computing the arithmetic mean of the two middle numbers.

   Example: The median of 6, 8, 10, 12, and 14 is 10.

   Example: The median of 6, 8, 10, 12, 14, and 16 is the arithmetic mean of 10 and 12.

   $$\frac{10 + 12}{2} = \frac{22}{2} = 11.$$

4. The **mode** of a group of numbers is the number that appears most often.

   Example: The mode of 10, 5, 7, 9, 12, 5, 10, 5 and 9 is 5.

5. To obtain the average of quantities that are weighted:

   a. Set up a table listing the quantities, their respective weights, and their respective values.

   b. Multiply the value of each quantity by its respective weight.

   c. Add up these products.

   d. Add up the weights.

   e. Divide the sum of the products by the sum of the weights.

70

Illustration: Assume that the weights for the following subjects are: English 3, History 2, Mathematics 2, Foreign Languages 2, and Art 1. What would be the average of a student whose marks are: English 80, History 85, Algebra 84, Spanish 82, and Art 90?

SOLUTION:

Subject	Weight	Mark
English	3	80
History	2	85
Algebra	2	84
Spanish	2	82
Art	1	90
English	$3 \times 80 = 240$	
History	$2 \times 85 = 170$	
Algebra	$2 \times 84 = 168$	
Spanish	$2 \times 82 = 164$	
Art	$1 \times 90 = 90$	
	832	

Sum of the weights: $3 + 2 + 2 + 2 + 1 = 10$
$$832 \div 10 = 83.2$$

Answer: Average = 83.2

Probability

6. The study of probability deals with predicting the outcome of chance events; that is, events in which one has no control over the results.

Examples: Tossing a coin, rolling dice, and drawing concealed objects from a bag are chance events.

7. The probability of a particular outcome is equal to the number of ways that outcome can occur, divided by the total number of possible outcomes.

Example: In tossing a coin, there are 2 possible outcomes: heads or tails. The probability that the coin will turn up heads is $1 \div 2$ or $\frac{1}{2}$.

Example: If a bag contains 5 balls of which 3 are red, the probability of drawing a red ball is $\frac{3}{5}$. The probability of drawing a non-red ball is $\frac{2}{5}$.

8. a. If an event is certain, its probability is 1.
Example: If a bag contains only red balls, the probability of drawing a red ball is 1.

b. If an event is impossible, its probability is 0.
Example: If a bag contains only red balls, the probability of drawing a green ball is 0.

9. Probability may be expressed in fractional, decimal, or percent form.

Example: An event having a probability of $\frac{1}{2}$ is said to be 50% probable.

10. A probability determined by random sampling of a group of items is assumed to apply to other items in that group and in other similar groups.

Illustration: A random sampling of 100 items produced in a factory shows that 7 are defective. How many items of the total production of 50,000 can be expected to be defective?

SOLUTION: The probability of an item's being defective is $\frac{7}{100}$, or 7%. Of the total production, 7% can be expected to be defective.

$$7\% \times 50,000 = .07 \times 50,000 = 3500$$

Answer: 3500 items

Practice Problems Involving Statistics and Probability

1. The arithmetic mean of 73.8, 92.2, 64.7, 43.8, 56.5, and 46.4 is
   (A) 60.6
   (B) 62.9
   (C) 64.48
   (D) 75.48

2. The median of the numbers 8, 5, 7, 5, 9, 9, 1, 8, 10, 5, and 10 is
   (A) 5
   (B) 7
   (C) 8
   (D) 9

3. The mode of the numbers 16, 15, 17, 12, 15, 15, 18, 19, and 18 is
   (A) 15
   (B) 16
   (C) 17
   (D) 18

4. A clerk filed 73 forms on Monday, 85 forms on Tuesday, 54 on Wednesday, 92 on Thursday, and 66 on Friday. What was the average number of forms filed per day?
   (A) 60
   (B) 72
   (C) 74
   (D) 92

5. The grades received on a test by twenty students were: 100, 55, 75, 80, 65, 65, 85, 90, 80, 45, 40, 50, 85, 85, 85, 80, 80, 70, 65, and 60. The average of these grades is
   (A) 70
   (B) 72
   (C) 77
   (D) 80

6. A buyer purchased 75 six-inch rulers costing 15¢ each, 100 one-foot rulers costing 30¢ each, and 50 one-yard rulers costing 72¢ each. What was the average price per ruler?
   (A) $26\frac{1}{8}$¢
   (B) $34\frac{1}{3}$¢
   (C) 39¢
   (D) 42¢

7. What is the average of a student who received 90 in English, 84 in Algebra, 75 in French, and 76 in Music, if the subjects have the following weights: English 4, Algebra 3, French 3, and Music 1?
   (A) 81
   (B) $81\frac{1}{2}$
   (C) 82
   (D) 83

Questions 8–11 refer to the following information:

A census shows that on a certain block the number of children in each family is 3, 4, 4, 0, 1, 2, 0, 2, and 2, respectively.

8. Find the average number of children per family.
   (A) 2
   (B) $2\frac{1}{2}$
   (C) 3
   (D) $3\frac{1}{2}$

9. Find the median number of children.
   (A) 1
   (B) 2
   (C) 3
   (D) 4

10. Find the mode of the number of children.
    (A) 0   (C) 2
    (B) 1   (D) 4

11. What is the probability that a family chosen at random on this block will have 4 children?
    (A) $\frac{4}{9}$ (C) $\frac{4}{7}$
    (B) $\frac{2}{9}$ (D) $\frac{2}{1}$

12. What is the probability that an even number will come up when a single die is thrown?
    (A) $\frac{1}{6}$ (C) $\frac{1}{2}$
    (B) $\frac{1}{3}$ (D) 1

13. A bag contains 3 black balls, 2 yellow balls, and 4 red balls. What is the probability of drawing a black ball?
    (A) $\frac{1}{2}$ (C) $\frac{2}{3}$
    (B) $\frac{1}{3}$ (D) $\frac{4}{9}$

14. In a group of 1000 adults, 682 are women. What is the probability that a person chosen at random from this group will be a man?
    (A) .318   (C) .5
    (B) .682   (D) 1

15. In a balloon factory, a random sampling of 100 balloons showed that 3 had pinholes in them. In a sampling of 2500 balloons, how many may be expected to have pinholes?
    (A) 30   (C) 100
    (B) 75   (D) 750

Statistics and Probability Problems — Correct Answers

1. **(B)**
2. **(C)**
3. **(A)**
4. **(C)**
5. **(B)**
6. **(B)**
7. **(D)**
8. **(A)**
9. **(B)**
10. **(C)**
11. **(B)**
12. **(C)**
13. **(B)**
14. **(A)**
15. **(B)**

Problem Solutions — Statistics and Probability

1. Find the sum of the values:

   $73.8 + 92.2 + 64.7 + 43.8 + 56.5 + 46.4 = 377.4$

   There are 6 values.

   $$\text{Arithmetic mean} = \frac{377.4}{6} = 62.9$$

   Answer: **(B)** 62.9

2. Arrange the numbers in order:

   1, 5, 5, 5, 7, 8, 8, 9, 9, 10, 10

   The middle number, or median, is 8.

   Answer: **(C)** 8

3. The mode is that number appearing most frequently. The number 15 appears three times.

   Answer: **(A)** 15

4. Average $= \dfrac{73 + 85 + 54 + 92 + 66}{5}$

   $= \dfrac{370}{5}$

   $= 74$

   Answer: **(C)** 74

5. Sum of the grades $= 1440$.

   $\dfrac{1440}{20} = 72$

   Answer: **(B)** 72

6. $\begin{array}{r} 75 \times 15\text{¢} = 1125\text{¢} \\ 100 \times 30\text{¢} = 3000\text{¢} \\ \underline{50 \times 72\text{¢}} = \underline{3600\text{¢}} \\ 225 \quad\quad 7725\text{¢} \end{array}$

   $\dfrac{7725\text{¢}}{225} = 34\tfrac{1}{3}\text{¢}$

   Answer: **(B)** $34\tfrac{1}{3}$¢

7.
Subject	Grade	Weight
English	90	4
Algebra	84	3
French	75	3
Music	76	1

   $(90 \times 4) + (84 \times 3) + (75 \times 3) + (76 \times 1)$
   $360 + 252 + 225 + 76 = 913$
   Weight $= 4 + 3 + 3 + 1 = 11$
   $913 \div 11 = 83$ average

   Answer: **(D)** 83

8. Average $= \dfrac{3+4+4+0+1+2+0+2+2}{9}$

   $= \dfrac{18}{9}$

   $= 2$

   Answer: **(A)** 2

9. Arrange the numbers in order:

   $$0, 0, 1, 2, 2, 2, 3, 4, 4$$

   Of the 9 numbers, the fifth (middle) number is 2.

   Answer: **(B)** 2

10. The number appearing most often is 2.

    Answer: **(C)** 2

11. There are 9 families, 2 of which have 4 children. The probability is $\tfrac{2}{9}$.

    Answer: **(B)** $\tfrac{2}{9}$

12. Of the 6 possible numbers, three are even (2, 4, and 6). The probability is $\tfrac{3}{6}$, or $\tfrac{1}{2}$.

    Answer: **(C)** $\tfrac{1}{2}$

13. There are 9 balls in all. The probability of drawing a black ball is $\tfrac{3}{9}$, or $\tfrac{1}{3}$.

    Answer: **(B)** $\tfrac{1}{3}$

14. If 682 people of the 1000 are women, $1000 - 682 = 318$ are men. The probability of choosing a man is $\tfrac{318}{1000} = .318$.

    Answer: **(A)** .318

15. There is a probability of $\tfrac{3}{100} = 3\%$ that a balloon may have a pinhole.

    $$3\% \times 2500 = 75.00$$

    Answer: **(B)** 75

GRAPHS

1. **Graphs** illustrate comparisons and trends in statistical information. The most commonly used graphs are **bar graphs**, **line graphs**, **circle graphs**, and **pictographs**.

Bar Graphs

2. **Bar graphs** are used to compare various quantities. Each bar may represent a single quantity or may be divided to represent several quantities.

3. Bar graphs may have horizontal or vertical bars.

Illustration:

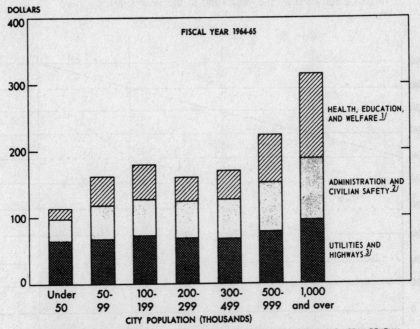

Municipal Expenditures, Per Capita

1/PUBLIC WELFARE, EDUCATION, HOSPITALS, HEALTH, LIBRARIES, AND HOUSING AND URBAN RENEWAL.
2/POLICE AND FIRE PROTECTION, FINANCIAL ADMINISTRATION, GENERAL CONTROL, GENERAL PUBLIC BUILDINGS, INTEREST ON GENERAL DEBT, AND OTHER.
3/HIGHWAYS, SEWERAGE, SANITATION, PARKS AND RECREATION, AND UTILITIES.
SOURCE: DEPARTMENT OF COMMERCE.

Question 1: What was the approximate municipal expenditure per capita in cities having populations of 200,000 to 299,000?

Answer: The middle bar of the seven shown represents cities having populations from 200,000 to 299,000. This bar reaches about halfway between 100 and 200. Therefore, the per capita expenditure was approximately $150.

Question 2: Which cities spent the most per capita on health, education, and welfare?

Answer: The bar for cities having populations of 1,000,000 and over has a larger striped section than the other bars. Therefore, those cities spent the most.

Question 3: Of the three categories of expenditures, which was least dependent on city size?

Answer: The expenditures for utilities and highways, the darkest part of each bar, varied least as city size increased.

Line Graphs

4. **Line graphs** are used to show trends, often over a period of time.

5. A line graph may include more than one line, with each line representing a different item.

Illustration:

The graph below indicates at 5 year intervals the number of citations issued for various offenses from the year 1960 to the year 1980.

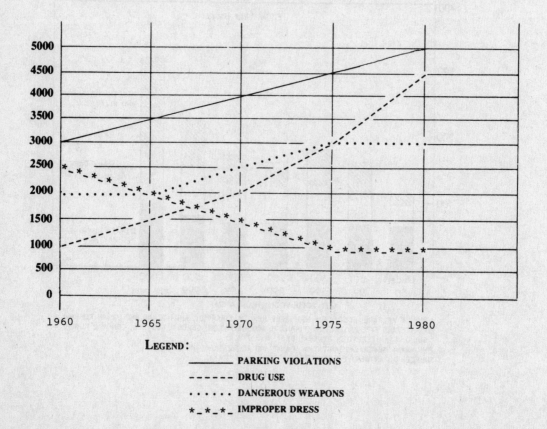

LEGEND:

——————— PARKING VIOLATIONS

– – – – – – DRUG USE

• • • • • • DANGEROUS WEAPONS

––*– IMPROPER DRESS

Question 4: Over the 20-year period, which offense shows an average rate of increase of more than 150 citations per year?

Answer: Drug use citations increased from 1000 in 1960 to 4500 in 1980. The average increase over the 20-year period is $\frac{3500}{20} = 175$.

Question 5: Over the 20-year period, which offense shows a constant rate of increase or decrease?

Answer: A straight line indicates a constant rate of increase or decrease. Of the four lines, the one representing parking violations is the only straight one.

Question 6: Which offense shows a total increase or decrease of 50% for the full 20-year period?

Answer: Dangerous weapons citations increased from 2000 in 1960 to 3000 in 1980, which is an increase of 50%.

Circle Graphs

6. **Circle graphs** are used to show the relationship of various parts of a quantity to each other and to the whole quantity.

7. Percents are often used in circle graphs. The 360 degrees of the circle represents 100%.

8. Each part of the circle graph is called a **sector**.

Illustration:

The following circle graph shows how the federal budget of $300.4 billion was spent.

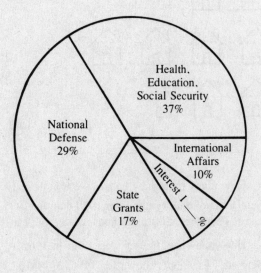

Question 7: What is the value of I?

Answer: There must be a total of 100% in a circle graph. The sum of the other sectors is:

$$17\% + 29\% + 37\% + 10\% = 93\%$$

Therefore, I = 100% − 93% = 7%.

Question 8: How much money was actually spent on national defense?

Answer: 29% × $300.4 billion = $87.116 billion
= $87,116,000,000

Question 9: How much more money was spent on state grants than on interest?

Answer: 17% − 7% = 10%
10% × $300.4 billion = $30.04 billion
= $30,040,000,000

Pictographs

9. **Pictographs** allow comparisons of quantities by using symbols. Each symbol represents a given number of a particular item.

Illustration:

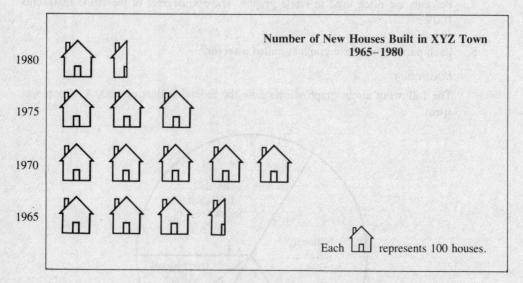

Number of New Houses Built in XYZ Town
1965–1980

Each represents 100 houses.

Question 10: How many more new houses were built in 1970 than in 1975?

Answer: There are two more symbols for 1970 than for 1975. Each symbol represents 100 houses. Therefore, 200 more houses were built in 1970.

Question 11: How many new houses were built in 1965?

Answer: There are $3\frac{1}{2}$ symbols shown for 1965; $3\frac{1}{2} \times 100 = 350$ houses.

Question 12: In which year were half as many houses built as in 1975?

Answer: In 1975, 3 × 100 = 300 houses were built. Half of 300, or 150, houses were built in 1980.

Practice Problems Involving Graphs

Questions 1–4 refer to the following graph:

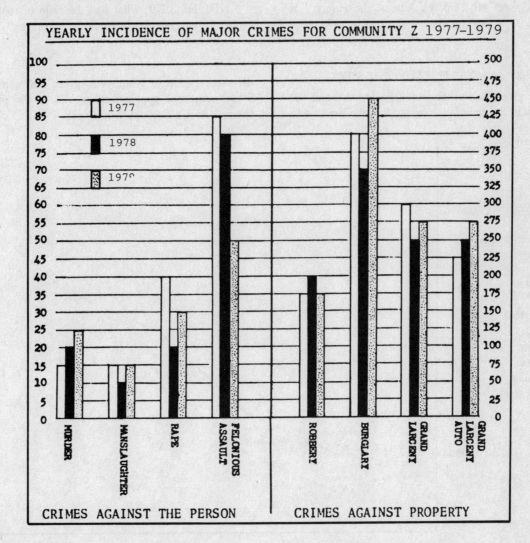

1. In 1979, the incidence of which of the following crimes was greater than in the previous two years?
   (A) grand larceny   (C) rape
   (B) murder   (D) robbery

2. If the incidence of burglary in 1980 had increased over 1979 by the same number as it had increased in 1979 over 1978, then the average

for this crime for the four-year period from 1977 through 1980 would be most nearly
   (A) 100   (C) 425
   (B) 400   (D) 440

3. The above graph indicates that the *percentage* increase in grand larceny auto from 1978 to 1979 was:
   (A) 5%   (C) 15%
   (B) 10%   (D) 20%

4. Which of the following cannot be determined because there is not enough information in the above graph to do so?

   (A) For the three-year period, what percentage of all "Crimes Against the Person" involved murders committed in 1978?

   (B) For the three-year period, what percentage of all "Major Crimes" was committed in the first six months of 1978?

   (C) Which major crimes followed a pattern of continuing yearly increases for the three-year period?

   (D) For 1979, what was the ratio of robbery, burglary, and grand larceny crimes?

Questions 5–7 refer to the following graph:

In the graph below, the lines labeled "A" and "B" represent the cumulative progress in the work of two file clerks, each of whom was given 500 consecutively numbered applications to file in the proper cabinets over a five-day work week.

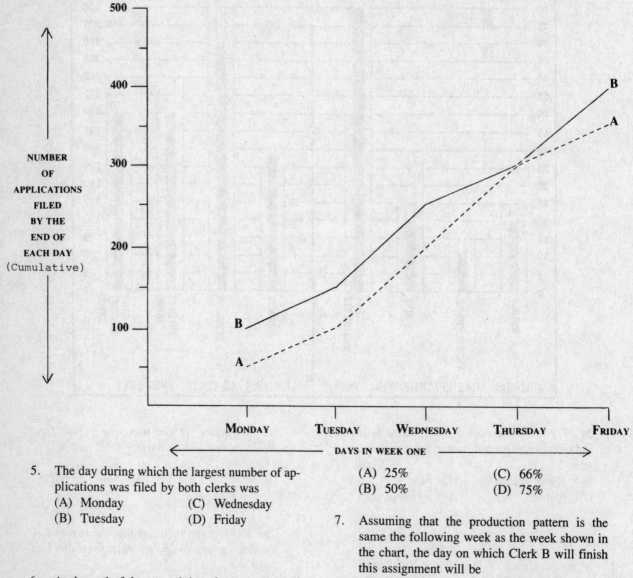

5. The day during which the largest number of applications was filed by both clerks was
   (A) Monday   (C) Wednesday
   (B) Tuesday   (D) Friday

6. At the end of the second day, the percentage of applications still to be filed was
   (A) 25%   (C) 66%
   (B) 50%   (D) 75%

7. Assuming that the production pattern is the same the following week as the week shown in the chart, the day on which Clerk B will finish this assignment will be
   (A) Monday   (C) Wednesday
   (B) Tuesday   (D) Friday

Questions 8–11 refer to the following graph:

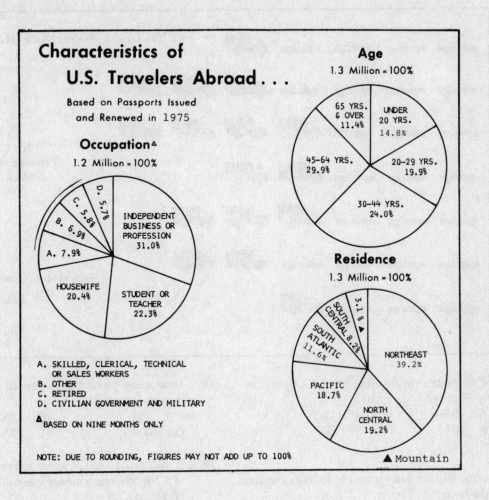

8. Approximately how many persons aged 29 or younger traveled abroad in 1975?
   - (A) 175,000
   - (B) 245,000
   - (C) 385,000
   - (D) 450,000

9. Of the people who did *not* live in the Northeast, what percent came from the North Central states?
   - (A) 19.2%
   - (B) 19.9%
   - (C) 26.5%
   - (D) 31.6%

10. The fraction of travelers from the four smallest occupation groups is most nearly equal to the fraction of travelers
   - (A) under age 20, and 65 and over, combined
   - (B) from the North Central and Mountain states
   - (C) between 45 and 64 years of age
   - (D) from the Housewife and Other categories

11. If the South Central, Mountain, and Pacific sections were considered as a single classification, how many degrees would its sector include?
   - (A) 30°
   - (B) 67°
   - (C) 108°
   - (D) 120°

Questions 12–15 refer to the following graph:

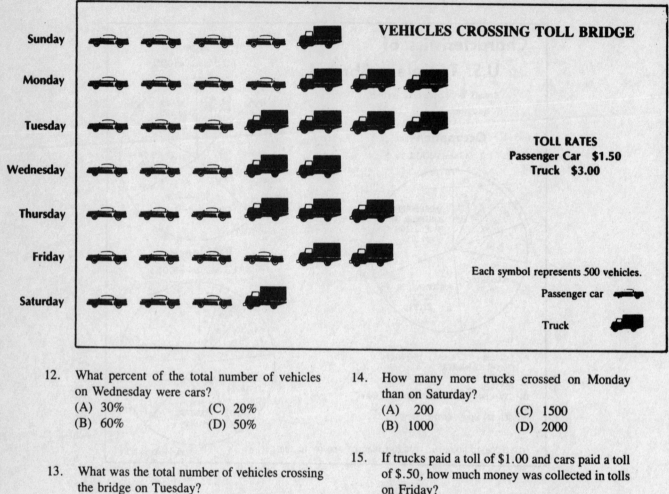

VEHICLES CROSSING TOLL BRIDGE

TOLL RATES
Passenger Car $1.50
Truck $3.00

Each symbol represents 500 vehicles.

Passenger car

Truck

12. What percent of the total number of vehicles on Wednesday were cars?
(A) 30%   (C) 20%
(B) 60%   (D) 50%

13. What was the total number of vehicles crossing the bridge on Tuesday?
(A) 7   (C) 1100
(B) 700   (D) 3500

14. How many more trucks crossed on Monday than on Saturday?
(A) 200   (C) 1500
(B) 1000   (D) 2000

15. If trucks paid a toll of $1.00 and cars paid a toll of $.50, how much money was collected in tolls on Friday?
(A) $400   (C) $2000
(B) $600   (D) $2500

Graphs — Correct Answers

1. **(B)**   6. **(D)**   11. **(C)**
2. **(D)**   7. **(A)**   12. **(B)**
3. **(B)**   8. **(D)**   13. **(D)**
4. **(B)**   9. **(D)**   14. **(B)**
5. **(C)**   10. **(A)**   15. **(C)**

Problem Solutions — Graphs

1. The incidence of murder increased from 15 in 1977 to 20 in 1978 to 25 in 1979.

   Answer: **(B)** murder

2. The incidence of burglary in 1977 was 400; in 1978 it was 350; and in 1979 it was 450. The increase from 1978 to 1979 was 100. An increase of 100 from 1979 gives 550 in 1980.
   The average of 400, 350, 450, and 550 is

   $$\frac{400 + 350 + 450 + 550}{4} = \frac{1750}{4}$$
   $$= 437.5$$

   Answer: **(D)** 440

3. The incidence of grand larceny auto went from 250 in 1978 to 275 in 1979, an increase of 25.
   The percent increase is

   $$\frac{25}{250} = .10 = 10\%$$

   Answer: **(B)** 10%

4. This graph gives information by year, not month. It is impossible to determine from the graph the percentage of crimes committed during the first six months of any year.

   Answer: **(B)**

5. For both A and B, the greatest increase in the cumulative totals occurred from the end of Tuesday until the end of Wednesday. Therefore, the largest number of applications was filed on Wednesday.

   Answer: **(C)** Wednesday

6. By the end of Tuesday, A had filed 100 applications and B had filed 150, for a total of 250. This left 750 of the original 1000 applications.

   $$\frac{750}{1000} = .75 = 75\%$$

   Answer: **(D)** 75%

7. During Week One, Clerk B files 100 applications on Monday, 50 on Tuesday, 100 on Wednesday, 50 on Thursday, and 100 on Friday, a total of 400. On Monday of Week Two, he will file numbers 401 to 500.

   Answer: **(A)** Monday

8.
20–29 yrs.:	19.9%
Under 20 yrs.:	+14.8%
	34.7%

   34.7% × 1.3 million = .4511 million
   = 451,100

   Answer: **(D)** 450,000

9. 100% − 39.2% = 60.8% did not live in Northeast.
   19.2% lived in North Central

   $$\frac{19.2}{60.8} = .316 \text{ approximately}$$

   Answer: **(D)** 31.6%

10. Four smallest groups of occupation:
    7.9 + 6.9 + 5.8 + 5.7 = 26.3
    Age groups under 20 and over 65:
    14.8 + 11.4 = 26.2

    Answer: **(A)**

11.
South Central:	8.2%
Mountain:	3.1%
Pacific:	18.7%
	30.0%

    30% × 360° = 108°

    Answer: **(C)** 108°

12. There are 5 vehicle symbols, of which 3 are cars. $\frac{3}{5}$ = 60%

    Answer: **(B)** 60%

13. On Tuesday, there were 3 × 500 = 1500 cars and 4 × 500 = 2000 trucks. The total number of vehicles was 3500.

    Answer: **(D)** 3500

14. The graph shows 2 more truck symbols on Monday than on Saturday. Each symbols represents 500 trucks, so there were 2 × 500 = 1000 more trucks on Monday.

    Answer: **(B)** 1000

15. On Friday there were

    $$4 \times 500 = 2000 \text{ cars}$$
    $$2 \times 500 = 1000 \text{ trucks}$$

Car tolls:	2000 × $ .50 =	$1000
Truck tolls:	1000 × $1.00 =	+ $1000
Total tolls:		$2000

    Answer: **(C)** $2000

RATIO AND PROPORTION

Ratio

1. A **ratio** expresses the relationship between two (or more) quantities in terms of numbers. The mark used to indicate ratio is the colon (:) and is read "to."

   Example: The ratio 2:3 is read "2 to 3."

2. A ratio also represents division. Therefore, any ratio of two terms may be written as a fraction, and any fraction may be written as a ratio.

   Example: $3:4 = \frac{3}{4}$
   $\frac{5}{6} = 5:6$

3. To simplify any complicated ratio of two terms containing fractions, decimals, or per cents:

   a. Divide the first term by the second.

   b. Write as a fraction in lowest terms.

   c. Write the fraction as a ratio.

   Illustration: Simplify the ratio $\frac{5}{6} : \frac{7}{8}$

   SOLUTION: $\frac{5}{6} \div \frac{7}{8} = \frac{5}{6} \times \frac{8}{7} = \frac{20}{21}$
   $\frac{20}{21} = 20:21$

   Answer: 20:21

4. To solve problems in which the ratio is given:

   a. Add the terms in the ratio.

   b. Divide the total amount that is to be put into a ratio by this sum.

   c. Multiply each term in the ratio by this quotient.

   Illustration: The sum of $360 is to be divided among three people according to the ratio 3:4:5. How much does each one receive?

   SOLUTION: $3 + 4 + 5 = 12$
   $\$360 \div 12 = \$30$
   $\$30 \times 3 = \$90$
   $\$30 \times 4 = \$120$
   $\$30 \times 5 = \$150$

   Answer: The money is divided thus: $90, $120, $150.

Proportion

5. a. A **proportion** indicates the equality of two ratios.

   Example: 2:4 = 5:10 is a proportion. This is read "2 is to 4 as 5 is to 10."

   b. In a proportion, the two outside terms are called the **extremes**, and the two inside terms are called the **means**.

   Example: In the proportion 2:4 = 5:10, 2 and 10 are the extremes, and 4 and 5 are the means.

   c. Proportions are often written in fractional form.

   Example: The proportion 2:4 = 5:10 may be written $\frac{2}{4} = \frac{5}{10}$.

   d. In any proportion, the product of the means equals the product of the extremes. If the proportion is in fractional form, the products may be found by cross-multiplication.

   Example: In $\frac{2}{4} = \frac{5}{10}$, $4 \times 5 = 2 \times 10$.

   e. The product of the extremes divided by one mean equals the other mean; the product of the means divided by one extreme equals the other extreme.

6. Many problems in which three terms are given and one term is unknown can be solved by using proportions. To solve such problems:

   a. Formulate the proportion very carefully according to the facts given. (If any term is misplaced, the solution will be incorrect.) Any symbol may be written in place of the missing term.

   b. Determine by inspection whether the means or the extremes are known. Multiply the pair that has both terms given.

   c. Divide this product by the third term given to find the unknown term.

   Illustration: The scale on a map shows that 2 cm represents 30 miles of actual length. What is the actual length of a road that is represented by 7 cm on the map?

   SOLUTION: The map lengths and the actual lengths are in proportion; that is, they have equal ratios. If m stands for the unknown length, the proportion is:

$$\frac{2}{7} = \frac{30}{m}$$

   As the proportion is written, m is an extreme and is equal to the product of the means, divided by the other extreme:

$$m = \frac{7 \times 30}{2}$$

$$m = \frac{210}{2}$$

$$m = 105$$

   Answer: 7 cm on the map represents 105 miles.

Illustration: If a money bag containing 500 nickels weighs 6 pounds, how much will a money bag containing 1600 nickels weigh?

SOLUTION: The weights of the bags and the number of coins in them are proportional. Suppose w represents the unknown weight. Then

$$\frac{6}{w} = \frac{500}{1600}$$

The unknown is a mean and is equal to the product of the extremes, divided by the other mean:

$$w = \frac{6 \times 1600}{500}$$
$$w = 19.2$$

Answer: A bag containing 1600 nickels weighs 19.2 pounds.

Practice Problems Involving Ratio and Proportion

1. The ratio of 24 to 64 is
   (A) 8:3
   (B) 24:100
   (C) 3:8
   (D) 64:100

2. The Baltimore Colts won 8 games and lost 3. The ratio of games won to games played is
   (A) 8:11
   (B) 3:11
   (C) 8:3
   (D) 3:8

3. The ratio of $\frac{1}{4}$ to $\frac{3}{5}$ is
   (A) 1 to 3
   (B) 3 to 20
   (C) 5 to 12
   (D) 3 to 4

4. If there are 16 boys and 12 girls in a class, the ratio of the number of girls to the number of children in the class is
   (A) 3 to 4
   (B) 3 to 7
   (C) 4 to 7
   (D) 4 to 3

5. 259 is to 37 as
   (A) 5 is to 1
   (B) 63 is to 441
   (C) 84 is to 12
   (D) 130 is to 19

6. 2 dozen cans of dog food at the rate of 3 cans for $1.45 would cost
   (A) $10.05
   (B) $11.20
   (C) $11.60
   (D) $11.75

7. A snapshot measures $2\frac{1}{2}$ inches by $1\frac{7}{8}$ inches. It is to be enlarged so that the longer dimension will be 4 inches. The length of the enlarged shorter dimension will be
   (A) $2\frac{1}{2}$ in
   (B) 3 in
   (C) $3\frac{3}{8}$ in
   (D) none of these

8. Men's white handkerchiefs cost $2.29 for 3. The cost per dozen handkerchiefs is
   (A) $27.48
   (B) $13.74
   (C) $9.16
   (D) $6.87

9. A certain pole casts a shadow 24 feet long. At the same time another pole 3 feet high casts a shadow 4 feet long. How high is the first pole, given that the heights and shadows are in proportion?
   (A) 18 ft
   (B) 19 ft
   (C) 20 ft
   (D) 21 ft

10. The actual length represented by $3\frac{1}{2}$ inches on a drawing having a scale of $\frac{1}{8}$ inch to the foot is
    (A) 3.75 ft
    (B) 28 ft
    (C) 360 ft
    (D) 120 ft

11. Aluminum bronze consists of copper and aluminum, usually in the ratio of 10:1 by weight. If an object made of this alloy weighs 77 lb, how many pounds of aluminum does it contain?
    (A) 7.7   (C) 70.0
    (B) 7.0   (D) 62.3

12. It costs 31 cents a square foot to lay vinyl flooring. To lay 180 square feet of flooring, it will cost
    (A) $16.20   (C) $55.80
    (B) $18.60   (D) $62.00

13. If a per diem worker earns $352 in 16 days, the amount that he will earn in 117 days is most nearly
    (A) $3050   (C) $2285
    (B) $2575   (D) $2080

14. Assuming that on a blueprint $\frac{1}{8}$ inch equals 12 inches of actual length, the actual length in inches of a steel bar represented on the blueprint by a line $3\frac{3}{4}$ inches long is
    (A) $3\frac{3}{4}$ (C) 450
    (B) 30   (D) 360

15. A, B, and C invested $9,000, $7,000 and $6,000, respectively. Their profits were to be divided according to the ratio of their investment. If B uses his share of the firm's profit of $825 to pay a personal debt of $230, how much will he have left?
    (A) $30.50   (C) $34.50
    (B) $32.50   (D) $36.50

Ratio and Proportion Problems — Correct Answers

1. **(C)**	6. **(C)**	11. **(B)**
2. **(A)**	7. **(B)**	12. **(C)**
3. **(C)**	8. **(C)**	13. **(B)**
4. **(B)**	9. **(A)**	14. **(D)**
5. **(C)**	10. **(B)**	15. **(B)**

Problem Solutions — Ratio and Proportion

1. The ratio 24 to 64 may be written 24:64 or $\frac{24}{64}$. In fraction form, the ratio can be reduced:

   $$\frac{24}{64} = \frac{3}{8} \text{ or } 3:8$$

   Answer: **(C)** 3:8

2. The number of games played was $3 + 8 = 11$. The ratio of games won to games played is 8:11.

   Answer: **(A)** 8:11

3. $\frac{1}{4}:\frac{3}{5} = \frac{1}{4} \div \frac{3}{5}$
   $= \frac{1}{4} \times \frac{5}{3}$
   $= \frac{5}{12}$
   $= 5:12$

   Answer: **(C)** 5 to 12

4. There are $16 + 12 = 28$ children in the class. The ratio of number of girls to number of children is 12:28.

   $$\frac{12}{28} = \frac{3}{7}$$

   Answer: **(B)** 3 to 7

5. The ratio $\frac{259}{37}$ reduces by 37 to $\frac{7}{1}$. The ratio $\frac{84}{12}$ also reduces to $\frac{7}{1}$. Therefore, $\frac{259}{37} = \frac{84}{12}$ is a proportion.

   Answer: **(C)** 84 is to 12

6. The number of cans are proportional to the price. Let p represent the unknown price:

   Then $\qquad \dfrac{3}{24} = \dfrac{1.45}{p}$

   $$p = \frac{1.45 \times 24}{3}$$

   $$p = \frac{34.80}{3}$$

   $$= \$11.60$$

   Answer: **(C)** $11.60

7. Let s represent the unknown shorter dimension:

   $$\frac{2\frac{1}{2}}{4} = \frac{1\frac{7}{8}}{s}$$

   $$s = \frac{4 \times 1\frac{7}{8}}{2\frac{1}{2}}$$

   $$= \frac{\overset{1}{\cancel{4}} \times \frac{15}{\cancel{8}}\,{}^{2}}{2\frac{1}{2}}$$

   $$= \tfrac{15}{2} \div 2\tfrac{1}{2}$$

   $$= \tfrac{15}{2} \div \tfrac{5}{2}$$

   $$= \tfrac{15}{2} \times \tfrac{2}{5}$$

   $$= 3$$

   Answer: **(B)** 3 in

8. If p is the cost per dozen (12):

   $$\frac{3}{12} = \frac{2.29}{p}$$

   $$p = \frac{\overset{4}{\cancel{12}} \times 2.29}{\underset{1}{\cancel{3}}}$$

   $$= 9.16$$

   Answer: **(C)** $9.16

9. If f is the height of the first pole, the proportion is:

   $$\frac{f}{24} = \frac{3}{4}$$

   $$f = \frac{\overset{6}{\cancel{24}} \times 3}{\underset{1}{\cancel{4}}}$$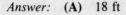

   $$= 18$$

   Answer: **(A)** 18 ft

10. If y is the unknown length:

    $$\frac{3\frac{1}{2}}{\frac{1}{8}} = \frac{y}{1}$$

    $$y = \frac{3\frac{1}{2} \times 1}{\frac{1}{8}}$$

    $$= 3\tfrac{1}{2} \div \tfrac{1}{8}$$

    $$= \tfrac{7}{2} \times \tfrac{8}{1}$$

    $$= 28$$

    Answer: **(B)** 28 ft

11. Since only two parts of a proportion are known (77 is total weight), the problem must be solved by the ratio method. The ratio 10:1 means that if the alloy were separated into equal parts, 10 of those parts would be copper and 1 would be aluminum, for a total of $10 + 1 = 11$ parts.

    $$77 \div 11 = 7 \text{ lb per part}$$

    The alloy has 1 part aluminum.

    $$7 \times 1 = 7 \text{ lb aluminum}$$

    Answer: **(B)** 7.0

12. The cost (c) is proportional to the number of square feet.

    $$\frac{\$.31}{c} = \frac{1}{180}$$

    $$c = \frac{\$.31 \times 180}{1}$$

    $$= \$55.80$$

    Answer: **(C)** $55.80

13. The amount earned is proportional to the number of days worked. If a is the unknown amount:

    $$\frac{\$352}{a} = \frac{16}{117}$$

    $$a = \frac{\$352 \times 117}{16}$$

    $$a = \$2574$$

    Answer: **(B)** $2575

14. If n is the unknown length:

$$\frac{\frac{1}{8}}{3\frac{3}{4}} = \frac{12}{n}$$

$$n = \frac{12 \times 3\frac{3}{4}}{\frac{1}{8}}$$

$$= \frac{\overset{3}{\cancel{12}} \times \frac{15}{\cancel{4}_1}}{\frac{1}{8}}$$

$$= \frac{45}{\frac{1}{8}}$$

$$= 45 \div \frac{1}{8}$$

$$= 45 \times \frac{8}{1}$$

$$= 360$$

Answer: **(D)** 360

15. The ratio of investment is:

$$9,000{:}7,000{:}6,000 \quad \text{or} \quad 9{:}7{:}6$$

$9 + 7 + 6 = 22$

$\$825 \div 22 = \$37.50$ each share of profit

$7 \times \$37.50 = \$262.50$ B's share of profit

$$\begin{array}{r} \$262.50 \\ -\ 230.00 \\ \hline \$\ 32.50 \end{array} \quad \text{amount B has left}$$

Answer: **(B)** $32.50

WORK AND TANK PROBLEMS

Work Problems

1. a. In work problems, there are three items involved: the number of people working, the time, and the amount of work done.

   b. The number of people working is directly proportional to the amount of work done; that is, the more people on the job, the more the work that will be done, and vice versa.

   c. The number of people working is inversely proportional to the time; that is, the more people on the job, the less time it will take to finish it, and vice versa.

   d. The time expended on a job is directly proportional to the amount of work done; that is, the more time expended on a job, the more work that is done, and vice versa.

Work at Equal Rates

2. a. When given the time required by a number of people working at equal rates to complete a job, multiply the number of people by their time to find the time required by one person to do the complete job.

   Example: If it takes 4 people working at equal rates 30 days to finish a job, then one person will take 30 × 4 or 120 days.

   b. When given the time required by one person to complete a job, to find the time required by a number of people working at equal rates to complete the same job, divide the time by the number of people.

   Example: If 1 person can do a job in 20 days, it will take 4 people working at equal rates 20 ÷ 4 or 5 days to finish the job.

3. To solve problems involving people who work at equal rates:

   a. Multiply the number of people by their time to find the time required by 1 person.

   b. Divide this time by the number of people required.

   Illustration: Four workers can do a job in 48 days. How long will it take 3 workers to finish the same job?

   SOLUTION: One worker can do the job in 48 × 4 or 192 days.
   3 workers can do the job in 192 ÷ 3 = 64 days.

   Answer: It would take 3 workers 64 days.

4. In some work problems, the rates, though unequal, can be equalized by comparison. To solve such problems:

   a. Determine from the facts given how many equal rates there are.

   b. Multiply the number of equal rates by the time given.

   c. Divide this by the number of equal rates.

   Illustration: Three workers can do a job in 12 days. Two of the workers work twice as fast as the third. How long would it take one of the faster workers to do the job himself?

   SOLUTION: There are two fast workers and one slow worker. Therefore, there are actually five slow workers working at equal rates.

   > 1 slow worker will take 12×5 or 60 days.
   > 1 fast worker = 2 slow workers; therefore, he will take $60 \div 2$ or 30 days to complete the job.

   Answer: It will take 1 fast worker 30 days to complete the job.

5. Unit time is time expressed in terms of 1 minute, 1 hour, 1 day, etc.

6. The rate at which a person works is the amount of work he can do in unit time.

7. If given the time it will take one person to do a job, then the reciprocal of the time is the part done in unit time.

   Example: If a worker can do a job in 6 days, then he can do $\frac{1}{6}$ of the work in 1 day.

8. The reciprocal of the work done in unit time is the time it will take to do the complete job.

   Example: If a worker can do $\frac{3}{7}$ of the work in 1 day, then he can do the whole job in $\frac{7}{3}$ or $2\frac{1}{3}$ days.

9. If given the various times in which each of a number of people can complete a job, to find the time it will take to do the job if all work together:

   a. Invert the time of each to find how much each can do in unit time.

   b. Add these reciprocals to find what part all working together can do in unit time.

   c. Invert this sum to find the time it will take all of them together to do the whole job.

   Illustration: If it takes A 3 days to dig a certain ditch, whereas B can dig it in 6 days, and C in 12, how long would it take all three to do the job?

   SOLUTION: A can do it in 3 days; therefore, he can do $\frac{1}{3}$ in one day. B can do it in 6 days; therefore, he can do $\frac{1}{6}$ in one day. C can do it in 12 days; therefore, he can do $\frac{1}{12}$ in one day.

   $$\tfrac{1}{3} + \tfrac{1}{6} + \tfrac{1}{12} = \tfrac{7}{12}$$

   A, B, and C can do $\frac{7}{12}$ of the work in one day; therefore, it will take them $\frac{12}{7}$ or $1\frac{5}{7}$ days to complete the job.

   Answer: A, B, and C, working together, can complete the job in $1\frac{5}{7}$ days.

10. If given the total time it requires a number of people working together to complete a job, and the times of all but one are known, to find the missing time:

   a. Invert the given times to find how much each can do in unit time.

   b. Add the reciprocals to find how much is done in unit time by those whose rates are known.

   c. Subtract this sum from the reciprocal of the total time to find the missing rate.

   d. Invert this rate to find the unknown time.

   Illustration: A, B, and C can do a job in 2 days. B can do it in 5 days, and C can do it in 4 days. How long would it take A to do it himself?

   SOLUTION: B can do it in 5 days; therefore, he can do $\frac{1}{5}$ in one day. C can do it in 4 days; therefore, he can do $\frac{1}{4}$ in one day. The part that can be done by B and C together in 1 day is:

   $$\tfrac{1}{5} + \tfrac{1}{4} = \tfrac{9}{20}$$

   The total time is 2 days; therefore, all can do $\frac{1}{2}$ in one day.

   $$\tfrac{1}{2} - \tfrac{9}{20} = \tfrac{1}{20}$$

   A can do $\frac{1}{20}$ in 1 day; therefore, he can do the whole job in 20 days.

   Answer: It would take A 20 days to complete the job himself.

11. In some work problems, certain values are given for the three factors — number of workers, the amount of work done, and the time. It is then usually required to find the changes that occur when one or two of the factors are given different values.

   One of the best methods of solving such problems is by directly making the necessary cancellations, divisions and multiplications.

   In this problem it is easily seen that more workers will be required since more houses are to be built in a shorter time.

   Illustration: If 60 workers can build 4 houses in 12 months, how many workers would be required to build 6 houses in 4 months?

   SOLUTION: To build 6 houses instead of 4 in the same amount of time, we would need $\frac{6}{4}$ of the number of workers.

   $$\tfrac{6}{4} \times 60 = 90$$

   Since we now have 4 months where previously we needed 12, we must triple the number of workers.

   $$90 \times 3 = 270$$

   Answer: 270 workers will be needed to build 6 houses in 4 months.

Tank Problems

12. The solution of tank problems is similar to that of work problems. Completely filling (or emptying) a tank may be thought of as completing a job.

13. a. If given the time it takes a pipe to fill or empty a tank, the reciprocal of the time will represent that part of the tank that is filled or emptied in unit time.

Example: If it takes a pipe 4 minutes to fill a tank, then $\frac{1}{4}$ of the tank is filled in one minute.

b. The amount that a pipe can fill or empty in unit time is its rate.

14. If given the part of a tank that a pipe or a combination of pipes can fill or empty in unit time, invert the part to find the total time required to fill or empty the whole tank.

Example: If a pipe can fill $\frac{2}{5}$ of a tank in 1 minute, then it will take $\frac{5}{2}$ or $2\frac{1}{2}$ minutes to fill the entire tank.

15. To solve tank problems in which only one action (filling or emptying) is going on:

a. Invert the time of each pipe to find how much each can do in unit time.

b. Add the reciprocals to find how much all can do in unit time.

c. Invert this sum to find the total time.

Illustration: Pipe A can fill a tank in 3 minutes whereas B can fill it in 4 minutes. How long would it take both pipes, working together, to fill it?

SOLUTION: Pipe A can fill it in 3 minutes; therefore, it can fill $\frac{1}{3}$ of the tank in one minute. Pipe B can fill it in 4 minutes; therefore, it can fill $\frac{1}{4}$ of the tank in one minute.

$$\tfrac{1}{3} + \tfrac{1}{4} = \tfrac{7}{12}$$

Pipe A and Pipe B can fill $\frac{7}{12}$ of the tank in one minute; therefore, they can fill the tank in $\frac{12}{7}$ or $1\frac{5}{7}$ minutes.

Answer: Pipes A and B, working together, can fill the tank in $1\frac{5}{7}$ minutes.

16. In problems in which both filling and emptying actions are occurring:

a. Determine which process has the faster rate.

b. The difference between the filling rate and the emptying rate is the part of the tank that is actually being filled or emptied in unit time. The fraction representing the slower action is subtracted from the fraction representing the faster process.

c. The reciprocal of this difference is the time it will take to fill or empty the tank.

Illustration: A certain tank can be filled by Pipe A in 12 minutes. Pipe B can empty the tank in 18 minutes. If both pipes are open, how long will it take to fill or empty the tank?

SOLUTION: Pipe A fills $\frac{1}{12}$ of the tank in 1 minute.
Pipe B empties $\frac{1}{18}$ of the tank in 1 minute.

$$\tfrac{1}{12} = \tfrac{3}{36}$$
$$\tfrac{1}{18} = \tfrac{2}{36}$$

Since $\frac{1}{12}$ is greater than $\frac{1}{18}$, the tank will ultimately be filled. In 1 minute, $\frac{3}{36} - \frac{2}{36} = \frac{1}{36}$ of the tank is actually filled. Therefore, the tank will be completely filled in 36 minutes.

Answer: It will take 36 minutes to fill the tank if both pipes are open.

Work and Tank Practice Problems

1. If 314 clerks filed 6594 papers in 10 minutes, what is the number filed per minute by the average clerk?
   (A) 2
   (B) 2.4
   (C) 2.1
   (D) 2.5

2. Four men working together can dig a ditch in 42 days. They begin, but one man works only half-days. How long will it take to complete the job?
   (A) 48 days
   (B) 45 days
   (C) 43 days
   (D) 44 days

3. A clerk is requested to file 800 cards. If he can file cards at the rate of 80 cards an hour, the number of cards remaining to be filed after 7 hours of work is
   (A) 140
   (B) 240
   (C) 260
   (D) 560

4. If it takes 4 days for 3 machines to do a certain job, it will take two machines
   (A) 6 days
   (B) $5\frac{1}{2}$ days
   (C) 5 days
   (D) $4\frac{1}{2}$ days

5. A stenographer has been assigned to place entries on 500 forms. She places entries on 25 forms by the end of half an hour, when she is joined by another stenographer. The second stenographer places entries at the rate of 45 an hour. Assuming that both stenographers continue to work at their respective rates of speed, the total number of hours required to carry out the entire assignment is
   (A) 5
   (B) $5\frac{1}{2}$
   (C) $6\frac{1}{2}$
   (D) 7

6. If in 5 days a clerk can copy 125 pages, 36 lines each, 11 words to the line, how many pages of 30 lines each and 12 words to the line can he copy in 6 days?
   (A) 145
   (B) 155
   (C) 160
   (D) 165

7. A and B do a job together in two hours. Working alone A does the job in 5 hours. How long will it take B to do the job alone?
   (A) $3\frac{1}{3}$ hr
   (B) $2\frac{1}{4}$ hr
   (C) 3 hr
   (D) 2 hr

8. A stenographer transcribes her notes at the rate of one line typed in ten seconds. At this rate, how long (in minutes and seconds) will it take her to transcribe notes, which will require seven pages of typing, 25 lines to the page?
   (A) 29 min 10 sec
   (B) 17 min 50 sec
   (C) 40 min 10 sec
   (D) 20 min 30 sec

9. A group of five clerks have been assigned to insert 24,000 letters into envelopes. The clerks perform this work at the following rates of speed: Clerk A, 1100 letters an hour; Clerk B, 1450 letters an hour; Clerk C, 1200 letters an hour; Clerk D 1300 letters an hour; Clerk E, 1250 letters an hour. At the end of two hours of work, Clerks C and D are assigned to another task. From the time that Clerks C and D were taken off the assignment, the number of hours required for the remaining clerks to complete this assignment is
   (A) less than 3 hr
   (B) 3 hr
   (C) more than 3 hr, but less than 4 hr
   (D) more than 4 hr

10. If a certain job can be performed by 18 workers in 26 days, the number of workers needed to perform the job in 12 days is
    (A) 24
    (B) 30
    (C) 39
    (D) 52

11. A steam shovel excavates 2 cubic yards every 40 seconds. At this rate, the amount excavated in 45 minutes is
    (A) 90 cu yd
    (B) 135 cu yd
    (C) 900 cu yd
    (D) 3600 cu yd

12. If a plant making bricks turns out 1250 bricks in 5 days, the number of bricks that can be made in 20 days is
    (A) 5000
    (B) 6250
    (C) 12,500
    (D) 25,000

13. A tank is $\frac{3}{4}$ full. Pipe A can fill the tank in 12 minutes. Pipe B can empty it in 8 minutes. If both pipes are open, how long will it take to empty the tank?
    (A) 14 min
    (B) 22 min
    (C) 16 min
    (D) 18 min

14. A tank that holds 400 gallons of water can be filled by one pipe in 15 minutes and emptied by another in 40 minutes. How long would it take to fill the tank if both pipes are open?
    (A) 20 min
    (B) 21 min
    (C) 23 min
    (D) 24 min

15. An oil burner in a housing development burns 76 gallons of fuel oil per hour. At 9 a.m. on a very cold day the superintendent asks the housing manager to put in an emergency order for more fuel oil. At that time, he reports that he has on hand 266 gallons. At noon, he again comes to the manager, notifying him that no oil has been delivered. The maximum amount of time that he can continue to furnish heat without receiving more oil is
    (A) $\frac{1}{2}$ hr
    (B) 1 hr
    (C) $1\frac{1}{2}$ hr
    (D) 2 hr

Work and Tank Problems — Correct Answers

1. **(C)**	6. **(D)**	11. **(B)**
2. **(A)**	7. **(A)**	12. **(A)**
3. **(B)**	8. **(A)**	13. **(D)**
4. **(A)**	9. **(B)**	14. **(D)**
5. **(B)**	10. **(C)**	15. **(A)**

Work and Tank Problem Solutions

1. 6594 papers ÷ 314 clerks = 21 papers per clerk in 10 minutes
   21 papers ÷ 10 minutes = 2.1 papers per minute filed by average clerk

   Answer: **(C)** 2.1

2. It would take 1 man 42 × 4 = 168 days to complete the job, working alone.
   If $3\frac{1}{2}$ men are working (one man works half-days, the other 3 work full days), the job would take 168 ÷ $3\frac{1}{2}$ = 48 days.

   Answer: **(A)** 48 days

3. In 7 hours the clerk files $7 \times 80 = 560$ cards. Since 800 cards must be filed, there are $800 - 560 = 240$ remaining.

Answer: **(B)** 240

4. It would take 1 machine $3 \times 4 = 12$ days to do the job. Two machines could do the job in $12 \div 2 = 6$ days.

Answer: **(A)** 6 days

5. At the end of the first half-hour, there are $500 - 25 = 475$ forms remaining. If the first stenographer completed 25 forms in half an hour, her rate is $25 \times 2 = 50$ forms per hour. The combined rate of the two stenographers is $50 + 45 = 95$ forms per hour. The remaining forms can be completed in $475 \div 95 = 5$ hours. Adding the first half-hour, the entire job requires $5\frac{1}{2}$ hours.

Answer: **(B)** $5\frac{1}{2}$

6. 36 lines $\times$ 11 words = 396 words on each page
125 pages $\times$ 396 words = 49,500 words in 5 days
$49,500 \div 5 = 9900$ words in 1 day
12 words $\times$ 30 lines = 360 words on each page
$9900 \div 360 = 27\frac{1}{2}$ pages in 1 day
$27\frac{1}{2} \times 6 = 165$ pages in 6 days.

Answer: **(D)** 165

7. If A can do the job alone in 5 hours, A can do $\frac{1}{5}$ of the job in 1 hour. Working together, A and B can do the job in 2 hours, therefore in 1 hour they do $\frac{1}{2}$ the job.
In 1 hour, B alone does
$$\frac{1}{2} - \frac{1}{5} = \frac{5}{10} - \frac{2}{10}$$
$$= \frac{3}{10} \text{ of the job.}$$
It would take B $\frac{10}{3}$ hours = $3\frac{1}{3}$ hours to do the whole job alone.

Answer: **(A)** $3\frac{1}{3}$ hr

8. She must type $7 \times 25 = 175$ lines. At the rate of 1 line per 10 seconds, it will take $175 \times 10 = 1750$ seconds.
$$1750 \text{ seconds} \div 60 = 29\frac{1}{6} \text{ minutes}$$
$$= 29 \text{ min } 10 \text{ sec}$$

Answer: **(A)** 29 min 10 sec

9.

Clerk	Number of letters per hr
A	1100
B	1450
C	1200
D	1300
E	+ 1250
Total =	6300

All 5 clerks working together process a total of 6300 letters per hour. After 2 hours, they have processed $6300 \times 2 = 12,600$. Of the original 24,000 letters there are

24,000
− 12,600
11,400 letters remaining

Clerks A, B, and E working together process a total of 3800 letters per hour. It will take them
$$11,400 \div 3800 = 3 \text{ hours}$$
to process the remaining letters.

Answer: **(B)** 3 hr

10. The job could be performed by 1 worker in 18×26 days = 468 days. To perform the job in 12 days would require $468 \div 12 = 39$ workers.

Answer: **(C)** 39

11. The shovel excavates 1 cubic yard in 20 seconds.
There are $45 \times 60 = 2700$ seconds in 45 minutes.
In 2700 seconds the shovel can excavate $2700 \div 20 = 135$ cubic yards.

Answer: **(B)** 135 cu yd

12. In 20 days the plant can produce four times as many bricks as in 5 days.
$$1250 \times 4 = 5000 \text{ bricks}$$

Answer: **(A)** 5000

13. Pipe A can fill the tank in 12 min or fill $\frac{1}{12}$ of the tank in 1 min. Pipe B can empty the tank in 8 min or empty $\frac{1}{8}$ of the tank in 1 min. In 1 minute, $\frac{1}{8} - \frac{1}{12}$ of the tank is emptied (since $\frac{1}{8}$ is greater than $\frac{1}{12}$).

$$
\begin{array}{r}
\frac{1}{8} \\
-\ \frac{1}{12}
\end{array} =
\begin{array}{r}
\frac{3}{24} \\
-\ \frac{2}{24} \\
\hline
\frac{1}{24}
\end{array}
$$ of the tank is emptied per minute

It would take 24 min to empty whole tank, but it is only $\frac{3}{4}$ full:

$$\frac{\overset{}{\cancel{3}}}{\underset{1}{\cancel{4}}} \times \overset{6}{\cancel{24}} = 18 \text{ minutes}$$

Answer: **(D)** 18 min

14. The first pipe can fill $\frac{1}{15}$ of the tank in 1 minute. The second pipe can empty $\frac{1}{40}$ of the tank in 1 minute. With both pipes open, $\frac{1}{15} - \frac{1}{40}$ of the tank will be filled per minute.

$$
\begin{array}{r}
\frac{1}{15} = \\
-\ \frac{1}{40} =
\end{array}
\begin{array}{r}
\frac{8}{120} \\
-\ \frac{3}{120} \\
\hline
\frac{5}{120} = \frac{1}{24}
\end{array}
$$

In 1 minute, $\frac{1}{24}$ of the tank is filled; therefore, it will take 24 minutes for the entire tank to be filled.

Answer: **(D)** 24 min

15. If 76 gallons are used per hour, it will take $266 \div 76 = 3\frac{1}{2}$ hours to use 266 gallons.

From 9 a.m. to noon is 3 hours; therefore, there is only fuel for $\frac{1}{2}$ hour more.

Answer: **(A)** $\frac{1}{2}$ hr

DISTANCE PROBLEMS

1. In distance problems, there are usually three quantities involved: the distance (in miles), the rate (in miles per hour — mph), and the time (in hours).

    a. To find the distance, multiply the rate by the time.

    Example: A man traveling 40 miles an hour for 3 hours travels 40 × 3 or 120 miles.

    b. The rate is the distance traveled in unit time. To find the rate, divide the distance by the time.

    Example: If a car travels 100 miles in 4 hours, the rate is 100 ÷ 4 or 25 miles an hour.

    c. To find the time, divide the distance by the rate.

    Example: If a car travels 150 miles at the rate of 30 miles an hour, the time is 150 ÷ 30 or 5 hours.

Combined Rates

2. a. When two people or objects are traveling towards each other, the rate at which they are approaching each other is the sum of their respective rates.

    b. When two people or objects are traveling in directly opposite directions, the rate at which they are separating is the sum of their respective rates.

3. To solve problems involving combined rates:

    a. Determine which of the three factors is to be found.

    b. Combine the rates and find the unknown factor.

    Illustration: A and B are walking towards each other over a road 120 miles long. A walks at a rate of 6 miles an hour, and B walks at a rate of 4 miles an hour. How soon will they meet?

    SOLUTION: The factor to be found is the time.

    Time = distance ÷ rate
    Distance = 120 miles
    Rate = 6 + 4 = 10 miles an hour
    Time = 120 ÷ 10 = 12 hours

    Answer: They will meet in 12 hours.

Illustration: Joe and Sam are walking in opposite directions. Joe walks at the rate of 5 miles an hour, and Sam walks at the rate of 7 miles an hour. How far apart will they be at the end of 3 hours?

SOLUTION: The factor to be found is distance.

$$\text{Distance} = \text{time} \times \text{rate}$$
$$\text{Time} = 3 \text{ hours}$$
$$\text{Rate} = 5 + 7 = 12 \text{ miles an hour}$$
$$\text{Distance} = 12 \times 3 = 36 \text{ miles}$$

Answer: They will be 36 miles apart at the end of 3 hours.

4. To find the time it takes a faster person or object to catch up with a slower person or object:

   a. Determine how far ahead the slower person or object is.

   b. Subtract the slower rate from the faster rate to find the gain in rate per unit time.

   c. Divide the distance that has been gained by the difference in rates.

Illustration: Two automobiles are traveling along the same road. The first one, which travels at the rate of 30 miles an hour, starts out 6 hours ahead of the second one, which travels at the rate of 50 miles an hour. How long will it take the second one to catch up with the first one?

SOLUTION: The first automobile starts out 6 hours ahead of the second. Its rate is 30 miles an hour. Therefore, it has traveled 6×30 or 180 miles by the time the second one starts. The second automobile travels at the rate of 50 miles an hour. Therefore, its gain is $50 - 30$ or 20 miles an hour. The second auto has to cover 180 miles. Therefore, it will take $180 \div 20$ or 9 hours to catch up with the first automobile.

Answer: It will take the faster auto 9 hours to catch up with the slower one.

Average of Two Rates

5. In some problems, two or more rates must be averaged. When the times are the same for two or more different rates, add the rates and divide by the number of rates.

Example: If a man travels for 2 hours at 30 miles an hour, at 40 miles an hour for the next 2 hours, and at 50 miles an hour for the next 2 hours, then his average rate for the 6 hours is $(30 + 40 + 50) \div 3 = 40$ miles an hour.

6. When the times are not the same, but the distances are the same:

   a. Assume the distance to be a convenient length.

   b. Find the time at the first rate.

   c. Find the time at the second rate.

   d. Find the time at the third rate, if any.

   e. Add up all the distances and divide by the total time to find the average rate.

Illustration: A boy travels a certain distance at the rate of 20 miles an hour and returns at the rate of 30 miles an hour. What is his average rate for both trips?

SOLUTION: The distance is the same for both trips. Assume that it is 60 miles. The time for the first trip is 60 ÷ 20 = 3 hours. The time for the second trip is 60 ÷ 30 = 2 hours. The total distance is 120 miles. The total time is 5 hours. Average rate is 120 ÷ 5 = 24 miles an hour.

Answer: The average rate is 24 miles an hour.

7. When the times are not the same and the distances are not the same:

   a. Find the time for the first distance.

   b. Find the time for the second distance.

   c. Find the time for the third distance, if any.

   d. Add up all the distances and divide by the total time to find the average rate.

   Illustration: A man travels 100 miles at 20 miles an hour, 60 miles at 30 miles an hour, and 80 miles at 10 miles an hour. What is his average rate for the three trips?

   SOLUTION: The time for the first trip is 100 ÷ 20 = 5 hours. The time for the second trip is 60 ÷ 30 = 2 hours. The time for the third trip is 80 ÷ 10 = 8 hours. The total distance is 240 miles. The total time is 15 hours. Average rate is 240 ÷ 15 = 16 hours.

   Answer: The average rate for the three trips is 16 miles an hour.

Gasoline Problems

8. Problems involving miles per gallon (mpg) of gasoline are solved in the same way as those involving miles per hour. The word "gallon" simply replaces the word "hour."

9. Miles per gallon = distance in miles ÷ no. of gallons

   Example: If a car can travel 100 miles using 4 gallons of gasoline, then its gasoline consumption is 100 ÷ 4, or 25 mpg.

Practice Problems Involving Distance

1. A ten-car train took 6 minutes to travel between two stations that are 3 miles apart. The average speed of the train was
   (A) 20 mph   (C) 30 mph
   (B) 25 mph   (D) 35 mph

2. A police car is ordered to report to the scene of a crime 5 miles away. If the car travels at an average rate of 40 miles per hour, the time it will take to reach its destination is
   (A) 3 min   (C) 10 min
   (B) 7.5 min   (D) 13.5 min

3. If the average speed of a train between two stations is 30 miles per hour and the two stations are $\frac{1}{2}$ mile apart, the time it takes the train to travel from one station to the other is
   (A) 1 min
   (B) 2 min
   (C) 3 min
   (D) 4 min

4. A car completes a 10-mile trip in 20 minutes. If it does one-half the distance at a speed of 20 miles an hour, its speed for the remainder of the distance must be
   (A) 30 mph
   (B) 40 mph
   (C) 50 mph
   (D) 60 mph

5. An express train leaves one station at 9:02 and arrives at the next station at 9:08. If the distance traveled is $2\frac{1}{2}$ miles, the average speed of the train (mph) is
   (A) 15 mph
   (B) 20 mph
   (C) 25 mph
   (D) 30 mph

6. A motorist averaged 60 miles per hour in going a distance of 240 miles. He made the return trip over the same distance in 6 hours. What was his average speed for the entire trip?
   (A) 40 mph
   (B) 48 mph
   (C) 50 mph
   (D) 60 mph

7. A city has been testing various types of gasoline for economy and efficiency. It has been found that a police radio patrol car can travel 18 miles on a gallon of Brand A gasoline, costing $1.30 a gallon, and 15 miles on a gallon of Brand B gasoline, costing $1.25 a gallon. For a distance of 900 miles, Brand B will cost
   (A) $10 more than Brand A
   (B) $10 less than Brand A
   (C) $100 more than Brand A
   (D) the same as Brand A

8. A suspect arrested in New Jersey is being turned over by New Jersey authorities to two New York City police officers for a crime committed in New York City. The New York officers receive their prisoner at a point $18\frac{1}{2}$ miles from their precinct station house, and travel directly toward their destination at an average speed of 40 miles an hour except for a delay of 10 minutes at one point because of a traffic tie-up. The time it should take the officers to reach their destination is, most nearly,
   (A) 18 min
   (B) 22 min
   (C) 32 min
   (D) 38 min

9. The Mayflower sailed from Plymouth, England, to Plymouth Rock, a distance of approximately 2800 miles, in 63 days. The average speed was closest to which one of the following?
   (A) $\frac{1}{2}$ mph
   (B) 1 mph
   (C) 2 mph
   (D) 3 mph

10. If a vehicle is to complete a 20-mile trip at an average rate of 30 miles per hour, it must complete the trip in
    (A) 20 min
    (B) 30 min
    (C) 40 min
    (D) 50 min

11. A car began a trip with 12 gallons of gasoline in the tank and ended with $7\frac{1}{2}$ gallons. The car traveled 17.3 miles for each gallon of gasoline. During the trip gasoline was bought for $10.00, at a cost of $1.25 per gallon. The total number of miles traveled during this trip was most nearly
    (A) 79
    (B) 196
    (C) 216
    (D) 229

12. A man travels a total of 4.2 miles each day to and from work. The traveling consumes 72 minutes each day. Most nearly how many hours would he save in 129 working days if he moved to another residence so that he would travel only 1.7 miles each day, assuming he travels at the same rate?
    (A) 92.11
    (B) 93.62
    (C) 95.35
    (D) 98.08

13. A man can travel a certain distance at the rate of 25 miles an hour by automobile. He walks back the same distance on foot at the rate of 10 miles an hour. What is his average rate for both trips?
    (A) $14\frac{2}{7}$ mph
    (B) $15\frac{1}{3}$ mph
    (C) $17\frac{1}{2}$ mph
    (D) 35 mph

14. Two trains running on the same track travel at the rates of 25 and 30 miles an hour. If the first train starts out an hour earlier, how long will it take the second train to catch up with it?
    (A) 2 hr   (C) 4 hr
    (B) 3 hr   (D) 5 hr

15. Two ships are 1550 miles apart sailing towards each other. One sails at the rate of 85 miles per day and the other at the rate of 65 miles per day. How far apart will they be at the end of 9 days?
    (A) 180 mi   (C) 220 mi
    (B) 200 mi   (D) 240 mi

Distance Problems — Correct Answers

1. **(C)**   6. **(B)**   11. **(C)**
2. **(B)**   7. **(A)**   12. **(A)**
3. **(A)**   8. **(D)**   13. **(A)**
4. **(D)**   9. **(C)**   14. **(D)**
5. **(C)**   10. **(C)**   15. **(B)**

Problem Solutions — Distance

1.
$$6 \text{ min} = \tfrac{6}{60} \text{ hr} = .1 \text{ hr}$$
Speed (rate) = distance ÷ time
$$\text{Speed} = 3 \div .1 = 30 \text{ mph}$$

Answer: **(C)** 30 mph

2.
Time = distance ÷ rate
Time = 5 ÷ 40 = .125 hr
.125 hr = .125 × 60 min
= 7.5 min

Answer: **(B)** 7.5 min

3.
Time = distance ÷ rate
Time = $\tfrac{1}{2}$ mi ÷ 30 mph
= $\tfrac{1}{60}$ hr
$\tfrac{1}{60}$ hr = 1 min

Answer: **(A)** 1 min

4.
First part of trip = $\tfrac{1}{2}$ of 10 miles = 5 miles
Time for first part = 5 ÷ 20
= $\tfrac{1}{4}$ hour
= 15 minutes

Second part of trip was 5 miles, completed in 20 − 15 minutes, or 5 minutes.
$$5 \text{ minutes} = \tfrac{1}{12} \text{ hour}$$
$$\text{Rate} = 5 \text{ mi} \div \tfrac{1}{12} \text{ hr}$$
$$= 60 \text{ mph}$$

Answer: **(D)** 60 mph

5. Time is 6 minutes, or .1 hour
$$\text{Speed} = \text{distance} \div \text{time}$$
$$= 2\tfrac{1}{2} \div .1$$
$$= 2.5 \div .1$$
$$= 25 \text{ mph}$$

Answer: **(C)** 25 mph

6.   Time for first 240 mi = 240 ÷ 60
                               = 4 hours
         Time for return trip = 6 hours
         Total time for round trip = 10 hours
         Total distance for round trip = 480 mi
                    Average rate = 480 mi ÷ 10 hr
                               = 48 mph

Answer: **(B)** 48 mph

7.   Brand A requires 900 ÷ 18 = 50 gal
         50 gal × $1.30 per gal = $65

         Brand B requires 900 ÷ 15 = 60 gal
         60 gal × $1.25 per gal = $75

Answer: **(A)** Brand B will cost $10 more than Brand A

8.   Time = distance ÷ rate
              Time = 18.5 mi ÷ 40 mph
                   = .4625 hours
                   = .4625 × 60 minutes
                   = 27.75 minutes
         27.75 + 10 = 37.75 minutes

Answer: **(D)** 38 min

9.   63 days = 63 × 24 hours
                 = 1512 hours
         Speed = 2800 mi ÷ 1512 hr
               = 1.85 mph

Answer: **(C)** 2 mph

10.   Time = 20 mi ÷ 30 mph
              = $\frac{2}{3}$ hr
         $\frac{2}{3}$ hr = $\frac{2}{3}$ × 60 min = 40 min

Answer: **(C)** 40 min

11. The car used
              $12 - 7\frac{1}{2} = 4\frac{1}{2}$ gal, plus
         $10.00 ÷ $1.25 = 8 gal,
         for a total of $12\frac{1}{2}$ gal, or 12.5 gal.
              12.5 gal × 17.3 mpg = 216.25 mi

Answer: **(C)** 216

12.   72 min = $\frac{72}{60}$ hr = 1.2 hr
         Rate = 4.2 mi ÷ 1.2 hr = 3.5 mph
         At this rate it would take 1.7 mi ÷ 3.5 mph =
         .486 hours (approx.) to travel 1.7 miles. The daily
         savings in time is 1.2 hr − .486 hr = .714 hr
              .714 hr × 129 days = 92.106 hr

Answer: **(A)** 92.11

13. Assume a convenient distance, say, 50 mi.
         Time by automobile = 50 mi ÷ 25 mph
                            = 2 hr
         Time walking = 50 mi ÷ 10 mph
                      = 5 hr
         Total time = 7 hours
         Total distance = 100 mi
         Average rate = 100 mi ÷ 7 hr
                      = $14\frac{2}{7}$ mph

Answer: **(A)** $14\frac{2}{7}$ mph

14.   30 mi − 25 mi = 5 mi gain per 1 hr
         During first hour, the first train travels 25 miles.
              25 mi ÷ 5 mph = 5 hr

Answer: **(D)** 5 hr

15.   85 mi × 9 da = 765 mi
             65 mi × 9 da = 585 mi
                            ⎯⎯⎯⎯
                            1350
         1550 mi − 1350 = 200 miles apart at end of
         9 days.

Answer: **(B)** 200 mi

INTEREST

1. **Interest (I)** is the price paid for the use of money. There are three items considered in interest:

   a. The **principal (p)**, which is the amount of money bearing interest.

   b. The **interest rate (r)**, expressed in percent on an annual basis.

   c. The **time (t)** during which the principal is used, expressed in terms of a year.

2. The basic formulas used in interest problems are:

   a. $I = prt$

   b. $p = \dfrac{I}{rt}$

   c. $r = \dfrac{I}{pt}$

   d. $t = \dfrac{I}{pr}$

3. a. For most interest problems, the year is considered to have 360 days. Months are considered to have 30 days, unless a particular month is specified.

   b. To use the interest formulas, time must be expressed as part of a year.

   Examples: 5 months $= \frac{5}{12}$ year

   36 days $= \frac{36}{360}$ year, or $\frac{1}{10}$ year

   1 year 3 months $= \frac{15}{12}$ year

   c. In reference to time, the prefix "semi" means "every half." The prefix "bi" means "every two."

   Examples: Semiannually means every half-year (every 6 months).
   Biannually means every 2 years.
   Semimonthly means every half-month (every 15 days, unless the month is specified).
   Biweekly means every 2 weeks (every 14 days).

4. There are two types of interest problems:

   a. **Simple interest**, in which the interest is calculated only once over a given period of time.

   b. **Compound interest**, in which interest is recalculated at given time periods based on previously earned interest.

Simple Interest

5. To find the interest when the principal, rate, and time are given:

    a. Change the rate of interest to a fraction.

    b. Express the time as a fractional part of a year.

    c. Multiply all three items.

    Illustration: Find the interest on $400 at $11\frac{1}{4}\%$ for 3 months and 16 days.

    SOLUTION:
    $$11\tfrac{1}{4}\% = \tfrac{45}{4}\% = \tfrac{45}{400}$$

    3 months and 16 days = 106 days
    (30 days per month)

    $$106 \text{ days} = \tfrac{106}{360} \text{ of a year} = \tfrac{53}{180} \text{ year}$$
    (360 days per year)

    $$\cancel{400} \times \frac{45}{\cancel{400}} \times \frac{53}{\cancel{180}} = \frac{53}{4}$$

    $$= 13.25$$

    Answer: Interest = $13.25

6. To find the principal if the interest, interest rate, and time are given:

    a. Change the interest rate to a fraction.

    b. Express the time as a fractional part of a year.

    c. Multiply the rate by the time.

    d. Divide the interest by this product.

    Illustration: What amount of money invested at 6% would receive interest of $18 over $1\frac{1}{2}$ years?

    SOLUTION:
    $$6\% = \tfrac{6}{100}$$
    $$1\tfrac{1}{2} \text{ years} = \tfrac{3}{2} \text{ years}$$
    $$\frac{\cancel{6}}{100} \times \frac{3}{\cancel{2}} = \frac{9}{100}$$
    $$\$18 \div \tfrac{9}{100} = \$\cancel{18} \times \frac{100}{\cancel{9}}$$
    $$= \$200$$

    Answer: Principal = $200

7. To find the rate if the principal, time, and interest are given:

    a. Change the time to a fractional part of a year.

    b. Multiply the principal by the time.

    c. Divide the interest by this product.

    d. Convert to a percent.

    Illustration: At what interest rate should $300 be invested for 40 days to accrue $2 in interest?

    SOLUTION: $\quad$ 40 days $= \frac{40}{360}$ of a year

    $$300 \times \frac{40}{360} = \frac{100}{3}$$

    $$\$2 \div \frac{100}{3} = 2 \times \frac{3}{100} = \frac{3}{50}$$

    $$\frac{3}{50} = 6\%$$

    Answer: Interest rate $= 6\%$

8. To find the time (in years) if the principal, interest, and interest rate are given:

    a. Change the interest rate to a fraction (or decimal).

    b. Multiply the principal by the rate.

    c. Divide the interest by this product.

    Illustration: Find the length of time for which $240 must be invested at 5% to accrue $16 in interest.

    SOLUTION: $\quad$ $5\% = .05$

    $$240 \times .05 = 12$$

    $$16 \div 12 = 1\tfrac{1}{3}$$

    Answer: Time $= 1\tfrac{1}{3}$ years

Compound Interest

9. Interest may be computed on a compound basis; that is, the interest at the end of a certain period (half year, full year, or whatever time stipulated) is added to the principal for the next period. The interest is then computed on the new increased principal, and for the next period the interest is again computed on the new increased principal. Since the principal constantly increases, compound interest yields more than simple interest.

10. To find the compound interest when given the principal, the rate, and time period:

   a. Calculate the interest as for simple interest problems, using the period of compounding for the time.

   b. Add the interest to the principal.

   c. Calculate the interest on the new principal over the period of compounding.

   d. Add this interest to form a new principal.

   e. Continue the same procedure until all periods required have been accounted for.

   f. Subtract the original principal from the final principal to find the compound interest.

Illustration: Find the amount that $200 will become if compounded semiannually at 8% for $1\frac{1}{2}$ years.

SOLUTION: Since it is to be compounded semiannually for $1\frac{1}{2}$ years, the interest will have to be computed 3 times:

Interest for the first period:	$.08 \times \frac{1}{2} \times \$200 = \$8$
First new principal:	$\$200 + \$8 = \$208$
Interest for the second period:	$.08 \times \frac{1}{2} \times \$208 = \$8.32$
Second new principal:	$\$208 + \$8.32 = \$216.32$
Interest for the third period:	$.08 \times \frac{1}{2} \times \$216.32 = \$8.6528$
Final principal:	$\$216.32 + \$8.6528 = \$224.9728$

Answer: $224.97 to the nearest cent

Bank Discounts

11. A promissory note is a committment to pay a certain amount of money on a given date, called the date of maturity.

12. When a promissory note is cashed by a bank in advance of its date of maturity, the bank deducts a discount from the principal and pays the rest to the depositor.

13. To find the bank discount:

    a. Find the time between the date the note is deposited and its date of maturity, and express this time as a fractional part of a year.

    b. Change the rate to a fraction.

    c. Multiply the principal by the time and the rate to find the bank discount.

    d. If required, subtract the bank discount from the original principal to find the amount the bank will pay the depositor.

Illustration: A $400 note drawn up on August 12, 1980, for 90 days is deposited at the bank on September 17, 1980. The bank charges a $6\frac{1}{2}\%$ discount on notes. How much will the depositor receive?

SOLUTION: From August 12, 1980, to September 17, 1980, is 36 days. This means that the note has 54 days to run.

$$54 \text{ days} = \tfrac{54}{360} \text{ of a year}$$
$$6\tfrac{1}{2}\% = \tfrac{13}{2}\% = \tfrac{13}{200}$$
$$\$400 \times \tfrac{13}{200} \times \tfrac{54}{360} = \tfrac{39}{10}$$
$$= \$3.90$$
$$\$400 - \$3.90 = \$396.10$$

Answer: The depositor will receive $396.10.

Practice Problems Involving Interest

1. What is the simple interest on $460 for 2 years at $8\frac{1}{2}\%$?
   (A) $46.00   (C) $78.20
   (B) $52.75   (D) $96.00

2. For borrowing $300 for one month, a man was charged $6.00. The rate of interest was
   (A) $\frac{1}{5}\%$ (C) 24%
   (B) 12%   (D) 2%

3. At a simple interest rate of 5% a year, the principal that will give $12.50 interest in 6 months is
   (A) $250   (C) $625
   (B) $500   (D) $650

4. Find the interest on $480 at $10\frac{1}{2}\%$ for 2 months and 15 days.
   (A) $ 9.50   (C) $13.25
   (B) $10.50   (D) $14.25

5. The interest on $300 at 6% for 10 days is
   (A) $ .50   (C) $2.50
   (B) $1.50   (D) $5.50

6. The scholarship board of a certain college lent a student $200 at an annual rate of 6% from September 30 until December 15. To repay the loan and accumulated interest the student must give the college an amount closest to which one of the following?
   (A) $202.50   (C) $203.50
   (B) $203.00   (D) $212.00

7. If $300 is invested at simple interest so as to yield a return of $18 in 9 months, the amount of money that must be invested at the same rate of interest so as to yield a return of $120 in 6 months is
   (A) $3000   (C) $2000
   (B) $3300   (D) $2300

8. When the principal is $600, the difference in one year between simple interest at 12% per annum and interest compounded semiannually at 12% per annum is
   (A) $ 2.16
   (B) $21.60
   (C) $ .22
   (D) $0.00

9. What is the compound interest on $600, compounded quarterly, at 6% for 9 months?
   (A) $27.38
   (B) $27.40
   (C) $27.41
   (D) $27.42

10. A 90-day note for $1200 is signed on May 12. Seventy-five days later the note is deposited at a bank that charges 8% discount on notes. The bank discount is
    (A) $8.40
    (B) $2.60
    (C) $2.00
    (D) $4.00

Interest Problems — Correct Answers

1.	(C)	6.	(A)
2.	(C)	7.	(A)
3.	(B)	8.	(A)
4.	(B)	9.	(C)
5.	(A)	10.	(D)

Problem Solutions — Interest

1. Principal = $460
    Rate = $8\frac{1}{2}\%$ = .085
    Time = 2 years
    Interest = $460 × .085 × 2
    = $78.20

    Answer: (C) $78.20

2. Principal = $300
    Interest = $6
    Time = $\frac{1}{12}$ year
    $300 × $\frac{1}{12}$ = $25
    $6 ÷ $25 = .24 = 24%

    Answer: (C) 24%

3. Rate = 5% = .05
    Interest = $12.50
    Time = $\frac{1}{2}$ year
    .05 × $\frac{1}{2}$ = .025
    $12.50 ÷ .025 = $500.00

    Answer: (B) $500

4. Time:
    2 months 15 days = 75 days or $\frac{75}{360}$ of a year
    Rate:
    $$10\frac{1}{2}\% = \frac{21}{2}\% = \frac{21}{200}$$
    Interest:
    $$480 × \frac{21}{200} × \frac{75}{360} = \frac{21}{2} = 10.50$$

    Answer: (B) $10.50

5.
$$\text{Principal} = \$300$$
$$\text{Rate} = .06 = \tfrac{6}{100}$$
$$\text{Time} = \tfrac{10}{360} = \tfrac{1}{36}$$
$$\text{Interest} = \$\overset{3}{\cancel{300}} \times \tfrac{\overset{1}{\cancel{6}}}{\cancel{100}} \times \tfrac{1}{\underset{6}{\cancel{36}}}$$
$$= \tfrac{3}{6} = \$.50$$

Answer: **(A)** $.50

6.
$$\text{Principal} = \$200$$
$$\text{Rate} = .06 = \tfrac{6}{100}$$

Time from Sept. 30 until Dec. 15 is 76 days. (31 days in October, 30 days in November, 15 days in December)

$$76 \text{ days} = \tfrac{76}{360} \text{ year}$$
$$\text{Interest} = \$\overset{2}{\cancel{200}} \times \tfrac{\overset{1}{\cancel{6}}}{\cancel{100}} \times \tfrac{76}{\underset{60}{\cancel{360}}}$$
$$= \$\tfrac{152}{60} = \$2.53$$
$$\$200 + \$2.53 = \$202.53$$

Answer: **(A)** closest to $202.50

7.
$$\text{Principal} = \$300$$
$$\text{Interest} = \$18$$
$$\text{Time} = \tfrac{9}{12} \text{ years} = \tfrac{3}{4} \text{ year}$$
$$\$300 \times \tfrac{3}{4} = \$225$$
$$\$18 \div \$225 = .08$$

Rate is 8%.

To yield $120 at 8% in 6 months,
$$\text{Interest} = \$120$$
$$\text{Rate} = .08$$
$$\text{Time} = \tfrac{1}{2} \text{ year}$$
$$.08 \times \tfrac{1}{2} = .04$$
$$\$120 \div .04 = \$3000 \text{ must be invested}$$

Answer: **(A)** $3000

8. Simple interest:
$$\text{Principal} = \$600$$
$$\text{Rate} = .12$$
$$\text{Time} = 1$$
$$\text{Interest} = \$600 \times .12 \times 1$$
$$= \$72.00$$

Compound interest:
$$\text{Principal} = \$600$$
$$\text{Period of compounding} = \tfrac{1}{2} \text{ year}$$
$$\text{Rate} = .12$$

For the first period,
$$\text{Interest} = \$600 \times .12 \times \tfrac{1}{2}$$
$$= \$36$$
$$\text{New principal} = \$600 + \$36$$
$$= \$636$$
For the second period,
$$\text{Interest} = \$636 \times .12 \times \tfrac{1}{2}$$
$$= \$38.16$$
$$\text{New principal} = \$636 + \$38.16$$
$$= \$674.16$$
$$\text{Total interest} = \$74.16$$
$$\text{Difference} = \$74.16 - 72.00$$
$$= \$2.16$$

Answer: **(A)** $2.16

9.
$$\text{Principal} = \$600$$
$$\text{Rate} = 6\% = \tfrac{6}{100}$$
$$\text{Time (period of compounding)} = \tfrac{3}{12} \text{ year} = \tfrac{1}{4} \text{ year}$$

In 9 months, the interest will be computed 3 times.

For first quarter,
$$\text{Interest} = \$600 \times \tfrac{6}{100} \times \tfrac{1}{4}$$
$$= \$9$$

New principal at end of first quarter:
$$\$600 + \$9 = \$609$$

For second quarter,
$$\text{Interest} = \$609 \times \tfrac{6}{100} \times \tfrac{1}{4}$$
$$= \$\tfrac{3654}{400} = \$9.135,$$
$$\text{or } \$9.14$$

New principal at end of second quarter:
$$\$609 + \$9.14 = \$618.14$$

For third quarter,
$$\text{Interest} = \$618.14 \times \tfrac{6}{100} \times \tfrac{1}{4}$$
$$= \$\tfrac{3708.84}{400}$$
$$= \$9.27$$

Total interest for the 3 quarters:
$$\$9 + \$9.14 + \$9.27 = \$27.41$$

Answer: **(C)** $27.41

10.
$$\text{Principal} = \$1200$$
$$\text{Time} = 90 \text{ days} - 75 \text{ days}$$
$$= 15 \text{ days}$$
$$15 \text{ days} = \tfrac{15}{360} \text{ year}$$
$$\text{Rate} = 8\% = \tfrac{8}{100}$$

$$\text{Bank discount} = \$\overset{12}{\cancel{1200}} \times \tfrac{\overset{1}{\cancel{8}}}{\cancel{100}} \times \tfrac{15}{\underset{45}{\cancel{360}}}$$
$$= \$\tfrac{180}{45} = \$4$$

Answer: **(D)** $4.00

TAXATION

1. The following facts should be taken into consideration when computing taxation problems:

   a. Taxes may be expressed as a percent or in terms of money based on a certain denomination.

   b. A **surtax** is an additional tax besides the regular tax rate.

2. In taxation, there are usually three items involved: the amount taxable, henceforth called the base, the tax rate, and the tax itself.

3. To find the tax when given the base and the tax rate in percent:

   a. Change the tax rate to a decimal.

   b. Multiply the base by the tax rate.

   Illustration: How much would be realized on $4000 if taxed 15%?

   SOLUTION: 15% = .15
                      $4000 × .15 = $600

   Answer: Tax = $600

4. To find the tax rate in percent form when given the base and the tax:

   a. Divide the tax by the base.

   b. Convert to a percent.

   Illustration: Find the tax rate at which $5600 would yield $784.

   SOLUTION: $784 ÷ $5600 = .14
                       .14 = 14%

   Answer: Tax rate = 14%

5. To find the base when given the tax rate and the tax:

   a. Change the tax rate to a decimal.

   b. Divide the tax by the tax rate.

   Illustration: What amount of money taxed 3% would yield $75?

   SOLUTION: 3% = .03
                      $75 ÷ .03 = $2500

   Answer: Base = $2500

6. When the tax rate is fixed and expressed in terms of money, take into consideration the denomination upon which it is based; that is, whether it is based on every $100, or $1000, etc.

7. To find the tax when given the base and the tax rate in terms of money:

   a. Divide the base by the denomination upon which the tax rate is based.

   b. Multiply this quotient by the tax rate.

   Illustration: If the tax rate is $3.60 per $1000, find the tax on $470,500.

   SOLUTION: $470,500 ÷ $1000 = 470.5
   $$470.5 \times \$3.60 = \$1,693.80$$

   Answer: $1,693.80

8. To find the tax rate based on a certain denomination when given the base and the tax derived:

   a. Divide the base by the denomination indicated.

   b. Divide the tax by this quotient.

   Illustration: Find the tax rate per $100 that would be required to raise $350,000 on $2,000,000 of taxable property.

   SOLUTION: $2,000,000 ÷ $100 = 20,000
   $$\$350,000 ÷ 20,000 = \$17.50$$

   Answer: Tax rate = $17.50 per $100

9. Since a surtax is an additional tax besides the regular tax, to find the total tax:

   a. Change the regular tax rate to a decimal.

   b. Multiply the base by the regular tax rate.

   c. Change the surtax rate to a decimal.

   d. Multiply the base by the surtax rate.

   e. Add both taxes.

   Illustration: Assuming that the tax rate is $2\frac{1}{3}\%$ on liquors costing up to $3.00, and 3% on those costing from $3.00 to $6.00, and $3\frac{1}{2}\%$ on those from $6.00 to $10.00, what would be the tax on a bottle costing $8.00 if there is a surtax of 5% on all liquors above $5.00?

   SOLUTION: An $8.00 bottle falls within the category of $6.00 to $10.00. The tax rate on such a bottle is

   $$3\tfrac{1}{2}\% = .035$$
   $$\$8.00 \times .035 = \$.28$$
   $$\text{surtax rate} = 5\% = .05$$
   $$\$8.00 \times .05 = \$.40$$
   $$\$.28 + \$.40 = \$.68$$

   Answer: Total tax = $.68

Practice Problems Involving Taxation

1. Mr. Jones' income for a year is $15,000. He pays $2250 for income taxes. The percent of his income that he pays for income taxes is
   (A) 9   (C) 15
   (B) 12   (D) 22

2. If the tax rate is $3\frac{1}{2}\%$ and the amount to be raised is $64.40, what is the base?
   (A) $1800   (C) $1850
   (B) $1840   (D) $1860

3. What is the tax rate per $1000 if a base of $338,500 would yield $616.07?
   (A) $1.80   (C) $1.95
   (B) $1.90   (D) $1.82

4. A man buys an electric light bulb for 54¢, which includes a 20% tax. What is the cost of the bulb without tax?
   (A) 43¢   (C) 45¢
   (B) 44¢   (D) 46¢

5. What tax rate on a base of $3650 would raise $164.25?
   (A) 4%   (C) $4\frac{1}{2}\%$
   (B) 5%   (D) $5\frac{1}{2}\%$

6. A piece of property is assessed at $22,850 and the tax rate is $4.80 per thousand. What is the amount of tax that must be paid on the property?
   (A) $109   (C) $109.68
   (B) $112   (D) $112.68

7. $30,000 worth of land is assessed at 120% of its value. If the tax rate is $5.12 per $1000 assessed valuation, the amount of tax to be paid is
   (A) $180.29   (C) $190.10
   (B) $184.32   (D) $192.29

8. Of the following real estate tax rates, which is the largest?
   (A) $31.25 per $1000   (C) 32¢ per $10
   (B) $3.45 per $100   (D) 3¢ per $1

9. A certain community needs $185,090.62 to cover its expenses. If its tax rate is $1.43 per $100 of assessed valuation, what must be the assessed value of its property?
   (A) $12,900,005   (C) $12,940,000
   (B) $12,943,400   (D) $12,840,535

10. A man's taxable income is $14,280. The state tax instructions tell him to pay 2% on the first $3000 of his taxable income, 3% on each of the second and third $3000, and 4% on the remainder. What is the total amount of income tax that he must pay?
    (A) $265.40   (C) $451.20
    (B) $309.32   (D) $454.62

Taxation Problems — Correct Answers

1. **(C)**
2. **(B)**
3. **(D)**
4. **(C)**
5. **(C)**

6. **(C)**
7. **(B)**
8. **(B)**
9. **(B)**
10. **(C)**

Problem Solutions — Taxation

1. $$\text{Tax} = \$2250$$
    $$\text{Base} = \$15,000$$
    $$\text{Tax rate} = \text{Tax} \div \text{Base}$$
    $$\text{Tax rate} = \$2250 \div \$15,000 = .15$$
    $$\text{Tax rate} = .15 = 15\%$$

    Answer: **(C)** 15

2. $$\text{Tax rate} = 3\tfrac{1}{2}\% = .035$$
    $$\text{Tax} = \$64.40$$
    $$\text{Base} = \text{Tax} \div \text{Tax rate}$$
    $$\text{Base} = \$64.40 \div .035$$
    $$= \$1840$$

    Answer: **(B)** $1840

3. $$\text{Base} = \$338,500$$
    $$\text{Tax} = \$616.07$$
    $$\text{Denomination} = \$1000$$
    $$\$338,500 \div \$1000 = 338.50$$
    $$\$616.07 \div 338.50 = \$1.82 \text{ per } \$1000$$

    Answer: **(D)** $1.82

4. 54¢ is 120% of the base (cost without tax)
    $$\text{Base} = 54 \div 120\%$$
    $$= 54 \div 1.20$$
    $$= 45$$

    Answer: **(C)** 45¢

5. $$\text{Base} = \$3650$$
    $$\text{Tax} = \$164.25$$
    $$\text{Tax rate} = \text{Tax} \div \text{Base}$$
    $$= \$164.25 \div \$3650$$
    $$= .045$$
    $$= 4\tfrac{1}{2}\%$$

    Answer: **(C)** $4\tfrac{1}{2}\%$

6. $$\text{Base} = \$22,850$$
    $$\text{Denomination} = \$1000$$
    $$\text{Tax rate} = \$4.80 \text{ per thousand}$$
    $$\frac{\$22,850}{\$1000} = 22.85$$
    $$22.85 \times \$4.80 = \$109.68$$

    Answer: **(C)** $109.68

7. $$\text{Base} = \text{Assessed valuation} = 120\% \text{ of } \$30,000$$
    $$= 1.20 \times \$30,000$$
    $$= \$36,000$$
    $$\text{Denomination} = \$1000$$
    $$\text{Tax rate} = \$5.12 \text{ per thousand}$$
    $$\frac{\$36,000}{\$1000} = 36$$
    $$36 \times \$5.12 = \$184.32$$

    Answer: **(B)** $184.32

8. Express each tax rate as a decimal:
    $$\$31.25 \text{ per } \$1000 = \frac{31.25}{1000} = .03125$$
    $$\$3.45 \text{ per } \$100 = \frac{3.45}{100} = .0345$$
    $$32\text{¢ per } \$10 = \frac{.32}{10} = .0320$$
    $$3\text{¢ per } \$1 = \frac{.03}{1} = .0300$$

    The largest decimal is .0345

    Answer: **(B)** $3.45 per $100

9. $$\text{Tax rate} = \$1.43 \text{ per } \$100$$
    $$= \frac{1.43}{100} = .0143$$
    $$= 1.43\%$$
    $$\text{Tax} = \$185,090.62$$
    $$\text{Base} = \text{Tax} \div \text{rate}$$
    $$= 185,090.62 \div .0143$$
    $$= \$12,943,400$$

    Answer: **(B)** $12,943,400

10. First $3000: $\quad$ .02 × $3000 = $ 60.00
    Second $3000: $\quad$ .03 × $3000 = $ 90.00
    Third $3000: $\quad$ .03 × $3000 = $ 90.00
    Remainder
    ($14,280 − $9000): .04 × $5280 = $211.20
    $$\text{Total tax} = \$451.20$$

    Answer: **(C)** $451.20

PROFIT AND LOSS

1. The following terms may be encountered in profit and loss problems:

   a. The **cost price** of an article is the price paid by a person who wishes to sell it again.

   b. There may be an **allowance** or **trade discount** on the cost price.

   c. The **list price** or **marked price** is the price at which the article is listed or marked to be sold.

   d. There may be a **discount** or **series of discounts** on the list price.

   e. The **selling price** or **sales price** is the price at which the article is finally sold.

   f. If the selling price is greater than the cost price, there has been a **profit**.

   g. If the selling price is lower than the cost price, there has been a **loss**.

   h. If the article is sold at the same price as the cost, there has been no loss or profit.

   i. Profit or loss may be based either on the cost price or on the selling price.

   j. Profit or loss may be stated in terms of dollars and cents, or in terms of percent.

   k. **Overhead** expenses include such items as rent, salaries, etc. Overhead expenses may be added to cost price to determine total cost when calculating profit or assigning selling price.

2. The basic formulas used in profit and loss problems are:

   $$\text{Selling price} = \text{cost price} + \text{profit}$$
   $$\text{Selling price} = \text{cost price} - \text{loss}$$

   Example: If the cost of an article is \$2.50, and the profit is \$1.50, then the selling price is \$2.50 + \$1.50 = \$4.00.

   Example: If the cost of an article is \$3.00, and the loss is \$1.20, then the selling price is \$3.00 − \$1.20 = \$1.80.

3. a. To find the profit in terms of money, subtract the cost price from the selling price, or selling price − cost price = profit.

   Example: If an article costing \$3.00 is sold for \$5.00, the profit is \$5.00 − \$3.00 = \$2.00.

   b. To find the loss in terms of money, subtract the selling price from the cost price, or: cost price − selling price = loss.

   Example: If an article costing \$2.00 is sold for \$1.50, the loss is \$2.00 − \$1.50 = \$.50.

4. To find the selling price if the profit or loss is expressed in percent based on cost price:

   a. Multiply the cost price by the percent of profit or loss to find the profit or loss in terms of money.

   b. Add this product to the cost price if a profit is involved, or subtract for a loss.

   Illustration: Find the selling price of an article costing $3.00 that was sold at a profit of 15% of the cost price.

   SOLUTION: 15% of $3.00 = .15 × $3.00
   $$= \$.45 \text{ profit}$$
   $$\$3.00 + \$.45 = \$3.45$$

   Answer: Selling price = $3.45

   Illustration: If an article costing $2.00 is sold at a loss of 5% of the cost price, find the selling price.

   SOLUTION: 5% of $2.00 = .05 × $2.00
   $$= \$.10 \text{ loss}$$
   $$\$2.00 - \$.10 = \$1.90$$

   Answer: Selling price = $1.90

5. To find the cost price when given the selling price and the percent of profit or loss based on the selling price:

   a. Multiply the selling price by the percent of profit or loss to find the profit or loss in terms of money.

   b. Subtract this product from the selling price if profit, or add the product to the selling price if a loss.

   Illustration: If an article sells for $12.00 and there has been a profit of 10% of the selling price, what is the cost price?

   SOLUTION: 10% of $12.00 = .10 × $12.00
   $$= \$1.20 \text{ profit}$$
   $$\$12.00 - \$1.20 = \$10.80$$

   Answer: Cost price = $10.80

   Illustration: What is the cost price of an article selling for $2.00 on which there has been a loss of 6% of the selling price?

   SOLUTION: 6% of $2.00 = .06 × $2.00
   $$= \$.12 \text{ loss}$$
   $$\$2.00 + \$.12 = \$2.12$$

   Answer: Cost price = $2.12

6. To find the percent of profit or percent of loss based on cost price:

   a. Find the profit or loss in terms of money.

   b. Divide the profit or loss by the cost price.

   c. Convert to a percent.

Illustration: Find the percent of profit based on cost price of an article costing $2.50 and selling for $3.00.

SOLUTION: $3.00 − $2.50 = $.50 profit

$$2.50 \overline{)\ .50} = 250 \overline{)\ 50.00}^{.20}$$
$$.20 = 20\%$$

Answer: Profit = 20%

Illustration: Find the percent of loss based on cost price of an article costing $5.00 and selling for $4.80.

SOLUTION: $5.00 − $4.80 = $.20 loss

$$5.00 \overline{)\ .20} = 500 \overline{)\ 20.00}^{.04}$$
$$.04 = 4\%$$

Answer: Loss = 4%

7. To find the percent of profit or percent of loss on selling price:

   a. Find the profit or loss in terms of money.

   b. Divide the profit or loss by the selling price.

   c. Convert to a percent.

   Illustration: Find the percent of profit based on the selling price of an article costing $2.50 and selling for $3.00.

   SOLUTION: $3.00 − $2.50 = $.50 profit
   $$3.00 \overline{)\ .50} = 300 \overline{)\ 50.00} = .16\tfrac{2}{3}$$
   $$= 16\tfrac{2}{3}\%$$

   Answer: Profit = $16\tfrac{2}{3}\%$

   Illustration: Find the percent of loss based on the selling price of an article costing $5.00 and selling for $4.80.

   SOLUTION: $5.00 − $4.80 = $.20 loss
   $$4.80 \overline{)\ .20} = 480 \overline{)\ 20.00} = .04\tfrac{1}{6}$$
   $$= 4\tfrac{1}{6}\%$$

   Answer: Loss = $4\tfrac{1}{6}\%$

8. To find the cost price when given the selling price and the percent of profit based on the cost price:

   a. Establish a relation between the selling price and the cost price.

   b. Solve to find the cost price.

   Illustration: An article is sold for $2.50, which is a 25% profit of the cost price. What is the cost price?

   SOLUTION: Since the selling price represents the whole cost price plus 25% of the cost price,

   $$2.50 = 125\% \text{ of the cost price}$$
   $$2.50 = 1.25 \text{ of the cost price}$$
   $$\text{Cost price} = 2.50 \div 1.25$$
   $$= 2.00$$

   Answer: Cost price = $2.00

9. To find the selling price when given the profit based on the selling price:

   a. Establish a relation between the selling price and the cost price.

   b. Solve to find the selling price.

   Illustration: A merchant buys an article for $27.00 and sells it at a profit of 10% of the selling price. What is the selling price?

   SOLUTION: $27.00 + profit = selling price
   Since the profit is 10% of the selling price, the cost price must be 90% of the selling price.

   $$27.00 = 90\% \text{ of the selling price}$$
   $$= .90 \text{ of the selling price}$$
   $$\text{Selling price} = 27.00 \div .90$$
   $$= 30.00$$

   Answer: Selling price = $30.00

Trade Discounts

10. A **trade discount**, usually expressed in percent, indicates the part that is to be deducted from the list price.

11. To find the selling price when given the list price and the trade discount:

    a. Multiply the list price by the percent of discount to find the discount in terms of money.

    b. Subtract the discount from the list price.

    Illustration: The list price of an article is $20.00. There is a discount of 5%. What is the selling price?

    SOLUTION: $20.00 × 5% = 20.00 × .05 = $1.00 discount
    $20.00 − $1.00 = $19.00

    Answer: Selling price = $19.00

    An alternate method of solving the above problem is to consider the list price to be 100%. Then, if the discount is 5%, the selling price is 100% − 5% = 95% of the list price. The selling price is
    $$95\% \text{ of } \$20.00 = .95 × \$20.00$$
    $$= \$19.00$$

Series of Discounts

12. There may be more than one discount to be deducted from the list price. These are called a **discount series**.

13. To find the selling price when given the list price and a discount series:

    a. Multiply the list price by the first percent of discount.

    b. Subtract this product from the list price.

c. Multiply the difference by the second discount.

d. Subtract this product from the difference.

e. Continue the same procedure if there are more discounts.

Illustration: Find the selling price of an article listed at $10.00 on which there are discounts of 20% and 10%.

SOLUTION: $10.00 \times 20\% = 10.00 \times .20 = \$2.00$
$\$10.00 - \$2.00 = \$8.00$
$\$8.00 \times 10\% = 8.00 \times .10 = \$.80$
$\$8.00 - \$.80 = \$7.20$

Answer: Selling price = $7.20

14. Instead of deducting each discount individually, it is often more practical to find the single equivalent discount first and then deduct. It does not matter in which order the discounts are taken.

15. The single equivalent discount may be found by assuming a list price of 100%. Leave all discounts in percent form.

a. Subtract the first discount from 100%, giving the net cost factor (NCF) had there been only one discount.

b. Multiply the NCF by the second discount. Subtract the product from the NCF, giving a second NCF that reflects both discounts.

c. If there is a third discount, multiply the second NCF by it and subtract the product from the second NCF, giving a third NCF that reflects all three discounts.

d. If there are more discounts, repeat the process.

e. Subtract the final NCF from 100% to find the single equivalent discount.

Illustration: Find the single equivalent discount of 20%, 25%, and 10%.

SOLUTION:

	100%	
	− 20%	first discount
	80%	first NCF
−25% of 80% =	20%	
	60%	second NCF
−10% of 60% =	6%	
	54%	third NCF

$100\% - 54\% = 46\%$ single equivalent discount

Answer: 46%

Illustration: An article lists at $750.00. With discounts of 20%, 25%, and 10%, what is the selling price of this article?

SOLUTION: As shown above, the single equivalent discount of 20%, 25%, and 10% is 46%.

$$46\% \text{ of } \$750 = .46 \times \$750$$
$$= \$345$$
$$\$750 - \$345 = \$405$$

Answer: Selling price = $405

Practice Problems Involving Profit and Loss

1. Dresses sold at $65.00 each. The dresses cost $50.00 each. The percentage of increase of the selling price over the cost is
   (A) 40
   (B) $33\frac{1}{3}$
   (C) $33\frac{1}{2}$
   (D) 30

2. A dealer bought a ladder for $27.00. What must it be sold for if he wishes to make a profit of 40% on the selling price?
   (A) $38.80
   (B) $43.20
   (C) $45.00
   (D) $67.50

3. A typewriter was listed at $120.00 and was bought for $96.00. What was the rate of discount?
   (A) $16\frac{2}{3}$%
   (B) 20%
   (C) 24%
   (D) 25%

4. A dealer sells an article at a loss of 50% of the cost. Based on the selling price, the loss is
   (A) 25%
   (B) 50%
   (C) 100%
   (D) none of these

5. What would be the marked price of an article if the cost was $12.60 and the gain was 10% of the cost price?
   (A) $11.34
   (B) $12.72
   (C) $13.86
   (D) $14.28

6. A stationer buys note pads at $.75 per dozen and sells them at 25 cents apiece. The profit based on the cost is
   (A) 50%
   (B) 300%
   (C) 200%
   (D) 100%

7. An article costing $18 is to be sold at a profit of 10% of the selling price. The selling price will be:
   (A) $19.80
   (B) $36.00
   (C) $18.18
   (D) $20.00

8. A calculating machine company offered to sell a city agency 4 calculating machines at a discount of 15% from the list price, and to allow the agency $85 for each of two old machines being traded in. The list price of the new machines is $625 per machine. If the city agency accepts this offer, the amount of money it will have to provide for the purchase of these 4 machines is
   (A) $1785
   (B) $2295
   (C) $1955
   (D) $1836

9. Pencils are purchased at $9 per gross and sold at 6 for 75 cents. The rate of profit based on the selling price is
   (A) 100%
   (B) 67%
   (C) 50%
   (D) 25%

10. The single equivalent discount of 20% and 10% is
    (A) 15%
    (B) 28%
    (C) 18%
    (D) 30%

Profit and Loss Problems — Correct Answers

1. **(D)**
2. **(C)**
3. **(B)**
4. **(C)**
5. **(C)**

6. **(B)**
7. **(D)**
8. **(C)**
9. **(C)**
10. **(B)**

Problem Solutions — Profit and Loss

1.   Selling price $-$ cost $= \$65 - \$50$
$$= \$15$$
$$\frac{\$15}{\$50} = .30 = 30\%$$

Answer: **(D)** 30%

2. Cost price $= 60\%$ of selling price, since the profit is 40% of the selling price, and the whole selling price is 100%.
$$\$27 = 60\% \text{ of selling price}$$
$$\text{Selling price} = \$27 \div 60\%$$
$$= \$27 \div .6$$
$$= \$45$$

Answer: **(C)** $45

3. The discount was $\$120 - \$96 = \$24$
$$\text{Rate of discount} = \frac{\$24}{\$120} = .20$$
$$= 20\%$$

Answer: **(B)** 20%

4. Loss $=$ cost $-$ selling price.
    Considering the cost to be 100% of itself, if the loss is 50% of the cost, the selling price is also 50% of the cost. (50% $= 100\% - 50\%$)
    Since the loss and the selling price are therefore the same, the loss is 100% of the selling price.

Answer: **(C)** 100%

5.   Gain (profit) $= 10\%$ of $12.60
$$= .10 \times \$12.60$$
$$= \$1.26$$
$$\text{Selling price} = \text{cost} + \text{profit}$$
$$= \$12.60 + \$1.26$$
$$= \$13.86$$

Answer: **(C)** $13.86

6. Each dozen note pads cost $.75 and are sold for
$$12 \times \$.25 = \$3.00$$
The profit is $\$3.00 - \$.75 = \$2.25$
$$\text{Profit based on cost} = \frac{\$2.25}{\$.75}$$
$$= 3$$
$$= 300\%$$

Answer: **(B)** 300%

7.   If profit $= 10\%$ of selling price, then cost $= 90\%$ of selling price
$$\$18 = 90\% \text{ of selling price}$$
$$\text{Selling price} = \$18 \div 90\%$$
$$= \$18 \div .90$$
$$= \$20$$

Answer: **(D)** $20.00

8. Discount for each new machine:
$$15\% \text{ of } \$625 = .15 \times \$625$$
$$= \$93.75$$
Each new machine will cost
$$\$625 - \$93.75 = \$531.25$$
Four new machines will cost
$$\$531.25 \times 4 = \$2125$$
But there is an allowance of $85 each for 2 old machines:
$$\$85 \times 2 = 170$$
Final cost to city:
$$\$2125 - \$170 = \$1955$$

Answer: **(C)** $1955

9.   1 gross $= 144$ units
Selling price for 6 pencils $= \$.75$
$$\text{Selling price for 1 pencil} = \frac{\$.75}{6}$$
$$\text{Selling price for 1 gross of pencils} = \frac{\$.75}{\overset{}{\underset{1}{6}}} \times \overset{24}{\cancel{144}}$$
$$= \$18.00$$
Cost for 1 gross of pencils $= \$9.00$
Profit for 1 gross of pencils $= \$18.00 - \$9.00$
$$= \$9.00$$
$$\frac{\text{profit}}{\text{selling price}} = \frac{\$9.00}{\$18.00}$$
$$= .5 = 50\%$$

Answer: **(C)** 50%

10.
$$
\begin{array}{r}
100\% \\
-\ \ 20\% \\
\hline
80\% \\
-10\% \text{ of } 80\% = -\ \ \ 8\% \\
\hline
72\% \\
\end{array}
$$
$100\% - 72\% = 28\%$ single equivalent discount

Answer: **(B)** 28%

PAYROLL

1. **Salaries** are computed over various time periods: hourly, daily, weekly, biweekly (every 2 weeks), semimonthly (twice each month), monthly, and yearly.

2. **Overtime** is usually computed as "time and a half"; that is, each hour in excess of the number of hours in the standard workday or workweek is paid at $1\frac{1}{2}$ time the regular hourly rate. Some companies pay "double time," twice the regular hourly rate, for work on Sundays and holidays.

   Illustration: An employee is paid weekly, based on a 40-hour workweek, with time and a half for overtime. If the employee's regular hourly rate is $4.50, how much will he earn for working 47 hours in one week?

   SOLUTION: Overtime hours = 47 − 40 = 7 hours
   Overtime pay = $1\frac{1}{2}$ × $4.50 = $6.75 per hour

   Overtime pay for 7 hours:
   7 × $6.75 = $47.25

   Regular pay for 40 hours:
   40 × $4.50 = $180.00
   Total pay = $47.25 + $180 = $227.25

   Answer: $227.25

3. a. In occupations such as retail sales, real estate, and insurance, earnings may be based on **commission**, which is a percent of the sales or a percent of the value of the transactions that are completed.

   b. Earnings may be from straight commission only, from salary plus commission, or from a commission that is graduated according to transaction volume.

   Illustration: A salesman earns a salary of $200 weekly, plus a commission based on sales volume for the week. The commission is 7% for the first $1500 of sales and 10% for all sales in excess of $1500. How much did he earn in a week in which his sales totaled $3200?

   SOLUTION: $3200 − $1500 = $1700   excess sales
   .07 × $1500 = $105   commission on first $1500
   .10 × $1700 = $170   commission on excess sales
   + $200   weekly salary
   $475   total earnings

   Answer: $475

4. **Gross pay** refers to the amount of money earned whether from salary, commission, or both, before any deductions are made.

5. **There are several deductions that are usually made from gross pay:**

a. **Withholding tax** is the amount of money withheld for income tax. It is based on wages, marital status, and number of exemptions (also called allowances) claimed by the employee. The withholding tax is found by referring to tables supplied by the federal, state or city governments.

Example:

Married Persons — Weekly Payroll Period

Wages		Number of withholding allowances claimed				
At least	But less than	0	1	2	3	4
		Amount of income tax to be withheld				
410	420	53	48	42	37	31
420	430	55	49	44	38	32
430	440	56	•51	45	40	34
440	450	58	52	47	41	35
450	460	59	54	48	43	37
460	470	61	55	50	44	38
470	480	62	57	51	46	40
480	490	64	58	53	47	41
490	500	65	60	54	49	43
500	510	67	61	56	50	44

Based on the above table, an employee who is married, claims three exemptions, and is paid a weekly wage of $434.50 will have $40.00 withheld for income tax. If the same employee earned $440 weekly it would be necessary to look on the next line for "at least $440 but less than $450" to find that $41.00 would be withheld.

b. The FICA (Federal Insurance Contribution Act) tax is also called the Social Security tax. In 1988, the FICA tax was 7.51% of the first $45,000 of annual wages; the wages in excess of $45,000 were not subject to the tax.

The FICA may be found by multiplying the wages up to and including $45,000 by .0751 or by using tables such as the one below.

Example:

Social Security Employee Tax Table

If wage payment is—	The employee tax to be deducted is—	If wage payment is—	The employee tax to be deducted is—
$61	$4.58	$81	$6.08
62	4.66	82	6.16
63	4.73	83	6.23
64	4.81	84	6.31
65	4.88	85	6.38
66	4.96	86	6.46
67	5.03	87	6.53
68	5.11	88	6.61
69	5.18	89	6.68
70	5.26	90	6.76
71	5.33	91	6.83
72	5.41	92	6.91
73	5.48	93	6.98
74	5.56	94	7.06
75	5.63	95	7.13
76	5.71	96	7.21
77	5.78	97	7.28
78	5.86	98	7.36
79	5.93	99	7.43
80	6.01	100	7.51

According to the table above, the Social Security tax, or FICA tax, on wages of $84.00 is $6.31. The FICA tax on $85.00 is $6.38.

Illustration: Based on 1988 tax figures, what is the total FICA tax on an annual salary of $30,000?

SOLUTION .0751 × $30,000 = $2253.00

Answer: $2253.00

c. Other deductions that may be made from gross pay are deductions for pension plans, loan payments, payroll savings plans, and union dues.

6. The **net pay**, or **take-home pay**, is equal to gross pay less the total deductions.

Illustration: Mr. Jay earns $550 salary per week with the following deductions: federal withholding tax, $106.70; FICA tax $41.31; state tax, $22.83; pension payment, $6.42; union dues, $5.84. How much take-home pay does he receive?

SOLUTION: Deductions $106.70
41.31
22.83
+   6.42
5.84
$183.10

Gross pay   = $550.00
Deductions   =   − 183.10
Net pay   = $366.90

Answer: His take-home pay is $366.90

Practice Problems Involving Payroll

1. Jane Rose's semimonthly salary is $750. Her yearly salary is
   (A) $9000
   (B) 12,500
   (C) $18,000
   (D) $19,500

2. John Doe earns $300 for a 40-hour week. If he receives time and a half for overtime, what is his hourly overtime wage?
   (A) $7.50
   (B) $9.25
   (C) $10.50
   (D) $11.25

3. Which salary is greater?
   (A) $350 weekly
   (B) $1378 monthly
   (C) $17,000 annually
   (D) $646 biweekly

4. A factory worker is paid on the basis of an 8-hour day, with an hourly rate of $3.50 and time and a half for overtime. Find his gross pay for a week in which he worked the following hours: Monday, 8; Tuesday, 9; Wednesday, 9½; Thursday, 8½; Friday, 9.
   (A) $140
   (B) $154
   (C) $161
   (D) $231

Questions 5 and 6 refer to the following table:

Single Persons — Weekly Payroll Period

Wages		Number of withholding allowances claimed				
At least	But less than	0	1	2	3	4
		Amount of income tax to be withheld				
340	350	49	43	37	32	26
350	360	50	45	39	33	28
360	370	52	46	40	35	29
370	380	55	48	42	36	31
380	390	58	49	43	38	32
390	400	60	51	45	39	34
400	410	63	53	46	41	35
410	420	66	55	48	42	37
420	430	69	58	49	44	38
430	440	72	61	51	45	40

5. If an employee is single and has one exemption, the income tax withheld from his weekly salary of $389.90 is
   (A) $51.00
   (B) $58.00
   (C) $49.00
   (D) $43.00

6. If a single person with two exemptions has $51.00 withheld for income tax, his weekly salary could *not* be
   (A) $430.00
   (B) $435.25
   (C) $437.80
   (D) $440.00

7. Sam Richards earns $1200 monthly. The following deductions are made from his gross pay monthly: federal withholding tax, $188.40; FICA tax, $84.60; state tax, $36.78; city tax, $9.24; savings bond, $37.50; pension plan, $5.32; repayment of pension loan, $42.30. His monthly net pay is
   (A) $795.86
   (B) $797.90
   (C) $798.90
   (D) $799.80

8. A salesman is paid a straight commission that is 23% of his sales. What is his commission on $1260 of sales?
   (A) $232.40
   (B) $246.80
   (C) $259.60
   (D) $289.80

9. Ann Johnson earns a salary of $150 weekly plus a commission of 9% of sales in excess of $500 for the week. For a week in which her sales were $1496, her earnings were
   (A) $223.64
   (B) $239.64
   (C) $253.64
   (D) $284.64

10. A salesperson is paid a 6% commission on the first $2500 of sales for the week, and 7½% on that portion of sales in excess of $2500. What is the commission earned in a week in which sales were $3280?
   (A) $196.80
   (B) $208.50
   (C) $224.30
   (D) $246.00

Payroll Problems — Correct Answers

1. (C)
2. (D)
3. (A)
4. (C)
5. (C)

6. (D)
7. (A)
8. (D)
9. (B)
10. (B)

Payroll Problems — Solutions

1. A semimonthly salary is paid twice a month. She receives $750 × 2 = $1500 each month, which is $1500 × 12 = $18,000 per year.

   Answer: **(C)** $18,000

2. The regular hourly rate is
   $$\$300 \div 40 = \$7.50$$
   The overtime rate is
   $$\$7.50 \times 1\tfrac{1}{2} = \$7.50 \times 1.5$$
   $$= \$11.25$$

   Answer: **(D)** $11.25

3. Write each salary as its yearly equivalent:
   $$\$350 \text{ weekly} = \$350 \times 52 \text{ yearly}$$
   $$= \$18,200 \text{ yearly}$$
   $$\$1378 \text{ monthly} = \$1378 \times 12 \text{ yearly}$$
   $$= \$16,536 \text{ yearly}$$
   $$\$17,000 \text{ annually} = \$17,000 \text{ yearly}$$
   $$\$646 \text{ biweekly} = \$646 \div 2 \text{ weekly}$$
   $$= \$323 \text{ weekly}$$
   $$= \$323 \times 52 \text{ yearly}$$
   $$= \$16,796 \text{ yearly}$$

   Answer: **(A)** $350 weekly

4. His overtime hours were:

Monday	0
Tuesday	1
Wednesday	$1\tfrac{1}{2}$
Thursday	$\tfrac{1}{2}$
Friday	1
Total	4 hours overtime

   Overtime rate per hour = $1\tfrac{1}{2} \times \$3.50$
   $$= 1.5 \times \$3.50$$
   $$= \$5.25$$
   Overtime pay = $4 \times \$5.25$
   $$= \$21$$

   Regular pay for 8 hours per day for 5 days or 40 hours.
   Regular pay = $40 \times \$3.50$
   $$= \$140$$
   Total wages = $\$140 + \$21$
   $$= \$161$$

   Answer: **(C)** $161

5. The correct amount is found on the line for wages of at least $380 but less than $390, and in the column under "1" withholding allowance. The amount withheld is $49.00

   Answer: **(C)** $49.00

6. In the column for 2 exemptions, or withholding allowances, $51.00 is found on the line for wages of at least $430, but less than $440. Choice (D) does not fall within that range.

   Answer: **(D)** $440

7. Deductions:
   $$\$188.40$$
   $$84.60$$
   $$36.78$$
   $$9.24$$
   $$37.50$$
   $$5.32$$
   $$+ \quad 42.30$$

Total	$404.14
Gross pay	= $1200.00
Total deductions	= − 404.14
	$ 795.86

   Answer: **(A)** $795.86

8. 23% of $1260 = .23 × $1260
   $$= \$289.80$$

   Answer: **(D)** $289.80

9. $1496 − 500 = $996 excess sales
   9% of $996 = .09 × $996
   $$= \$89.64 \text{ commission}$$

$150.00	salary
+ 89.64	commission
$239.64	total earnings

   Answer: **(B)** $239.64

10. $3280 − $2500 = $780 excess sales
    Commission on $2500:
    .06 × $2500 = $150.00
    Commission on $780:
    .075 × $780 = + 58.50
    Total = $208.50

    Answer: **(B)** $208.50

SEQUENCES

1. A **sequence** is a list of numbers based on a certain pattern. There are three main types of sequences:

   a. If each term in a sequence is being increased or diminished by the same number to form the next term, then it is an **arithmetic sequence**. The number being added or subtracted is called the **common difference**.

   Examples: 2, 4, 6, 8, 10 . . . is an arithmetic sequence in which the common difference is 2.

   14, 11, 8, 5, 2 . . . is an arithmetic sequence in which the common difference is 3.

   b. If each term of a sequence is being multiplied by the same number to form the next term, then it is a **geometric sequence**. The number multiplying each term is called the **common ratio**.

   Examples: 2, 6, 18, 54 . . . is a geometric sequence in which the common ratio is 3.

   64, 16, 4, 1 . . . is a geometric sequence in which the common ratio is $\frac{1}{4}$.

   c. If the sequence is neither arithmetic nor geometric, it is a **miscellaneous sequence**. Such a sequence may have each term a square or a cube, or the difference may be squares or cubes; or there may be a varied pattern in the sequence that must be determined.

2. A sequence may be ascending, that is, the numbers increase; or descending, that is, the numbers decrease.

3. To determine whether the sequence is arithmetic:

   a. If the sequence is ascending, subtract the first term from the second, and the second term from the third. If the difference is the same in both cases, the sequence is arithmetic.

   b. If the sequence is descending, subtract the second term from the first, and the third term from the second. If the difference is the same in both cases, the sequence is arithmetic.

4. To determine whether the sequence is geometric, divide the second term by the first, and the third term by the second. If the ratio is the same in both cases, the sequence is geometric.

5. To find a missing term in an arithmetic sequence that is ascending:

   a. Subtract any term from the one following it to find the common difference.

   b. Add the common difference to the term preceding the missing term.

   c. If the missing term is the first term, it may be found by subtracting the common difference from the second term.

   Illustration: What number follows $16\frac{1}{3}$ in this sequence:

   $$3,\ 6\frac{1}{3},\ 9\frac{2}{3},\ 13,\ 16\frac{1}{3} \ldots$$

   SOLUTION: $6\frac{1}{3} - 3 = 3\frac{1}{3},\ 9\frac{2}{3} - 6\frac{1}{3} = 3\frac{1}{3}$
   The sequence is arithmetic; the common difference is $3\frac{1}{3}$.

   $$16\frac{1}{3} + 3\frac{1}{3} = 19\frac{2}{3}$$

   Answer: The missing term, which is the term following $16\frac{1}{3}$, is $19\frac{2}{3}$.

6. To find a missing term in an arithmetic sequence that is descending:

   a. Subtract any term from the one preceding it to find the common difference.

   b. Subtract the common difference from the term preceding the missing term.

   c. If the missing term is the first term, it may be found by adding the common difference to the second term.

   Illustration: Find the first term in the sequence:

   $$\text{——},\ 16,\ 13\frac{1}{2},\ 11,\ 8\frac{1}{2},\ 6 \ldots$$

   SOLUTION: $16 - 13\frac{1}{2} = 2\frac{1}{2},\ 13\frac{1}{2} - 11 = 2\frac{1}{2}$
   The sequence is arithmetic; the common difference is $2\frac{1}{2}$.

   $$16 + 2\frac{1}{2} = 18\frac{1}{2}$$

   Answer: The term preceding 16 is $18\frac{1}{2}$.

7. To find a missing term in a geometric sequence:

   a. Divide any term by the one preceding it to find the common ratio.

   b. Multiply the term preceding the missing term by the common ratio.

   c. If the missing term is the first term, it may be found by dividing the second term by the common ratio.

   Illustration: Find the missing term in the sequence:

   $$2,\ 6,\ 18,\ 54,\ \text{——}$$

   SOLUTION: $6 \div 2 = 3,\ 18 \div 6 = 3$
   The sequence is geometric; the common ratio is 3.

   $$54 \times 3 = 162$$

   Answer: The missing term is 162.

Illustration: Find the missing term in the sequence:

$$———, 32, 16, 8, 4, 2$$

SOLUTION: $16 \div 32 = \frac{1}{2}$ (common ratio)

$$32 \div \frac{1}{2} = 32 \times \frac{2}{1}$$
$$= 64$$

Answer: The first term is 64.

8. If, after trial, a sequence is neither arithmetic nor geometric, it must be one of a miscellaneous type. Test to see whether it is a sequence of squares or cubes or whether the difference is the square or the cube of the same number; or the same number may be first squared, then cubed, etc.

Practice Problems Involving Sequences

Find the missing term in each of the following sequences:

1. ———, 7, 10, 13

2. 5, 10, 20, ———, 80

3. 49, 45, 41, ———, 33, 29

4. 1.002, 1.004, 1.006, ———

5. 1, 4, 9, 16, ———

6. 10, $7\frac{7}{8}$, $5\frac{3}{4}$, $3\frac{5}{8}$, ———

7. ———, 3, $4\frac{1}{2}$, $6\frac{3}{4}$

8. 55, 40, 28, 19, 13, ———

9. 9, 3, 1, $\frac{1}{3}$, $\frac{1}{9}$, ———

10. 1, 3, 7, 15, 31, ———

Sequence Problems — Correct Answers

1. 4
2. 40
3. 37
4. 1.008
5. 25

6. $1\frac{1}{2}$
7. 2
8. 10
9. $\frac{1}{27}$
10. 63

Problem Solutions — Sequences

1. This is an ascending arithmetic sequence in which the common difference is $10 - 7$, or 3. The first term is $7 - 3 = 4$.

2. This is a geometric sequence in which the common ratio is $10 \div 5$, or 2. The missing term is $20 \times 2 = 40$.

3. This is a descending arithmetic sequence in which the common difference is $49 - 45$, or 4. The missing term is $41 - 4 = 37$.

4. This is an ascending arithmetic sequence in which the common difference is $1.004 - 1.002$, or .002. The missing term is $1.006 + .002 = 1.008$.

5. This sequence is neither arithmetic nor geometric. However, if the numbers are rewritten as 1^2, 2^2, 3^2, and 4^2, it is clear that the next number must be 5^2, or 25.

6. This is a descending arithmetic sequence in which the common difference is $10 - 7\frac{7}{8} = 2\frac{1}{8}$. The missing term is $3\frac{5}{8} - 2\frac{1}{8} = 1\frac{4}{8}$, or $1\frac{1}{2}$.

7. This is a geometric sequence in which the common ratio is:

$$4\tfrac{1}{2} \div 3 = \tfrac{9}{2} \times \tfrac{1}{3}$$
$$= \tfrac{3}{2}$$

The first term is $3 \div \tfrac{3}{2} = 3 \times \tfrac{2}{3}$
$$= 2$$

Therefore, the missing term is 2.

8. There is no common difference and no common ratio in this sequence. However, note the differences between terms:

$$55 \underbrace{} 40 \underbrace{} 28 \underbrace{} 19 \underbrace{} 13$$

15	12	9	6
5×3	4×3	3×3	2×3

The differences are multiples of 3. Following the same pattern, the difference between 13 and the next term must be 1×3, or 3. The missing term is then $13 - 3 = 10$.

9. This is a geometric sequence in which the common ratio is $3 \div 9 = \tfrac{1}{3}$. The missing term is $\tfrac{1}{9} \times \tfrac{1}{3} = \tfrac{1}{27}$.

10. This sequence is neither arithmetic nor geometric. However, note the difference between terms:

$$1 \underbrace{} 3 \underbrace{} 7 \underbrace{} 15 \underbrace{} 31$$

2	4	8	16
2^1	2^2	2^3	2^4

The difference between 31 and the next term must be 2^5, or 32. The missing term is $31 + 32 = 63$.

ARITHMETIC REVIEW

Exam 1

1. A cashier can count 1286 coins in one hour. How many coins can he count in $3\frac{1}{2}$ hours?
   (A) 3880
   (B) 3902
   (C) 4253
   (D) 4501

2. Jane has two pieces of ribbon. One piece is $2\frac{3}{4}$ yards; the other, $2\frac{2}{3}$ yards. To make the two pieces equal she must cut off from the longer piece
   (A) 9 in
   (B) 8 in
   (C) 6 in
   (D) 3 in

3. The present size of a dollar bill in the United States is 2.61 inches by 6.14 inches. The number of square inches of paper used for one bill is closest to which one of the following?
   (A) 160 sq in
   (B) 17.8 sq in
   (C) 16.0 sq in
   (D) 1.78 sq in

4. The total saving in purchasing 30 fifty-cent rulers at a reduced rate of $4.60 per dozen is
   (A) $1.20
   (B) $1.75
   (C) $2.85
   (D) $3.50

5. The difference between one hundred five thousand eighty-four and ninety-three thousand seven hundred nine is
   (A) 11,375
   (B) 12,131
   (C) 56,294
   (D) 56,375

6. If a distance estimated at 150 feet is really 140 feet, the percent of error in this estimate is
   (A) $6\frac{2}{3}\%$
   (B) $7\frac{1}{7}\%$
   (C) 10%
   (D) 14%

7. Assuming that on a blueprint $\frac{1}{4}$ inch equals 12 inches, the actual length in feet of a steel bar represented on the blueprint by a line $3\frac{3}{8}$ inches long is
   (A) $3\frac{3}{8}$
   (B) $6\frac{3}{4}$
   (C) $12\frac{1}{2}$
   (D) $13\frac{1}{2}$

8. The difference between 320 centimeters and 3 meters is
   (A) 20 cm
   (B) 31.7 cm
   (C) 200 cm
   (D) 317 cm

9. If two of the angles of a triangle are 30° and 60°, the triangle is
   (A) right
   (B) acute
   (C) obtuse
   (D) equilateral

10. If A can do a job in three hours and B can do the same job in five hours, then both working together can finish the job in
    (A) 4 hr
    (B) $1\frac{3}{8}$ hr
    (C) $2\frac{1}{8}$ hr
    (D) $1\frac{7}{8}$ hr

11. If Mrs. Jones bought $3\frac{3}{4}$ yards of muslin at $1.16 per yard and $4\frac{3}{8}$ yards of polyester at $3.87 per yard, the amount of change she received from $25 is
    (A) $2.12
    (B) $2.28
    (C) $2.59
    (D) $2.63

12. A piece of cardboard in the shape of a 15-inch square is rolled so as to form a cylindrical surface, without overlapping. The number of inches in the diameter of the cylinder is approximately
    (A) 45
    (B) 23
    (C) 5
    (D) 2.5

13. The single commercial discount that is equivalent to successive discounts of 10% and 10% is
    (A) 20%
    (B) 19%
    (C) 17%
    (D) 15%

14. The value of forty thousand nickels is
    (A) $20
    (B) $200
    (C) $2000
    (D) $20,000

15. A checking account has a balance of $253.36. If deposits of $36.95, $210.23 and $7.34, and withdrawals of $117.35, $23.37 and $15.98 are made, what is the new balance of the account?
    (A) $155.54   (C) $364.58
    (B) $351.18   (D) $664.58

16. Six-percent simple annual interest on $2436.18 is most nearly
    (A) $145.08   (C) $146.08
    (B) $145.17   (D) $146.17

17. Suppose that a pile of 96 file cards measures one inch in height and that it takes you $\frac{1}{2}$ hour to file these cards. If you are given three piles of cards that measure $2\frac{1}{2}$ inches high, $1\frac{3}{4}$ inches high, and $3\frac{3}{8}$ inches high respectively, the time it would take to file the cards is most nearly
    (A) 2 hr 30 min   (C) 6 hr 45 min
    (B) 3 hr 50 min   (D) 8 hr 15 min

18. If the sum of the edges of a cube is 48 inches, the volume of the cube is
    (A) 512 cu in   (C) 64 cu in
    (B) 96 cu in   (D) 12 cu in

19. If the sum of 42.83 and 72.9 is subtracted from 200, the result is
    (A) 230.07   (C) 48.12
    (B) 169.93   (D) 84.27

20. Assuming there are 28.4 grams per ounce, the number of kilograms in 3 pounds is closest to
    (A) .45   (C) 1.36
    (B) .85   (D) 1.92

21. A drawer contains 5 red pens and 3 blue pens. If Mr. Jones takes a pen from the drawer without looking, what is the probability that he will take a blue pen?
    (A) $\frac{3}{8}$ (C) $\frac{3}{5}$
    (B) $\frac{1}{2}$ (D) $\frac{5}{8}$

22. If $1\frac{1}{2}$ cups of cereal are used with $4\frac{1}{2}$ cups of water, the amount of water needed with $\frac{3}{4}$ of a cup of cereal is:
    (A) 2 cups   (C) $2\frac{1}{8}$ cups
    (B) $2\frac{1}{4}$ cups   (D) $2\frac{1}{2}$ cups

23. The regular price of a TV set that sold for $118.80 at a 20% reduction sale is
    (A) $148.50   (C) $138.84
    (B) $142.60   (D) $ 95.04

24. The number missing in the sequence 2, 6, 12, 20, ?, 42, 56, 72 is
    (A) 30   (C) 36
    (B) 40   (D) 28

25. A rectangular flower bed whose dimensions are 16 yards by 12 yards is surrounded by a walk 3 yards wide. The area of the walk is
    (A) 93 sq yd   (C) 96 sq yd
    (B) 204 sq yd   (D) 150 sq yd

26. John drives 60 miles to his destination at an average speed of 40 mph and makes the return trip at an average speed of 30 mph. His average speed for the entire trip is
    (A) $32\frac{1}{5}$ mph   (C) 35 mph
    (B) $34\frac{2}{7}$ mph   (D) $43\frac{1}{3}$ mph

Questions 27 and 28 refer to the following graph:

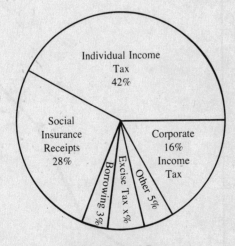

Origin of Federal Revenues of $352.6 Billion

27. The total individual income tax and corporate income tax revenues were, to the nearest billion dollars,
    (A) 58   (C) 205
    (B) 123   (D) 256

28. The revenue from excise tax was, to the nearest billion dollars,
    (A) 14   (C) 20
    (B) 15   (D) 21

29. If a worker has completed ⅝ of a job, what percent of the job remains to be completed?
    (A) 37½%   (C) 60%
    (B) 40%   (D) 62½%

30. After one year, Mr. Richards paid back a total of $1695.00 as payment for a $1500.00 loan. All the money paid over $1500.00 was simple interest. The interest charge was most nearly
    (A) 13%   (C) 9%
    (B) 11%   (D) 7%

31. If the City Department of Purchase bought 190 manual typewriters for $79.35 each and 208 manual typewriters for $83.99 each, the total price paid for these manual typewriters is
    (A) $31,581.30   (C) $33,427.82
    (B) $32,546.42   (D) $33,586.30

Questions 32 and 33 refer to the following graph:

32. In which month were approximately 270 cases investigated?
    (A) May   (C) July
    (B) June   (D) August

33. The total number of cases investigated during the first three months of the year was approximately
    (A) 330   (C) 390
    (B) 350   (D) 430

34. Twelve clerks are assigned to enter certain data on index cards. This number of clerks could perform the task in 18 days. After these clerks have worked on this assignment for 6 days, 4 more clerks are added to the staff to do this work. Assuming that all the clerks work at the same rate of speed, the entire task, instead of taking 18 days, will be performed in
    (A) 9 days   (C) 15 days
    (B) 12 days   (D) 16 days

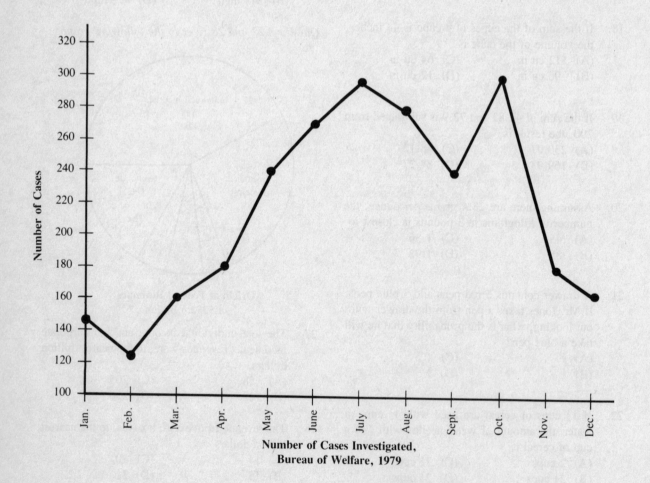

**Number of Cases Investigated,
Bureau of Welfare, 1979**

35. Smith earns $7.20 per hour for a 40-hour week, with time and a half for overtime hours. In a week in which he worked 46 hours, he earned
    (A) $331.20   (C) $424.20
    (B) $352.80   (D) $496.80

36. In the Fahrenheit scale, the temperature that is equivalent to 50° Celsius is
    (A) 122°   (C) 106°
    (B) 90°   (D) 87°

37. The circumference of a circle is 10π. The area of the same circle is
    (A) 5π (C) 25π
    (B) 10π (D) 100π

38. A bank pays 6% interest, compounded quarterly, on savings accounts. How much interest will $300 earn in 9 months?
    (A) $13.31   (C) $13.71
    (B) $13.51   (D) $13.91

39. A champion runner ran the 100-yard dash in three track meets. The first time, he ran it in 10.2 seconds; the second in 10.4 seconds; and the third time in 10 seconds. What was his average time?
    (A) 10.2 sec   (C) 10.35 sec
    (B) 10.3 sec   (D) 10.4 sec

40. Joshua Howard is paid a yearly salary of $18,000. His monthly paycheck shows the following deductions: federal income tax, $292.20; FICA, $91.95; state tax, $42.45; pension, $4.32. What is his yearly take-home pay?
    (A) $12,828.96   (C) $15,238.42
    (B) $13,366.53   (D) $17,569.08

41. Two adjacent walls of a 40′ by 35′ office are to be painted. The walls are 8′ high and include no doors or windows. If each gallon of the paint to be used covers 450 square feet, how many gallons are needed?
    (A) $1\frac{1}{3}$ (C) $2\frac{1}{3}$
    (B) $1\frac{1}{2}$ (D) $2\frac{1}{2}$

42. Mr. Harvey receives a salary of $300 per week plus 2% commission on sales. What were his total earnings for a week in which his sales were $5846?
    (A) $406.92   (C) $426.92
    (B) $416.92   (D) $436.92

43. If real estate tax is $1.62 per $100 assessed valuation, the tax that must be paid on property assessed at $82,200 is closest to
    (A) $152   (C) $1086
    (B) $694   (D) $1332

44. A worked 5 days on overhauling an old car. B worked 4 days more to finish the job. After the sale of the car, the net profit was $243. They wanted to divide the profit on the basis of the time spent by each. A's share of the profit was
    (A) $108   (C) $127
    (B) $135   (D) $143

45. Mr. Jones wishes to purchase one item priced at $24.50 and another item priced at $43.28. If sales tax is 5%, what is the total amount he must pay for the two items?
    (A) $67.78   (C) $71.16
    (B) $67.79   (D) $71.17

46. A circle graph shows that 32% of the tourists to a city are German, 28% are Spanish, 20% are English, 10% are miscellaneous, and the rest are French. How many degrees of the circle should be devoted to the French?
    (A) 12   (C) 30
    (B) 24   (D) 36

47. Of the following readings of rainfall — 1.2 inches, 2.4 inches, 2.2 inches, 3.5 inches, 4.3 inches, 2.3 inches, 4.2 inches, 3.9 inches, 3.0 inches, 3.3 inches, 2.9 inches, 3.6 inches, 4.5 inches, 4.7 inches, and 4.6 inches — the median is
    (A) 3.5   (C) 3.4
    (B) 3.3   (D) 3.0

48. If FICA tax is 6.13%, the FICA tax on wages of $450.70 is closest to
    (A) $27.60   (C) $27.80
    (B) $27.70   (D) $27.90

49. In a particular company, two employees received hourly wages of $4.50, three employees received hourly wages of $4.15, and five employees received hourly wages of $4.75. The average hourly wage of this group of employees is
   (A) $4.37
   (B) $4.47
   (C) $4.52
   (D) $4.63

50. A invested $7000 in a business venture and his partner, B, invested $8000. They agreed to share the profits in the same ratio. What was A's share of a profit of $2250?
   (A) $1050.00
   (B) $1200.00
   (C) $1968.75
   (D) $2250.00

Exam 1 — Correct Answers

1. **(D)**	11. **(C)**	21. **(A)**	31. **(B)**	41. **(A)**
2. **(D)**	12. **(C)**	22. **(B)**	32. **(B)**	42. **(B)**
3. **(C)**	13. **(B)**	23. **(A)**	33. **(D)**	43. **(D)**
4. **(D)**	14. **(C)**	24. **(A)**	34. **(C)**	44. **(B)**
5. **(A)**	15. **(B)**	25. **(B)**	35. **(B)**	45. **(D)**
6. **(B)**	16. **(D)**	26. **(B)**	36. **(A)**	46. **(D)**
7. **(D)**	17. **(B)**	27. **(C)**	37. **(C)**	47. **(A)**
8. **(A)**	18. **(C)**	28. **(D)**	38. **(C)**	48. **(A)**
9. **(A)**	19. **(D)**	29. **(A)**	39. **(A)**	49. **(C)**
10. **(D)**	20. **(C)**	30. **(A)**	40. **(A)**	50. **(A)**

Exam 1 — Solutions

1.
$$1286 \times 3\tfrac{1}{2} = \overset{643}{\cancel{1286}} \times \frac{7}{\cancel{2}}$$
$$= 4501$$

Answer: **(D)** 4501

2.
$$2\tfrac{3}{4} = 2\tfrac{9}{12}$$
$$- 2\tfrac{2}{3} = - 2\tfrac{8}{12}$$
$$\overline{\tfrac{1}{12}}$$

One piece is $\frac{1}{12}$ of a yard longer than the other.

$$1 \text{ yd} = 36 \text{ in}$$
$$\tfrac{1}{12} \text{ of } 1 \text{ yd} = \tfrac{1}{12} \times 36 \text{ in}$$
$$= 3 \text{ in}$$

Answer: **(D)** 3 in

3.
$$
\begin{array}{r}
2.61 \text{ in} \\
\times\ 6.14 \text{ in} \\
\hline
1044 \\
261 \\
15\ 66 \\
\hline
16.0254 \text{ sq in}
\end{array}
$$

Answer: **(C)** 16.0 sq in

4. $30 = 2\frac{1}{2}$ dozen

At the reduced rate, $2\frac{1}{2}$ dozen rulers will cost

$$2\frac{1}{2} \times \$4.60 = \$11.50$$

At the regular rate, 30 rulers will cost

$$30 \times \$.50 = \$15.00$$

The saving is
$$\begin{array}{r} \$15.00 \\ -\ 11.50 \\ \hline \$\ 3.50 \end{array}$$

Answer: **(D)** $3.50

5. $\begin{array}{r} 105,084 \\ -\ 93,709 \\ \hline 11,375 \end{array}$

Answer: **(A)** 11,375

6.   Estimated distance $= 150$ ft
Real distance $\quad= 140$ ft
Amount of error $\ \ = \ \ 10$ ft

$$10 \text{ ft} \div 140 \text{ ft} = .07\frac{1}{7} = 7\frac{1}{7}\%$$

Answer: **(B)** $7\frac{1}{7}\%$

7.   12 inches $= 1$ foot, therefore each $\frac{1}{4}$ inch on the blueprint represents 1 foot. $3\frac{3}{8}$ inches represent:

$$3\frac{3}{8} \text{ in} \div \frac{1}{4} \text{ in} = \frac{27}{\cancel{8}_2} \times \frac{\cancel{4}^1}{1} \text{ ft}$$
$$= \frac{27}{2} \text{ ft}$$
$$= 13\frac{1}{2} \text{ feet}$$

Answer: **(D)** $13\frac{1}{2}$

8.   3 meters $= 300$ centimeters
320 cm $-$ 300 cm $= 20$ cm

Answer: **(A)** 20 cm

9.   The sum of the angles of a triangle is 180°. If two of the angles are 30° and 60°, the third must be 90°, which is a right angle. The triangle is therefore a right triangle.

Answer: **(A)** right

10.   A can do $\frac{1}{3}$ of the job in 1 hour. B can do $\frac{1}{5}$ of the job in 1 hour. In each hour working together, they will complete $\frac{1}{3} + \frac{1}{5}$ of the job.

$$\frac{1}{3} + \frac{1}{5} = \frac{5}{15} + \frac{3}{15}$$
$$= \frac{8}{15}$$

It will take $\frac{15}{8}$ hours to complete the job working together.

$$\frac{15}{8} = 1\frac{7}{8}$$

Answer: **(D)** $1\frac{7}{8}$ hr

11.   Cost of muslin:
$$\$1.16 \times 3\frac{3}{4} = \$1.16 \times \frac{15}{4} = \$4.35$$
Cost of polyester:
$$\$3.87 \times 4\frac{2}{3} = \$3.87 \times \frac{14}{3} = \$18.06$$
Total cost:
$$\$4.35 + \$18.06 = \$22.41$$
Change:
$$\$25.00 - \$22.41 = \$2.59$$

Answer: **(C)** $2.59

12.   The circumference is 15″. To find the diameter, divide the circumference by π $\left(\frac{22}{7}\right)$.

$$15 \div \frac{22}{7} = 15 \times \frac{7}{22}$$
$$= \frac{105}{22}$$
$$= 4\frac{17}{22}$$

The diameter is approximately 5″.

Answer: **(C)** 5

13. $\begin{array}{r} 100\% \\ -\ 10\% \\ \hline 90\% \end{array}$

-10% of $90\% = \begin{array}{r} -\ 9\% \\ \hline 81\% \end{array}$

$100\% - 81\% = 19\%$ single equivalent discount

Answer: **(B)** 19%

14. $\begin{array}{r} 40,000 \\ \times \quad .05 \\ \hline 2000.00 \end{array}$

Answer: **(C)** $2000

15. Deposits:

$$\begin{array}{r} \$\ 36.95 \\ 210.23 \\ +\quad 7.34 \\ \hline \$254.52 \end{array}$$

Withdrawals:

$$\begin{array}{r} \$117.35 \\ 23.37 \\ +\quad 15.98 \\ \hline \$156.70 \end{array}$$

$$\begin{array}{rl} \$253.36 & \text{original balance} \\ +\quad 254.52 & \text{deposits} \\ \hline 507.88 & \\ -\quad 156.70 & \text{withdrawals} \\ \hline \$351.18 & \text{new balance} \end{array}$$

Answer: **(B)** $351.18

16.

$$\begin{array}{r} \$2436.18 \\ \times\qquad .06 \\ \hline \$146.1708 \end{array}$$

Answer: **(D)** $146.17

17. The cards total $2\frac{1}{2} + 1\frac{3}{4} + 3\frac{3}{8}$ inches in height.

$$\begin{array}{rl} 2\frac{1}{2} = & 2\frac{4}{8} \\ 1\frac{3}{4} = & 1\frac{6}{8} \\ +\ 3\frac{3}{8} = & +\ 3\frac{3}{8} \\ \hline & 6\frac{13}{8} = 7\frac{5}{8} \end{array}$$

Each inch takes $\frac{1}{2}$ hour to file. $7\frac{5}{8}$ inches will take $7\frac{5}{8} \times \frac{1}{2}$ hours.

$$7\frac{5}{8} \times \frac{1}{2} = \frac{61}{8} \times \frac{1}{2}$$
$$= \frac{61}{16}$$
$$= 3\frac{13}{16}$$

1 hour = 60 min

$$\frac{13}{16} \text{ of } 60 = \frac{13}{16} \times \overset{15}{\underset{4}{60}} = \frac{195}{4} = 48\frac{3}{4} \text{ min}$$

Notice that only one of the answer choices is possible once you know that it will take 3+ hours. It is not really necessary to solve for the exact number of minutes.

Answer: **(B)** 3 hr 50 min

18.

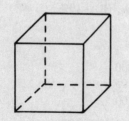

A cube has 12 edges. If the sum of the edges is 48 inches, each edge is 48 inches ÷ 12 = 4 inches.

The volume of the cube = 4^3 cubic inches
$$= 4 \times 4 \times 4 \text{ cu in}$$
$$= 64 \text{ cu in}$$

Answer: **(C)** 64 cu in

19. Sum of 42.83 and 72.9:

$$\begin{array}{r} 42.83 \\ +\ 72.9 \\ \hline 115.73 \end{array}$$

Subtract from 200:

$$\begin{array}{r} 200.00 \\ -\ 115.73 \\ \hline 84.27 \end{array}$$

Answer: **(D)** 84.27

20.

$$\begin{array}{rl} 3 \text{ pounds} = & 3 \times 16 \text{ ounces} \\ = & 48 \text{ ounces} \end{array}$$

$$48 \text{ ounces} \times \begin{array}{c} 28.4 \text{ grams} \\ \text{per ounce} \end{array} = 1363.2 \text{ grams}$$

$$1 \text{ kilogram} = 1000 \text{ grams}$$
$$1363.2 \text{ grams} \div 1000 = 1.3632 \text{ kilograms}$$

Answer: **(C)** 1.36

21. There is a total of 8 pens in the drawer. The probability that the pen is blue is $\frac{3}{8}$.

Answer: **(A)** $\frac{3}{8}$

22. If the cereal is reduced by half, the water must also be halved.

$$\frac{1}{2} \times 4\frac{1}{2} = \frac{1}{2} \times \frac{9}{2}$$
$$= \frac{9}{4}$$
$$= 2\frac{1}{4}$$

Answer: **(B)** $2\frac{1}{4}$ cups

23. The sale price is 80% of the regular price.

$$\begin{array}{rl} \text{Regular price} = & \$118.80 \div .80 \\ = & \$148.50 \end{array}$$

Answer: **(A)** $148.50

24. Find the differences between the terms of the sequence:

    2___6___12___20 — 42___56___72
      4   6   8   14   16

    The difference between 20 and the missing term must be 10. Therefore, the missing term is 30.

    Answer: **(A)** 30

25.

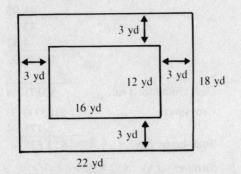

    Area of large rectangle is:
    $$22 \text{ yd} \times 18 \text{ yd} = 396 \text{ sq yd}$$
    Area of small rectangle is:
    $$16 \text{ yd} \times 12 \text{ yd} = 192 \text{ sq yd}$$
    Area of path is difference or
    $$396 \text{ sq yd} - 192 \text{ sq yd} = 204 \text{ sq yd}$$

    Answer: **(B)** 204 sq yd

26. It took 60 miles $\div$ 40 mph = $1\frac{1}{2}$ hours to drive to his destination. The return trip took 60 miles $\div$ 30 mph = 2 hours. The entire trip was 120 miles and took $3\frac{1}{2}$ hours.

    $$\begin{aligned}
    \text{Average speed} &= 120 \text{ mi} \div 3\frac{1}{2} \text{ hr} \\
    &= 120 \div \tfrac{7}{2} \text{ mph} \\
    &= 120 \times \tfrac{2}{7} \text{ mph} \\
    &= \tfrac{240}{7} \text{ mph} \\
    &= 34\tfrac{2}{7} \text{ mph}
    \end{aligned}$$

    Answer: **(B)** $34\frac{2}{7}$ mph

27. Individual income tax =   42%
    Corporate income tax = + 16%
         Total   =   58%

    58% of $352.6 billion = .58 × $352.6 billion
                          = $204.508 billion

    Answer: **(C)** 205

28. The total of all of the sectors of the graph except excise tax is 94%. Therefore, excise tax revenues are 100% − 94%, or 6%.

    6% of $352.6 billion = .06 × $352.6 billion
                         = $21.156 billion

    Answer: **(D)** 21

29. The whole job is $\frac{8}{8}$. If $\frac{5}{8}$ is completed, $\frac{3}{8}$ remains.
    $$\begin{aligned}
    \tfrac{3}{8} &= 3 \div 8 \\
    &= .375 \\
    &= 37\tfrac{1}{2}\%
    \end{aligned}$$

    Answer: **(A)** $37\frac{1}{2}\%$

30. $1695.00
     − 1500.00
       $ 195.00   interest
                 Rate of interest $= \frac{195}{1500}$
                                  $= .13$
                                  $= 13\%$

    Answer: **(A)** 13%

31.   190 × $79.35 = $15,076.50
         208 × $83.99 = + $17,469.92
                  Total = $32,546.42

    Answer: **(B)** $32,546.42

32. The dot for June lies between 260 and 280.

    Answer: **(B)** June

33. January: 145
    February: 125
    March: + 160
                    430

    Answer: **(D)** 430

34. The first 12 clerks complete $\frac{6}{18}$, or $\frac{1}{3}$ of the job in 6 days, leaving $\frac{2}{3}$ of the job to be completed.

    One clerk would require $12 \times 18 = 216$ days to complete the job, working alone. Sixteen clerks require $216 \div 16$, or $13\frac{1}{2}$ days for the entire job. But only $\frac{2}{3}$ of the job remains. To do $\frac{2}{3}$ of the job, sixteen clerks require
    $$\begin{aligned}
    \tfrac{2}{3} \times 13\tfrac{1}{2} &= \tfrac{2}{3} \times \tfrac{27}{2} \\
    &= 9 \text{ days}
    \end{aligned}$$

    The entire job takes 6 days + 9 days = 15 days.

    Answer: **(C)** 15 days

35. Smith worked $46 - 40 = 6$ hours overtime. For each overtime hour he earned:

$$1\tfrac{1}{2} \times \$7.20 = \$10.80$$

Overtime pay: $\$10.80 \times 6 = \$\ 64.80$
Regular pay: $\$7.20 \times 40 = \underline{\ 288.00}$
                                         $\$352.80$

Answer: **(B)** $352.80

36. The formula for changing Celsius to Fahrenheit is

$$F = \tfrac{9}{5}°C + 32$$
$$F = \tfrac{9}{\cancel{5}_{1}} \times \cancel{50}^{10} + 32$$
$$F = 90 + 32 = 122$$

Answer: **(A)** 122°

37. If the circumference of a circle is 10π, its diameter is 10 and its radius is $10 \div 2$, or 5.

$$\text{The area of a circle} = \pi r^2$$
$$= \pi \times 5^2$$
$$= 25\pi$$

Answer: **(C)** 25π

38. The interest will be compounded three times in 9 months.

First period:

$$.06 \times \tfrac{1}{\cancel{4}_{1}} \times \cancel{\$300}^{75} = \$4.50 \text{ interest}$$

New principal:
     $\$300 + \$4.50 = \$304.50$

Second period:
     $.06 \times \tfrac{1}{4} \times \$304.50 = \$4.57 \text{ interest}$

New principal:
     $\$304.50 + \$4.57 = \$309.07$

Third period:
     $.06 \times \tfrac{1}{4} \times \$309.07 = \$4.64 \text{ interest}$

Total interest: $\$\ 4.50$
                       4.57
               $+\ \underline{\ \ 4.64}$
                  $\$13.71$

Answer: **(C)** $13.71

39.   10.2
      10.4
      $\underline{10}$
      30.6

$$\text{Average} = 30.6 \div 3 = 10.2$$

Answer: **(A)** 10.2 seconds

40. Monthly deductions: $\$\ 292.20$
                               91.95
                               42.45
                          $+\ \underline{\ \ 4.32}$
                             $\$\ 430.92$
                          $\times\ \underline{\ \ \ \ \ 12}$

Deductions for year: $\$5171.04$
Gross pay: $\$18,000.00$
                       $-\ \underline{\ 5,171.04}$
Take-home pay: $\$12,828.96$

Answer: **(A)** $12,828.96

41. Area of 40' wall $= 40' \times 8' = $ 320 sq ft
   Area of 35' wall $= 35' \times 8' = \underline{+\ 280}$ sq ft
                    Total area $= $ 600 sq ft

$$600 \div 450 = 1\tfrac{1}{3} \text{ gallons}$$

Answer: **(A)** $1\tfrac{1}{3}$

42.   Commission $= 2\%$ of $5846
                        $= .02 \times \$5846$
                        $= \$116.92$
  Salary + commission $= \$300 + \$116.92$
                             $= \$416.92$

Answer: **(B)** $416.92

43. $\dfrac{1.62}{100} = 1.62\%$

   1.62% of $\$82,200 = .0162 \times \$82,200$
                           $= \$1331.64$

Answer: **(D)** $1332

44. A and B worked in the ratio 5:4.
               $5 + 4 = 9$
           $\$243 \div 9 = \$27$
         A's share $= 5 \times \$27 = \$135$

Answer: **(B)** $135

45. The two items together cost

$$\begin{array}{r} \$24.50 \\ +\ 43.28 \\ \hline \$67.78 \end{array}$$

5% tax
$$\begin{array}{r} \times\quad .05 \\ \hline \$3.3890 \end{array}$$

The tax for the two items was $3.39.

Total cost:
$$\begin{array}{r} \$67.78 \\ +\ \ 3.39 \\ \hline \$71.17 \end{array}$$

Answer: **(D)** $71.17

46. The total must be 100%.

German	32%
Spanish	28%
English	20%
Miscellaneous	10%
	90%

Therefore, 100% − 90%, or 10% are French.

A circle contains 360°.
$$10\% \text{ of } 360° = .10 \times 360°$$
$$= 36°$$

Answer: **(D)** 36

47. To find the median, arrange the values in order:

$$1.2,\ 2.2,\ 2.3,\ 2.4,\ 2.9,$$
$$3.0,\ 3.3,\ 3.5,\ 3.6,\ 3.9,$$
$$4.2,\ 4.3,\ 4.5,\ 4.6,\ 4.7$$

The median is the middle value in the list, or 3.5.

Answer: **(A)** 3.5

48.
$$\begin{array}{r} \$450.70 \\ \times\quad .0613 \\ \hline 135210 \\ 45070\ \ \\ 24\ 0420\ \ \ \ \\ \hline \$27.627910 \end{array}$$

Answer: **(A)** $27.60

49.
$$\begin{array}{r} 2 \times \$4.50 = \quad\$\ 9.00 \\ 3 \times \$4.15 = \quad 12.45 \\ 5 \times \$4.75 = +\ 23.75 \\ \hline \$45.20 \end{array}$$

$$\$45.20 \div 10 = \$4.52$$

Answer: **(C)** $4.52

50. A and B invested in the ratio 7:8.
$$7 + 8 = 15$$
$$\$2250 \div 15 = \$150$$
$$\text{A's share} = 7 \times \$150 = \$1050.$$

Answer: **(A)** $1050.00

ARITHMETIC REVIEW

Exam 2

1. A bag of nickels and dimes contains $11.50. If there are 73 dimes, how many nickels are there?
   (A) 78　　　　　(C) 82
   (B) 80　　　　　(D) 84

2. A shipment consists of 340 ten-foot pieces of conduit with a coupling on each piece. If the conduit weighs 0.85 lb per foot and each coupling weighs 0.15 lb, the total weight of the shipment is
   (A) 340 lb　　　　(C) 2941 lb
   (B) 628 lb　　　　(D) 3400 lb

3. A carton contains 9 dozen file folders. If a clerk removes 53 folders, how many are left in the carton?
   (A) 37　　　　　(C) 55
   (B) 44　　　　　(D) 62

4. What tax rate on a base of $4782 would yield $286.92?
   (A) 6%　　　　　(C) 12%
   (B) $8\frac{1}{4}$%　　　　(D) $16\frac{2}{3}$%

5. A can type 500 form letters in five hours. B can type 400 of these form letters in five hours. If A and B are to work together, the number of hours it will take them to type 540 form letters is most nearly
   (A) 2　　　　　(C) 4
   (B) 3　　　　　(D) 5

6. The difference between one tenth of 2000 and one-tenth percent of 2000 is
   (A) 0　　　　　(C) 180
   (B) 18　　　　　(D) 198

7. If the fractions $\frac{5}{7}$, $\frac{1}{2}$, $\frac{3}{5}$, and $\frac{2}{3}$ are arranged in ascending order of size, the result is
   (A) $\frac{1}{2}, \frac{3}{5}, \frac{2}{3}, \frac{5}{7}$　　　(C) $\frac{1}{2}, \frac{2}{3}, \frac{3}{5}, \frac{5}{7}$
   (B) $\frac{3}{5}, \frac{5}{7}, \frac{2}{3}, \frac{1}{2}$　　　(D) $\frac{5}{7}, \frac{2}{3}, \frac{3}{5}, \frac{1}{2}$

8. An employee has $\frac{2}{9}$ of his salary withheld for income tax. The percent of his salary that is withheld is most nearly
   (A) 16%　　　　(C) 20%
   (B) 18%　　　　(D) 22%

9. Frank and John repaired an old auto and sold it for $900. Frank worked on it 10 days and John worked 8 days. They divided the money in the ratio of the time spent on the work. Frank received
   (A) $400　　　　(C) $500
   (B) $450　　　　(D) $720

10. A driver traveled 100 miles at the rate of 40 mph, then traveled 80 miles at 60 mph. The total number of hours for the entire trip was
    (A) $1\frac{3}{20}$　　　　(C) $2\frac{1}{4}$
    (B) $1\frac{3}{4}$　　　　(D) $3\frac{5}{6}$

11. On a house plan on which 2 inches represents 5 feet, the length of a room measures $7\frac{1}{2}$ inches. The actual length of the room is
    (A) $12\frac{1}{2}$ ft　　　(C) $17\frac{1}{2}$ ft
    (B) $15\frac{3}{4}$ ft　　　(D) $18\frac{3}{4}$ ft

12. The ratio between .01% and .1 is
    (A) 1 to 10　　　(C) 1 to 1000
    (B) 1 to 100　　　(D) 1 to 10,000

13. After an article is discounted at 25%, it sells for $375. The original price of the article was
    (A) $93.75　　　(C) $375
    (B) $350　　　　(D) $500

142

14. If Mr. Mitchell has $627.04 in his checking account and then writes three checks for $241.75, $13.24, and $102.97, what will be his new balance?
    (A) $257.88   (C) $357.96
    (B) $269.08   (D) $369.96

15. If erasers cost 8¢ each for the first 250, 7¢ each for the next 250, and 5¢ for every eraser thereafter, how many erasers may be purchased for $50?
    (A) 600   (C) 850
    (B) 750   (D) 1000

16. Assume that it is necessary to partition a room measuring 40 feet by 20 feet into eight smaller rooms of equal size. Allowing no room for aisles, the *minimum* amount of partitioning that would be needed is
    (A) 90 ft   (C) 110 ft
    (B) 100 ft   (D) 140 ft

17. As a result of reports received by the Housing Authority concerning the reputed ineligibility of 756 tenants because of above-standard incomes, an intensive check of their employers has been ordered. Four housing assistants have been assigned to this task. At the end of 6 days at 7 hours each, they have checked on 336 tenants. In order to speed up the investigation, two more housing assistants are assigned at this point. If they worked at the same rate, the number of additional 7-hour days it would take to complete the job is, most nearly,
    (A) 1   (C) 5
    (B) 3   (D) 7

18. A bird flying 400 miles covers the first 100 at the rate of 100 miles an hour, the second 100 at the rate of 200 miles an hour, the third 100 at the rate of 300 miles an hour, and the last 100 at the rate of 400 miles an hour. The average speed was
    (A) 192 mph   (C) 250 mph
    (B) 212 mph   (D) 150 mph

19. At 5 o'clock the smaller angle between the hands of the clock is
    (A) 5°   (C) 120°
    (B) 75°   (D) 150°

20. $7\frac{2}{3}$% of $1200 is
    (A) $87   (C) $112
    (B) $92   (D) $920

21. A certain family spends 30% of its income for food, 8% for clothing, 25% for shelter, 4% for recreation, 13% for education, and 5% for miscellaneous items. The weekly earnings are $500. What is the number of weeks it would take this family to save $15,000?
    (A) 100   (C) 175
    (B) 150   (D) 200

22. A 12-gallon mixture of antifreeze and water is 25% antifreeze. If 3 gallons of water are added to it, the strength of the mixture is now
    (A) 12%   (C) 20%
    (B) $16\frac{2}{3}$%   (D) 35%

23. A cab driver works on a commission basis, receiving $42\frac{1}{2}$% of the fares. In addition, his earnings from tips are valued at 29% of the commissions. If his average weekly fares equal $520, then his monthly earnings are
    (A) between $900 and $1000
    (B) between $1000 and $1100
    (C) between $1100 and $1200
    (D) over $1200

Questions 24 and 25 refer to the following graph:

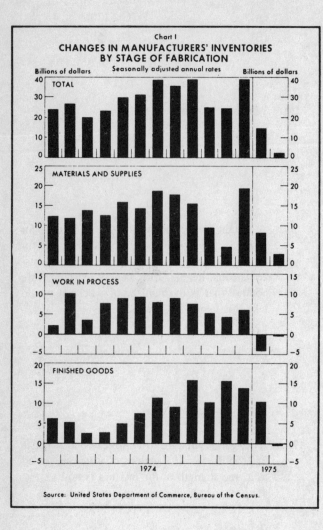

Chart I
CHANGES IN MANUFACTURERS' INVENTORIES
BY STAGE OF FABRICATION
Seasonally adjusted annual rates
Billions of dollars   Billions of dollars

TOTAL

MATERIALS AND SUPPLIES

WORK IN PROCESS

FINISHED GOODS

1974   1975

Source: United States Department of Commerce, Bureau of the Census.

24. For how many months are materials and supplies inventories over $15 billion at the same time that finished goods inventories are over $10 billion?
(A) one   (C) three
(B) two   (D) four

25. In June 1974 the ratio of finished goods to work in process was approximately
(A) 7:16   (C) 9:7
(B) 9:16   (D) 7:9

26. An employee's net pay is equal to his total earnings less all deductions. If an employee's total earnings in a pay period are $497.05, what is his net pay if he has the following deductions: federal income tax, $90.32; FICA, $28.74; state tax, $18.79; city tax, $7.25; pension, $1.88?
(A) $351.17   (C) $350.17
(B) $351.07   (D) $350.07

27. Assume that two types of files have been ordered: 200 of type A and 100 of type B. When the files are delivered, the buyer discovers that 25% of each type is damaged. Of the remaining files, 20% of type A and 40% of type B are the wrong color. The total number of files that are the wrong color is
(A) 30   (C) 50
(B) 40   (D) 60

28. A parade is marching up an avenue for 60 city blocks. A sample count of the number of people watching the parade is taken, first in a block near the end of the parade, and then in a block at the middle. The former count is 4000, the latter is 6000. If the average for the entire parade is assumed to be the average of the two samples, then the estimated number of persons watching the entire parade is most nearly
(A) 240,000   (C) 480,000
(B) 300,000   (D) 600,000

29. If A takes 6 days to do a task and B takes 3 days to do the same task, working together they should do the same task in
(A) $2\frac{2}{3}$ days   (C) $2\frac{1}{3}$ days
(B) 2 days   (D) $2\frac{1}{2}$ days

30. The total length of fencing needed to enclose a rectangular area 46 feet by 34 feet is
(A) 26 yd 1 ft   (C) 52 yd 2 ft
(B) $26\frac{2}{3}$ yd   (D) $53\frac{1}{3}$ yd

31. Find the length of time it would take $432 to yield $74.52 in interest at $5\frac{3}{4}$% per annum.
(A) 2 yr 10 mo   (C) 3 yr 10 mo
(B) 3 yr   (D) 4 yr

32. The price of a radio is $31.29, which includes a 5% sales tax. What was the price of the radio before the tax was added?
(A) $29.80   (C) $29.90
(B) $29.85   (D) $29.95

33. If a person in the 19% income tax bracket pays $3515 in income taxes, his taxable income was
    (A) $18,500
    (B) $32,763
    (C) $53,800
    (D) $67,785

34. Of the numbers 6, 5, 3, 3, 6, 3, 4, 3, 4, 3, the mode is
    (A) 3
    (B) 4
    (C) 5
    (D) 6

35. In a circle graph a sector of 108 degrees is shaded to indicate the overhead in doing $150,000 gross business. The overhead amounts to
    (A) $1200
    (B) $4500
    (C) $12,000
    (D) $45,000

36. Two people start at the same point and walk in opposite directions. If one walks at the rate of 2 miles per hour and the other walks at the rate of 3 miles per hour, in how many hours will they be 20 miles apart?
    (A) 2
    (B) 3
    (C) 4
    (D) 5

37. In a group of 100 people, 37 wear glasses. What is the probability that a person chosen at random from this group does *not* wear glasses?
    (A) .37
    (B) .50
    (C) .63
    (D) 1.00

38. The interest on $148.00 at 6% for 60 days is
    (A) $8.88
    (B) $2.96
    (C) $14.80
    (D) $ 1.48

39. A man bought a camera that was listed at $160. He was given successive discounts of 20% and 10%. The price he paid was
    (A) $112.00
    (B) $115.20
    (C) $119.60
    (D) $129.60

40. The water level of a swimming pool measuring 75 feet by 42 feet is to be raised four inches. If there are 7.48 gallons in a cubic foot, the number of gallons of water that will be needed is
    (A) 140
    (B) 31,500
    (C) 7854
    (D) 94,500

41. A salesman is paid $4\frac{1}{2}$% commission on his first $7000 of sales and $5\frac{1}{2}$% commission on all sales in excess of $7000. If his sales were $9600, how much commission did he earn?
    (A) $432
    (B) $458
    (C) $480
    (D) $528

42. How many boxes 3 inches by 4 inches by 5 inches can fit into a carton 3 feet by 4 feet by 5 feet?
    (A) 60
    (B) 144
    (C) 1728
    (D) 8640

43. The value of 32 nickels, 73 quarters, and 156 dimes is
    (A) $26.10
    (B) $31.75
    (C) $35.45
    (D) $49.85

44. The area of the shaded figure is
    (A) 4π
    (B) 5π
    (C) 16π
    (D) 21π

45. The wage rate in a certain trade is $8.60 an hour for a 40-hour week and $1\frac{1}{2}$ times the base pay for overtime. An employee who works 48 hours in a week earns
    (A) $447.20
    (B) $498.20
    (C) $582.20
    (D) $619.20

46. Jane Michaels borrowed $200 on March 31 at the simple interest rate of 8% per year. If she wishes to repay the loan and the interest on May 15, what is the total amount she must pay?
    (A) $201
    (B) $202
    (C) $203
    (D) $204

47. How many decigrams are in .57 kilograms?
    (A) 57
    (B) 570
    (C) 5700
    (D) 57,000

48. If candies are bought at $1.10 per dozen and sold at 3 for 55 cents, the total profit on $5\frac{1}{2}$ dozen is
    (A) $5.55
    (B) $6.05
    (C) $6.55
    (D) $7.05

49. The number missing in the sequence 2, 5, 10, 17, ——, 37, 50, 65 is
   (A) 22
   (C) 26
   (B) 24
   (D) 27

50. A cylindrical container has a diameter of 14 inches and a height of 6 inches. If one gallon equals 231 cubic inches, the capacity of the tank is approximately
   (A) $2\frac{2}{7}$ gal
   (C) $1\frac{1}{7}$ gal
   (B) 4 gal
   (D) 3 gal

Exam 2 — Correct Answers

1. (D)	11. (D)	21. (D)	31. (B)	41. (B)
2. (C)	12. (C)	22. (C)	32. (A)	42. (C)
3. (C)	13. (D)	23. (C)	33. (A)	43. (C)
4. (A)	14. (B)	24. (C)	34. (A)	44. (D)
5. (B)	15. (B)	25. (D)	35. (D)	45. (A)
6. (D)	16. (B)	26. (D)	36. (C)	46. (B)
7. (A)	17. (C)	27. (D)	37. (C)	47. (C)
8. (D)	18. (A)	28. (B)	38. (D)	48. (B)
9. (C)	19. (D)	29. (B)	39. (B)	49. (C)
10. (D)	20. (B)	30. (D)	40. (C)	50. (B)

Exam 2 — Solutions

1.
$$73 \text{ dimes} = 73 \times \$.10$$
$$= \$7.30$$
$$\$11.50 - \$7.30 = \$4.20$$

There is $4.20 worth of nickels in the bag.
$$\$4.20 \div \$.05 = 84 \text{ nickels}$$

Answer: **(D)** 84

2. Each 10-foot piece weighs
$$10 \times .85 \text{ lb} = \begin{array}{r} 8.5 \text{ lb} \\ + .15 \text{ lb} \\ \hline 8.65 \text{ lb} \end{array}$$

The entire shipment weighs
$$340 \times 8.65 \text{ lb} = 2941 \text{ lb}$$

Answer: **(C)** 2941 lb

3. The carton contains $9 \times 12 = 108$ folders.

$$108 - 53 = 55 \text{ remain in carton}$$

Answer: **(C)** 55

4.
$$\text{Rate} = \text{tax} \div \text{base}$$

$$286.92 \div 4782 = 4782\overline{)\begin{array}{r} .06 \\ 286.92 \\ \underline{286\ 92} \end{array}}$$

$$.06 = 6\%$$

Answer: **(A)** 6%

5. A can type $500 \div 5 = 100$ letters per hour.
    B can type $400 \div 5 = \ \ 80$ letters per hour.
    Together they type 180 letters per hour.
    $$540 \div 180 = 3$$
    It will take 3 hours to type 540 letters.

    Answer: **(B)** 3

6. $$\tfrac{1}{10} \text{ of } 2000 = \tfrac{1}{10} \times 2000 = 200$$
    $$\tfrac{1}{10}\% \text{ of } 2000 = .001 \times 2000 = 2$$
    The difference is $200 - 2 = 198$.

    Answer: **(D)** 198

7. To compare the fractions, change them to fractions having the same denominator.
    $$\text{L.C.D.} = 7 \times 2 \times 5 \times 3 = 210$$
    $$\tfrac{5}{7} = \tfrac{150}{210}$$
    $$\tfrac{1}{2} = \tfrac{105}{210}$$
    $$\tfrac{3}{5} = \tfrac{126}{210}$$
    $$\tfrac{2}{3} = \tfrac{140}{210}$$
    The correct order is $\tfrac{105}{210}, \tfrac{126}{210}, \tfrac{140}{210}, \tfrac{150}{210}$.

    Answer: **(A)** $\tfrac{1}{2}, \tfrac{3}{5}, \tfrac{2}{3}, \tfrac{5}{7}$

8.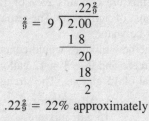

    $.22\tfrac{2}{9} = 22\%$ approximately

    Answer: **(D)** 22%

9. Frank and John worked in the ratio 10:8.
    $$10 + 8 = 18$$
    $$\$900 \div 18 = \$50$$
    $$\text{Frank's share} = 10 \times \$50 = \$500$$

    Answer: **(C)** \$500

10. The first part of the trip took
    $$100 \text{ mi} \div 40 \text{ mph} = 2\tfrac{1}{2} \text{ hours}$$
    The second part of the trip took
    $$80 \text{ mi} \div 60 \text{ mph} = 1\tfrac{1}{3} \text{ hours}$$
    $$2\tfrac{1}{2} = \ \ 2\tfrac{3}{6}$$
    $$\underline{+ \ 1\tfrac{1}{3} = + \ 1\tfrac{2}{6}}$$
    $$3\tfrac{5}{6}$$

    Answer: **(D)** $3\tfrac{5}{6}$

11. Let f represent the actual number of feet. The plan lengths and the actual lengths are in proportion. Therefore,
    $$\frac{2}{7\tfrac{1}{2}} = \frac{5}{f}$$
    and
    $$f = \frac{5 \times 7\tfrac{1}{2}}{2}$$
    $$= \frac{5 \times \tfrac{15}{2}}{2}$$
    $$= \tfrac{75}{2} \div 2$$
    $$= \tfrac{75}{2} \times \tfrac{1}{2}$$
    $$= \tfrac{75}{4} = 18\tfrac{3}{4}$$

    Answer: **(D)** $18\tfrac{3}{4}$ ft

12. $$\frac{.01\%}{.1} = \frac{.0001}{.1}$$
    $$= .001$$
    $$= \frac{1}{1000}$$

    Answer: **(C)** 1 to 1000

13. \$375 is 75% of the original price.
    $$\text{The original price} = \$375 \div 75\%$$
    $$= \$375 \div .75$$
    $$= \$500$$

    Answer: **(D)** \$500

14. Total of checks: \$241.75
    $$13.24$$
    $$\underline{+ \quad 102.97}$$
    $$\$357.96$$

    $$\$627.04 \quad \text{old balance}$$
    $$\underline{- \quad 357.96 \quad \text{checks}}$$
    $$\$269.08 \quad \text{new balance}$$

    Answer: **(B)** \$269.08

15. First 250 erasers:
    $$250 \times \$.08 = \$20.00$$
    Next 250 erasers:
    $$250 \times \$.07 = \$17.50$$
    Total for 500 erasers:
    $$\$20.00 + \$17.50 = \$37.50$$
    $$\$50.00 - \$37.50 = \$12.50$$
    \$12.50 remains for 5¢ erasers:
    $$\$12.50 \div \$.05 = 250 \text{ erasers}$$
    $$500 + 250 = 750$$

    Answer: **(B)** 750

16. The room may be partitioned as is shown below:

The total amount of partitioning is 100 feet.

Answer: **(B)** 100 ft

17. Four assistants completed 336 cases in 42 hours (6 days at 7 hours per day). Therefore, each assistant completed 336 ÷ 4, or 84 cases in 42 hours, for a rate of 2 cases per hour per assistant.

After the first 6 days, the number of cases remaining is

$$756 - 336 = 420$$

It will take 6 assistants, working at the rate of 2 cases per hour per assistant $420 \div 12$ or 35 hours to complete the work. If each workday has 7 hours, then $35 \div 7$ or 5 days are needed.

Answer: **(C)** 5

18. At 100 mph, 100 miles will take 1 hour. At 200 mph, 100 miles will take $\frac{1}{2}$ hour. At 300 mph, 100 miles will take $\frac{1}{3}$ hour. At 400 mph, 100 miles will take $\frac{1}{4}$ hour.

Total time:
$$
\begin{aligned}
1 &= \tfrac{12}{12}\\
\tfrac{1}{2} &= \tfrac{6}{12}\\
\tfrac{1}{3} &= \tfrac{4}{12}\\
+\ \tfrac{1}{4} &= +\tfrac{3}{12}\\
\hline
&\ \tfrac{25}{12} = 2\tfrac{1}{12}\ \text{hours}
\end{aligned}
$$

400 miles $\div\ 2\tfrac{1}{12}$ hours $= 400 \div \tfrac{25}{12}$ mph

$= \overset{16}{\cancel{400}} \times \tfrac{12}{\underset{1}{\cancel{25}}}$ mph

$= 192$ mph

Answer: **(A)** 192 mph

19. Each hour is represented by
$$360° \div 12 = 30°$$

The smaller angle formed by the hands
$$= 5 \times 30°$$
$$= 150°$$

12   5

Answer: **(D)** 150°

20. $7\tfrac{2}{3}\%$ of $1200 = \tfrac{23}{3}\%$ of $1200

To change $\tfrac{23}{3}\%$ to a fraction, divide by 100:

$$\tfrac{23}{3} \div 100 = \tfrac{23}{3} \times \tfrac{1}{100} = \tfrac{23}{300}$$

$$\underset{1}{\cancel{\tfrac{23}{300}}} \times \overset{4}{\cancel{\$1200}} = \$92$$

Answer: **(B)** $92

21. The family spends a total of 85% of its income. Therefore, 100% − 85%, or 15%, remains for savings.

$$15\% \text{ of } \$500 = .15 \times \$500$$
$$= \$75 \text{ per week}$$
$$\$15,000 \div \$75 = 200 \text{ weeks}$$

Answer: **(D)** 200

22. The mixture contains 25% of 12 gallons, or 3 gallons, of antifreeze. The remaining 9 gallons must be water.

The new mixture would contain 3 gallons of antifreeze and $9 + 3 = 12$ gallons of water, for a total of 15 gallons. The strength would be

$$\frac{3 \text{ gal. antifreeze}}{15 \text{ gal. mixture}} = .20$$
$$= 20\%$$

Answer: **(C)** 20%

23. Commission $= 42\tfrac{1}{2}\%$ of fares
$$42\tfrac{1}{2}\% \text{ of } \$520 = .425 \times \$520$$
$$= \$221 \text{ commission}$$

Tips $= 29\%$ of commission
$$29\% \text{ of } \$221 = .29 \times \$221$$
$$= \$64.09 \text{ tips}$$

Weekly earnings:
$$
\begin{aligned}
\$221.00\\
+\ \ \ 64.09\\
\hline
\$285.09
\end{aligned}
$$

Monthly earnings:
$$
\begin{aligned}
\$285.09\\
\times \ \ \ \ \ \ \ 4\\
\hline
\$1140.36
\end{aligned}
$$

Answer: **(C)** between $1100 and $1200

24. Finished goods inventories are over $10 billion in July, September, October, November, and December of 1974 and January of 1975. Of those months, materials and supplies inventories are over $15 billion in July, September, and December.

Answer: **(C)** three

25. In June 1974 finished goods inventories were approximately $7 billion, and work in process inventories were approximately $9 billion. The ratio is 7:9.

Answer: **(D)** 7:9

26.

$$\begin{array}{r} \$\ 90.32 \\ 28.74 \\ 18.79 \\ 7.25 \\ \underline{1.88} \end{array}$$

Total deductions $146.98

$$\begin{array}{r} \$497.05 \quad \text{total earnings} \\ \underline{-\ 146.98} \quad \text{deductions} \\ \$350.07 \quad \text{net pay} \end{array}$$

Answer: **(D)** $350.07

27. If 25% are damaged, then 75% are not damaged.

Type A: 75% of 200 = .75 × 200
                      = 150

        20% of 150 are wrong color
        20% of 150 = .20 × 150
                   = 30

Type B: 75% of 100 = .75 × 100
                      = 75

        40% of 75 are wrong color
        40% of 75 = .40 × 75
                  = 30

Total wrong color = 30 + 30 = 60

Answer: **(D)** 60

28. Average is $\dfrac{4000 + 6000}{2}$ = 5000 per block.

If there are 60 blocks, there are
        60 × 5000 = 300,000 people

Answer: **(B)** 300,000

29. A can do $\frac{1}{6}$ of the task in 1 day, and B can do $\frac{1}{3}$ in 1 day.

Together, in 1 day they can do

$$\begin{array}{r} \frac{1}{6} = \quad \frac{1}{6} \\ \underline{+\ \frac{1}{3}} = \underline{+\ \frac{2}{6}} \\ \frac{3}{6} = \frac{1}{2} \text{ of the job} \end{array}$$

It will take 2 days to complete the job if they work together.

Answer: **(B)** 2 days

30.

Perimeter = 46′ + 34′ + 46′ + 34′
          = 160′

160 ft ÷ 3 ft per yd = $\frac{160}{3}$ yd
                     = $53\frac{1}{3}$ yd

Answer: **(D)** $53\frac{1}{3}$ yd

31.

$$\$432 \times 5\frac{3}{4}\% = \overset{108}{\cancel{432}} \times \frac{23}{\underset{100}{\cancel{100}}}$$
$$= \$\frac{2484}{100}$$
$$= \$24.84$$

$74.52 ÷ $24.84 = 3

Answer: **(B)** 3 yr

32. $31.29 = 105% of price before tax

Price before tax = $31.29 ÷ 105%
                 = $31.29 ÷ 1.05
                 = $29.80

Answer: **(A)** $29.80

33. $3515 = 19% of taxable income

Taxable income = $3515 ÷ 19%
               = $3515 ÷ .19
               = $18,500

Answer: **(A)** $18,500

34. The mode is the value appearing most frequently. For the list given, the mode is 3.

Answer: **(A)** 3

35. A sector of 108° is
$$\frac{108°}{360°} = \frac{3}{10} \text{ of the circle}$$

$$\frac{3}{10} \times \$150,000 = \$45,000$$

Answer: **(D)** $45,000

36. In 1 hour they are 5 miles apart.
$$20 \text{ mi} \div 5 \text{ mi} = 4 \text{ hr}$$
It will take 4 hours to be 20 miles apart.

Answer: **(C)** 4

37. If 37 wear glasses, 100 − 37, or 63 do not wear glasses.
The probability is $\frac{63}{100} = .63$

Answer: **(C)** .63

38.
$$60 \text{ days} = \tfrac{60}{360} \text{ year}$$
$$= \tfrac{1}{6} \text{ year}$$
$$\text{Interest} = \$148 \times .06 \times \tfrac{1}{6}$$
$$= \$1.48$$

Answer: **(D)** $1.48

39. First discount:
$$20\% \text{ of } \$160 = .20 \times \$160 = \$32$$
$$\$160 - \$32 = \$128$$
Second discount:
$$10\% \text{ of } \$128 = .10 \times \$128 = \$12.80$$
$$\$128.00 - \$12.80 = \$115.20$$

Answer: **(B)** $115.20

40.
$$4 \text{ in} = \tfrac{1}{3} \text{ ft}$$
$$\text{Volume to be added} = \overset{25}{\cancel{75}} \times 42 \times \tfrac{1}{\underset{1}{\cancel{3}}}$$
$$= 1050 \text{ cu ft}$$
$$= 1050 \times 7.48 \text{ gal}$$
$$= 7854 \text{ gal}$$

Answer: **(C)** 7854

41. Commission on first $7000:
$$4\tfrac{1}{2}\% \text{ of } \$7000 = .045 \times \$7000$$
$$= \$315$$
Commission on remainder:
$$\$9600 - \$7000 = \$2600$$
$$5\tfrac{1}{2}\% \text{ of } \$2600 = .055 \times \$2600$$
$$= \$143$$
$$\text{Total commission} = \$315 + \$143$$
$$= \$458$$

Answer: **(B)** $458

42. Volume of the carton = 3 ft × 4 ft × 5 ft
$$= 36 \text{ in} \times 48 \text{ in} \times 60 \text{ in}$$
$$= 103,680 \text{ cu in}$$
Volume of each box = 3 in × 4 in × 5 in
$$= 60 \text{ cu in}$$
$$103,680 \div 60 = 1728$$

Answer: **(C)** 1728

43.
$$\begin{aligned}
32 \text{ nickels} &= 32 \times \$.05 = \$\ 1.60 \\
73 \text{ quarters} &= 73 \times \$.25 = \ 18.25 \\
156 \text{ dimes} &= 156 \times \$.10 = \ 15.60 \\
&\qquad\qquad\text{Total} = \$35.45
\end{aligned}$$

Answer: **(C)** $35.45

44. The area of the shaded figure equals the area of the larger circle minus the area of the smaller circle.
$$\begin{aligned}
\text{Area of larger circle} &= 5^2\pi = 25\pi \\
- \text{ Area of smaller circle} &= 2^2\pi = \ \ 4\pi \\
\text{Area of shaded figure} &= 21\pi
\end{aligned}$$

Answer: **(D)** 21π

45.
$$48 - 40 = 8 \text{ hours overtime}$$
Salary for 8 hours overtime:
$$1\tfrac{1}{2} \times \$8.60 \times 8 = \tfrac{3}{\underset{1}{\cancel{2}}} \times \$8.60 \times \overset{4}{\cancel{8}}$$
$$= \$103.20$$
Salary for 40 hours regular time:
$$\$8.60 \times 40 = \$344.00$$
$$\text{Total salary} = \$344.00 + \$103.20$$
$$= \$447.20$$

Answer: **(A)** $447.20

46. From March 31 to May 15 is 45 days, which is $\frac{45}{360}$ of a year.

$$\text{Interest} = \$200 \times .08 \times \frac{\overset{1}{\cancel{\frac{45}{360}}}}{8}$$

$$= \$\frac{16}{8}$$

$$= \$2$$

She must pay $200 + $2 = $202.

Answer: **(B)** $202

47.
$$.57 \text{ kilograms} = .57 \times 1000 \text{ grams}$$
$$= 570 \text{ grams}$$
$$= 570 \div .10 \text{ decigrams}$$
$$= 5700 \text{ decigrams}$$

Answer: **(C)** 5700

48. The cost of $5\frac{1}{2}$ dozen is
$$5\frac{1}{2} \times \$1.10 = 5.5 \times \$1.10$$
$$= \$6.05$$

The candies sell at 3 for $.55. A dozen sell for $4 \times \$.55$, or $2.20. The selling price of $5\frac{1}{2}$ dozen is
$$5\frac{1}{2} \times \$2.20 = 5.5 \times \$2.20$$
$$= \$12.10$$
$$\text{Profit} = \$12.10 - \$6.05$$
$$= \$6.05$$

Answer: **(B)** $6.05

49. Find the differences between terms:

$$2 \underset{3}{\underbrace{\quad}} 5 \underset{5}{\underbrace{\quad}} 10 \underset{7}{\underbrace{\quad}} 17 \underline{\quad\quad} 37 \underset{13}{\underbrace{\quad}} 50 \underset{15}{\underbrace{\quad}} 65$$

The difference between 17 and the missing term must be 9. The missing term is
$$17 + 9 = 26$$

Answer: **(C)** 26

50. The volume of a cylinder $= \pi r^2 h$. If the diameter is 14, the radius is 7. Using $\pi = \frac{22}{7}$,

$$\text{Volume} = \frac{22}{\cancel{7}} \times \overset{7}{\cancel{49}} \times 6$$
$$\underset{1}{}$$

$$= 924 \text{ cubic inches}$$
$$924 \div 231 = 4 \text{ gallons}$$

Answer: **(B)** 4 gal

Using the Answer Sheet

The answer sheet for your civil service exam is machine scored. You cannot give any explanations to the machine, so you must fill out the answer sheet clearly and correctly.

1. Blacken your answer space firmly and completely. ● is the only correct way to mark the answer sheet. ◑, ⊗, ⊙, and ∅ are all unacceptable. The machine might not read them at all.

2. Mark only one answer for each question. If you mark more than one answer you will be considered wrong even if one of the answers is correct.

3. If you change your mind, you must erase your mark. Attempting to cross out an incorrect answer like this ● will not work. You must erase any incorrect answer completely. An incomplete erasure might be read as a second answer.

4. All of your answering should be in the form of blackened spaces. The machine cannot read English. Do not write any notes in the margins.

5. MOST IMPORTANT: Answer each question in the right place. Question 1 must be answered in space 1; question 12 in space 12. If you should skip an answer space and mark a series of answers in the wrong places, you must erase all those answers and do the questions over, marking your answers in the proper places. You cannot afford to use the limited time in this way. Therefore, as you answer *each* question, look at its number and check that you are marking your answer in the space with the same number.

TEAR HERE

Arithmetic Review—Exam 1

1 Ⓐ Ⓑ Ⓒ Ⓓ	11 Ⓐ Ⓑ Ⓒ Ⓓ	21 Ⓐ Ⓑ Ⓒ Ⓓ	31 Ⓐ Ⓑ Ⓒ Ⓓ	41 Ⓐ Ⓑ Ⓒ Ⓓ
2 Ⓐ Ⓑ Ⓒ Ⓓ	12 Ⓐ Ⓑ Ⓒ Ⓓ	22 Ⓐ Ⓑ Ⓒ Ⓓ	32 Ⓐ Ⓑ Ⓒ Ⓓ	42 Ⓐ Ⓑ Ⓒ Ⓓ
3 Ⓐ Ⓑ Ⓒ Ⓓ	13 Ⓐ Ⓑ Ⓒ Ⓓ	23 Ⓐ Ⓑ Ⓒ Ⓓ	33 Ⓐ Ⓑ Ⓒ Ⓓ	43 Ⓐ Ⓑ Ⓒ Ⓓ
4 Ⓐ Ⓑ Ⓒ Ⓓ	14 Ⓐ Ⓑ Ⓒ Ⓓ	24 Ⓐ Ⓑ Ⓒ Ⓓ	34 Ⓐ Ⓑ Ⓒ Ⓓ	44 Ⓐ Ⓑ Ⓒ Ⓓ
5 Ⓐ Ⓑ Ⓒ Ⓓ	15 Ⓐ Ⓑ Ⓒ Ⓓ	25 Ⓐ Ⓑ Ⓒ Ⓓ	35 Ⓐ Ⓑ Ⓒ Ⓓ	45 Ⓐ Ⓑ Ⓒ Ⓓ
6 Ⓐ Ⓑ Ⓒ Ⓓ	16 Ⓐ Ⓑ Ⓒ Ⓓ	26 Ⓐ Ⓑ Ⓒ Ⓓ	36 Ⓐ Ⓑ Ⓒ Ⓓ	46 Ⓐ Ⓑ Ⓒ Ⓓ
7 Ⓐ Ⓑ Ⓒ Ⓓ	17 Ⓐ Ⓑ Ⓒ Ⓓ	27 Ⓐ Ⓑ Ⓒ Ⓓ	37 Ⓐ Ⓑ Ⓒ Ⓓ	47 Ⓐ Ⓑ Ⓒ Ⓓ
8 Ⓐ Ⓑ Ⓒ Ⓓ	18 Ⓐ Ⓑ Ⓒ Ⓓ	28 Ⓐ Ⓑ Ⓒ Ⓓ	38 Ⓐ Ⓑ Ⓒ Ⓓ	48 Ⓐ Ⓑ Ⓒ Ⓓ
9 Ⓐ Ⓑ Ⓒ Ⓓ	19 Ⓐ Ⓑ Ⓒ Ⓓ	29 Ⓐ Ⓑ Ⓒ Ⓓ	39 Ⓐ Ⓑ Ⓒ Ⓓ	49 Ⓐ Ⓑ Ⓒ Ⓓ
10 Ⓐ Ⓑ Ⓒ Ⓓ	20 Ⓐ Ⓑ Ⓒ Ⓓ	30 Ⓐ Ⓑ Ⓒ Ⓓ	40 Ⓐ Ⓑ Ⓒ Ⓓ	50 Ⓐ Ⓑ Ⓒ Ⓓ

Arithmetic Review—Exam 2

1 Ⓐ Ⓑ Ⓒ Ⓓ	11 Ⓐ Ⓑ Ⓒ Ⓓ	21 Ⓐ Ⓑ Ⓒ Ⓓ	31 Ⓐ Ⓑ Ⓒ Ⓓ	41 Ⓐ Ⓑ Ⓒ Ⓓ
2 Ⓐ Ⓑ Ⓒ Ⓓ	12 Ⓐ Ⓑ Ⓒ Ⓓ	22 Ⓐ Ⓑ Ⓒ Ⓓ	32 Ⓐ Ⓑ Ⓒ Ⓓ	42 Ⓐ Ⓑ Ⓒ Ⓓ
3 Ⓐ Ⓑ Ⓒ Ⓓ	13 Ⓐ Ⓑ Ⓒ Ⓓ	23 Ⓐ Ⓑ Ⓒ Ⓓ	33 Ⓐ Ⓑ Ⓒ Ⓓ	43 Ⓐ Ⓑ Ⓒ Ⓓ
4 Ⓐ Ⓑ Ⓒ Ⓓ	14 Ⓐ Ⓑ Ⓒ Ⓓ	24 Ⓐ Ⓑ Ⓒ Ⓓ	34 Ⓐ Ⓑ Ⓒ Ⓓ	44 Ⓐ Ⓑ Ⓒ Ⓓ
5 Ⓐ Ⓑ Ⓒ Ⓓ	15 Ⓐ Ⓑ Ⓒ Ⓓ	25 Ⓐ Ⓑ Ⓒ Ⓓ	35 Ⓐ Ⓑ Ⓒ Ⓓ	45 Ⓐ Ⓑ Ⓒ Ⓓ
6 Ⓐ Ⓑ Ⓒ Ⓓ	16 Ⓐ Ⓑ Ⓒ Ⓓ	26 Ⓐ Ⓑ Ⓒ Ⓓ	36 Ⓐ Ⓑ Ⓒ Ⓓ	46 Ⓐ Ⓑ Ⓒ Ⓓ
7 Ⓐ Ⓑ Ⓒ Ⓓ	17 Ⓐ Ⓑ Ⓒ Ⓓ	27 Ⓐ Ⓑ Ⓒ Ⓓ	37 Ⓐ Ⓑ Ⓒ Ⓓ	47 Ⓐ Ⓑ Ⓒ Ⓓ
8 Ⓐ Ⓑ Ⓒ Ⓓ	18 Ⓐ Ⓑ Ⓒ Ⓓ	28 Ⓐ Ⓑ Ⓒ Ⓓ	38 Ⓐ Ⓑ Ⓒ Ⓓ	48 Ⓐ Ⓑ Ⓒ Ⓓ
9 Ⓐ Ⓑ Ⓒ Ⓓ	19 Ⓐ Ⓑ Ⓒ Ⓓ	29 Ⓐ Ⓑ Ⓒ Ⓓ	39 Ⓐ Ⓑ Ⓒ Ⓓ	49 Ⓐ Ⓑ Ⓒ Ⓓ
10 Ⓐ Ⓑ Ⓒ Ⓓ	20 Ⓐ Ⓑ Ⓒ Ⓓ	30 Ⓐ Ⓑ Ⓒ Ⓓ	40 Ⓐ Ⓑ Ⓒ Ⓓ	50 Ⓐ Ⓑ Ⓒ Ⓓ

Sentence Completions—Test 1

1 Ⓐ Ⓑ Ⓒ Ⓓ	6 Ⓐ Ⓑ Ⓒ Ⓓ	11 Ⓐ Ⓑ Ⓒ Ⓓ	16 Ⓐ Ⓑ Ⓒ Ⓓ	21 Ⓐ Ⓑ Ⓒ Ⓓ
2 Ⓐ Ⓑ Ⓒ Ⓓ	7 Ⓐ Ⓑ Ⓒ Ⓓ	12 Ⓐ Ⓑ Ⓒ Ⓓ	17 Ⓐ Ⓑ Ⓒ Ⓓ	22 Ⓐ Ⓑ Ⓒ Ⓓ
3 Ⓐ Ⓑ Ⓒ Ⓓ	8 Ⓐ Ⓑ Ⓒ Ⓓ	13 Ⓐ Ⓑ Ⓒ Ⓓ	18 Ⓐ Ⓑ Ⓒ Ⓓ	23 Ⓐ Ⓑ Ⓒ Ⓓ
4 Ⓐ Ⓑ Ⓒ Ⓓ	9 Ⓐ Ⓑ Ⓒ Ⓓ	14 Ⓐ Ⓑ Ⓒ Ⓓ	19 Ⓐ Ⓑ Ⓒ Ⓓ	24 Ⓐ Ⓑ Ⓒ Ⓓ
5 Ⓐ Ⓑ Ⓒ Ⓓ	10 Ⓐ Ⓑ Ⓒ Ⓓ	15 Ⓐ Ⓑ Ⓒ Ⓓ	20 Ⓐ Ⓑ Ⓒ Ⓓ	25 Ⓐ Ⓑ Ⓒ Ⓓ

Sentence Completions—Test 2

1 Ⓐ Ⓑ Ⓒ Ⓓ	6 Ⓐ Ⓑ Ⓒ Ⓓ	11 Ⓐ Ⓑ Ⓒ Ⓓ	16 Ⓐ Ⓑ Ⓒ Ⓓ	21 Ⓐ Ⓑ Ⓒ Ⓓ
2 Ⓐ Ⓑ Ⓒ Ⓓ	7 Ⓐ Ⓑ Ⓒ Ⓓ	12 Ⓐ Ⓑ Ⓒ Ⓓ	17 Ⓐ Ⓑ Ⓒ Ⓓ	22 Ⓐ Ⓑ Ⓒ Ⓓ
3 Ⓐ Ⓑ Ⓒ Ⓓ	8 Ⓐ Ⓑ Ⓒ Ⓓ	13 Ⓐ Ⓑ Ⓒ Ⓓ	18 Ⓐ Ⓑ Ⓒ Ⓓ	23 Ⓐ Ⓑ Ⓒ Ⓓ
4 Ⓐ Ⓑ Ⓒ Ⓓ	9 Ⓐ Ⓑ Ⓒ Ⓓ	14 Ⓐ Ⓑ Ⓒ Ⓓ	19 Ⓐ Ⓑ Ⓒ Ⓓ	24 Ⓐ Ⓑ Ⓒ Ⓓ
5 Ⓐ Ⓑ Ⓒ Ⓓ	10 Ⓐ Ⓑ Ⓒ Ⓓ	15 Ⓐ Ⓑ Ⓒ Ⓓ	20 Ⓐ Ⓑ Ⓒ Ⓓ	25 Ⓐ Ⓑ Ⓒ Ⓓ

Sentence Completions—Test 3

1 Ⓐ Ⓑ Ⓒ Ⓓ	6 Ⓐ Ⓑ Ⓒ Ⓓ	11 Ⓐ Ⓑ Ⓒ Ⓓ	16 Ⓐ Ⓑ Ⓒ Ⓓ	21 Ⓐ Ⓑ Ⓒ Ⓓ
2 Ⓐ Ⓑ Ⓒ Ⓓ	7 Ⓐ Ⓑ Ⓒ Ⓓ	12 Ⓐ Ⓑ Ⓒ Ⓓ	17 Ⓐ Ⓑ Ⓒ Ⓓ	22 Ⓐ Ⓑ Ⓒ Ⓓ
3 Ⓐ Ⓑ Ⓒ Ⓓ	8 Ⓐ Ⓑ Ⓒ Ⓓ	13 Ⓐ Ⓑ Ⓒ Ⓓ	18 Ⓐ Ⓑ Ⓒ Ⓓ	23 Ⓐ Ⓑ Ⓒ Ⓓ
4 Ⓐ Ⓑ Ⓒ Ⓓ	9 Ⓐ Ⓑ Ⓒ Ⓓ	14 Ⓐ Ⓑ Ⓒ Ⓓ	19 Ⓐ Ⓑ Ⓒ Ⓓ	24 Ⓐ Ⓑ Ⓒ Ⓓ
5 Ⓐ Ⓑ Ⓒ Ⓓ	10 Ⓐ Ⓑ Ⓒ Ⓓ	15 Ⓐ Ⓑ Ⓒ Ⓓ	20 Ⓐ Ⓑ Ⓒ Ⓓ	25 Ⓐ Ⓑ Ⓒ Ⓓ

Sentence Completions—Test 4

1 Ⓐ Ⓑ Ⓒ Ⓓ	6 Ⓐ Ⓑ Ⓒ Ⓓ	11 Ⓐ Ⓑ Ⓒ Ⓓ	16 Ⓐ Ⓑ Ⓒ Ⓓ	21 Ⓐ Ⓑ Ⓒ Ⓓ
2 Ⓐ Ⓑ Ⓒ Ⓓ	7 Ⓐ Ⓑ Ⓒ Ⓓ	12 Ⓐ Ⓑ Ⓒ Ⓓ	17 Ⓐ Ⓑ Ⓒ Ⓓ	22 Ⓐ Ⓑ Ⓒ Ⓓ
3 Ⓐ Ⓑ Ⓒ Ⓓ	8 Ⓐ Ⓑ Ⓒ Ⓓ	13 Ⓐ Ⓑ Ⓒ Ⓓ	18 Ⓐ Ⓑ Ⓒ Ⓓ	23 Ⓐ Ⓑ Ⓒ Ⓓ
4 Ⓐ Ⓑ Ⓒ Ⓓ	9 Ⓐ Ⓑ Ⓒ Ⓓ	14 Ⓐ Ⓑ Ⓒ Ⓓ	19 Ⓐ Ⓑ Ⓒ Ⓓ	24 Ⓐ Ⓑ Ⓒ Ⓓ
5 Ⓐ Ⓑ Ⓒ Ⓓ	10 Ⓐ Ⓑ Ⓒ Ⓓ	15 Ⓐ Ⓑ Ⓒ Ⓓ	20 Ⓐ Ⓑ Ⓒ Ⓓ	25 Ⓐ Ⓑ Ⓒ Ⓓ

Sentence Completions—Test 5

1 Ⓐ Ⓑ Ⓒ Ⓓ	6 Ⓐ Ⓑ Ⓒ Ⓓ	11 Ⓐ Ⓑ Ⓒ Ⓓ	16 Ⓐ Ⓑ Ⓒ Ⓓ	21 Ⓐ Ⓑ Ⓒ Ⓓ
2 Ⓐ Ⓑ Ⓒ Ⓓ	7 Ⓐ Ⓑ Ⓒ Ⓓ	12 Ⓐ Ⓑ Ⓒ Ⓓ	17 Ⓐ Ⓑ Ⓒ Ⓓ	22 Ⓐ Ⓑ Ⓒ Ⓓ
3 Ⓐ Ⓑ Ⓒ Ⓓ	8 Ⓐ Ⓑ Ⓒ Ⓓ	13 Ⓐ Ⓑ Ⓒ Ⓓ	18 Ⓐ Ⓑ Ⓒ Ⓓ	23 Ⓐ Ⓑ Ⓒ Ⓓ
4 Ⓐ Ⓑ Ⓒ Ⓓ	9 Ⓐ Ⓑ Ⓒ Ⓓ	14 Ⓐ Ⓑ Ⓒ Ⓓ	19 Ⓐ Ⓑ Ⓒ Ⓓ	24 Ⓐ Ⓑ Ⓒ Ⓓ
5 Ⓐ Ⓑ Ⓒ Ⓓ	10 Ⓐ Ⓑ Ⓒ Ⓓ	15 Ⓐ Ⓑ Ⓒ Ⓓ	20 Ⓐ Ⓑ Ⓒ Ⓓ	25 Ⓐ Ⓑ Ⓒ Ⓓ

Sentence Completions—Test 6

1 Ⓐ Ⓑ Ⓒ Ⓓ	6 Ⓐ Ⓑ Ⓒ Ⓓ	11 Ⓐ Ⓑ Ⓒ Ⓓ	16 Ⓐ Ⓑ Ⓒ Ⓓ	21 Ⓐ Ⓑ Ⓒ Ⓓ
2 Ⓐ Ⓑ Ⓒ Ⓓ	7 Ⓐ Ⓑ Ⓒ Ⓓ	12 Ⓐ Ⓑ Ⓒ Ⓓ	17 Ⓐ Ⓑ Ⓒ Ⓓ	22 Ⓐ Ⓑ Ⓒ Ⓓ
3 Ⓐ Ⓑ Ⓒ Ⓓ	8 Ⓐ Ⓑ Ⓒ Ⓓ	13 Ⓐ Ⓑ Ⓒ Ⓓ	18 Ⓐ Ⓑ Ⓒ Ⓓ	23 Ⓐ Ⓑ Ⓒ Ⓓ
4 Ⓐ Ⓑ Ⓒ Ⓓ	9 Ⓐ Ⓑ Ⓒ Ⓓ	14 Ⓐ Ⓑ Ⓒ Ⓓ	19 Ⓐ Ⓑ Ⓒ Ⓓ	24 Ⓐ Ⓑ Ⓒ Ⓓ
5 Ⓐ Ⓑ Ⓒ Ⓓ	10 Ⓐ Ⓑ Ⓒ Ⓓ	15 Ⓐ Ⓑ Ⓒ Ⓓ	20 Ⓐ Ⓑ Ⓒ Ⓓ	25 Ⓐ Ⓑ Ⓒ Ⓓ

TEAR HERE

Part II
VOCABULARY

Introduction

This part is designed to help you expand your vocabulary quickly and to give you practice answering vocabulary questions. It is divided into four chapters. The first, Study Aids, includes an etymology chart and a word list. Read this material through carefully, but don't try to absorb it all at once. Follow the study suggestions. To study effectively, you might divide the word list into sections. Learn each section thoroughly before going on to the next one. For maximum concentration, a half-hour of studying at a time is plenty.

The second chapter gives you practical advice on how to answer synonym questions and has a series of twelve practice tests on synonyms. The third chapter teaches you how to answer antonym questions (opposites) and offers a series of eight practice tests on antonyms. The fourth chapter gives instruction on answering sentence completion questions and concludes with a series of six practice tests. Each chapter is followed immediately by its own answer key. To get the maximum benefit from the practice tests, space them out. You might take one or two at the beginning to see how well you do. Then spend some time on the Study Aids before taking another round of tests. Keep an honest record of your scores so you'll know how much you've improved.

Use the practice tests to help you learn. When you check your answers, make sure you understand why they were wrong, or right. Make a note of any words you're not sure of and look them up in the word list or in a dictionary. The correct answers to the sentence completion practice tests are followed by explanations which clarify the reasoning behind each correct answer choice.

This part is only a starting place. There are many ways you can work to increase your vocabulary. Make it a habit to read as much as you can, and read many different kinds of things. You will not only learn the meanings of new words from their context in actual writing; you will also learn how they are used. If you come across a word that seems mysterious, look it up in a dictionary. Then look again at how it is being used in context. Practice using it yourself so that it becomes a permanent part of your vocabulary. With a little effort you will find your knowledge of words steadily growing.

STUDY AIDS

Etymology

Etymology is the study of how words are formed. Many English words, especially the longer and more difficult ones, are built up out of basic parts or roots. One of the most efficient ways of increasing your vocabulary is to learn some of these parts. Once you know some basic building blocks, you will find it easier to remember words you've learned and to puzzle out unfamiliar ones.

Let's look at some examples. The word *biography* is made up of two important parts. *Graphy* comes from a Greek word meaning "writing." Many English words use this root. *Graphology*, for instance, is the study of handwriting. *Graphite* is the carbon material used in "lead" pencils. The *telegraph* is a device for writing at a distance. The *bio* part, also from Greek, means "life." It too is at the root of many English words, such as *biology* (the study of life) and *biochemistry* (the chemistry of life). When we put *bio* and *graphy* together, we get a word meaning "writing about a person's life." We can make another word by adding another part, this time from the Latin word *auto*, meaning "self." An *autobiography* is "a person's written account of his own life."

Looking over the last two paragraphs, can you guess the meaning of the word part *-logy* ?

Any good dictionary will give you the etymology of words. When you look up an unfamiliar word, make it a habit to look at how it was formed. Does the word have a root that helps to explain its meaning? Is it related to other words that you already know?

The following chart lists over 150 common word parts. Each part is defined and an example is given of a word in which it appears. Don't try to memorize the whole chart at once. If you study only a small section at a time, you'll get better results. When you've learned one of the building blocks, remember to look for it in your reading. See if you can think of other words in which the word part appears. Use the dictionary to check your guesses. To help you apply what you've learned, there are seven etymology exercises following the chart.

WORD PART	MEANING	EXAMPLE
a, ab, abs	from, away	*abrade* — to wear off
		absent — away, not present
act, ag	do, act, drive	*action* — a doing
		agent — one who acts for another
alter, altr	other, change	*alternate* — to switch back and forth
am, ami	love, friend	*amorous* — loving
anim	mind, life, spirit	*animated* — spirited
annu, enni	year	*annual* — yearly

156

WORD PART	MEANING	EXAMPLE
ante	before	*antediluvian* — before the Flood
anthrop	man	*anthropology* — study of mankind
anti	against	*antiwar* — against war
arbit	judge	*arbiter* — a judge
arch	first, chief	*archetype* — first model
aud, audit, aur	hear	*auditorium* — place where performances are heard
auto	self	*automobile* — self-moving vehicle
bell	war	*belligerent* — warlike
bene, ben	good, well	*benefactor* — one who does good deeds
bi	two	*bilateral* — two-sided
bibli	book	*bibliophile* — book lover
bio	life	*biology* — study of life
brev	short	*abbreviate* — to shorten
cad, cas, cid	fall	*casualty* — one who has fallen
cede, ceed, cess	go, yield	*exceed* — go beyond *recession* — a going backwards
cent	hundred	*century* — hundred years
chrom	color	*monochrome* — having one color
chron	time	*chronology* — time order
cide, cis	cut, kill	*suicide* — a self-killing *incision* — a cutting into
circum	around	*circumnavigate* — to sail around
clam, claim	shout	*proclaim* — to declare loudly
clin	slope, lean	*decline* — to slope downward
cogn	know	*recognize* — to know
com, co, col, con, cor	with, together	*concentrate* — to bring closer together *cooperate* — to work with *collapse* — to fall together
contra, contro, counter	against	*contradict* — to speak against *counterclockwise* — against the clock's direction
corp	body	*incorporate* — to bring into a body *corpse* — dead body
cosm	order, world	*cosmos* — universe
cre, cresc, cret	grow	*increase* — to grow *accretion* — growth by addition
cred	trust, believe	*incredible* — unbelievable
culp	blame	*culprit* — one who is to blame

WORD PART	MEANING	EXAMPLE
cur, curr, curs	run, course	*current* — presently running
de	away from, down, opposite	*detract* — to draw away from
dec	ten	*decade* — ten years
dem	people	*democracy* — rule by the people
dic, dict	say, speak	*dictation* — a speaking
		predict — to say in advance, to foretell
dis, di	not, away from	*dislike* — to not like
		digress — to turn away from the subject
doc, doct	teach, prove	*indoctrinate* — to teach
domin	rule	*domineer* — to rule over
du	two	*duo* — a couple
duc, duct	lead	*induct* — to lead in
dur	hard, lasting	*durable* — able to last
equ	equal	*equivalent* — of equal value
ev	time, age	*longevity* — age, length of life
ex, e, ef	from, out	*expatriate* — one who lives outside his native country
		emit — to send out
extra	outside, beyond	*extraterrestrial* — from beyond the earth
fac, fact, fect, fic	do, make	*factory* — place where things are made
		fictitious — made up or imaginary
fer	bear, carry	*transfer* — to carry across
fid	belief, faith	*fidelity* — faithfulness
fin	end, limit	*finite* — limited
flect, flex	bend	*reflect* — to bend back
flu, fluct, flux	flow	*fluid* — flowing substance
		influx — a flowing in
fore	in front of, previous	*forecast* — to tell ahead of time
		foreleg — front leg
form	shape	*formation* — shaping
fort	strong	*fortify* — to strengthen
frag, fract	break	*fragile* — easily broken
		fracture — a break
fug	flee	*fugitive* — one who flees
gen	birth, kind, race	*engender* — to give birth to
geo	earth	*geology* — study of the earth
grad, gress	step, go	*progress* — to go forward
graph	writing	*autograph* — to write one's own name
her, hes	stick, cling	*adhere* — to cling
		cohesive — sticking together

WORD PART	MEANING	EXAMPLE
homo	same, like	*homophonic* — sounding the same
hyper	too much, over	*hyperactive* — overly active
in, il, ig, im, ir	not	*incorrect* — not correct
		ignorant — not knowing
		illogical — not logical
		irresponsible — not responsible
in, il, im, ir	on, into, in	*impose* — to place on
		invade — to go into
inter	between, among	*interplanetary* — between planets
intra, intro	within, inside	*intrastate* — within a state
ject	throw	*reject* — to throw back
junct	join	*juncture* — place where things join
leg	law	*legal* — lawful
leg, lig, lect	choose, gather, read	*legible* — readable
		eligible — able to be chosen
		select — to choose
lev	light, rise	*alleviate* — to make lighter
liber	free	*liberation* — a freeing
loc	place	*location* — place
log	speech, study	*dialogue* — speech for two characters
		psychology — study of the mind
luc, lum	light	*translucent* — allowing light to pass through
		luminous — shining
magn	large, great	*magnify* — to make larger
mal, male	bad, wrong, poor	*maladjusted* — poorly adjusted
		malevolent — ill-wishing
mar	sea	*marine* — sea-dwelling
ment	mind	*demented* — out of one's mind
meter, metr, mens	measure	*chronometer* — time-measuring device
		commensurate — of equal measure
micr	small	*microwave* — small wave
min	little	*minimum* — least
mis	badly, wrongly	*misunderstand* — to understand wrongly
mit, miss	send	*remit* — to send back
		mission — a sending
mono	single, one	*monorail* — train that runs on a single track
morph	shape	*anthropomorphic* — man-shaped

WORD PART	MEANING	EXAMPLE
mov, mob, mot	move	*removal* — a moving away
		mobile — able to move
multi	many	*multiply* — to become many
mut	change	*mutation* — change
nasc, nat	born	*innate* — inborn
		native — belonging by or from birth
neg	deny	*negative* — no, not
neo	new	*neologism* — new word
nom	name	*nomenclature* — system of naming
		nominate — to name for office
non	not	*nonentity* — a nobody
nov	new	*novice* — newcomer, beginner
		innovation — something new
omni	all	*omnipresent* — present in all places
oper	work	*operate* — to work
		cooperation — a working together
path, pat, pass	feel, suffer	*patient* — suffering
		compassion — a feeling with
ped, pod	foot	*pedestrian* — one who goes on foot
pel, puls	drive, push	*impel* — to push
phil	love	*philosophy* — love of wisdom
phob	fear	*phobic* — irrationally fearing
phon	sound	*symphony* — a sounding together
phot	light	*photosynthesis* — synthesis of chemical compounds in plants with the aid of light
		photon — light particle
poly	many	*polygon* — many-sided figure
port	carry	*import* — to carry into a country
		portable — able to be carried
pot	power	*potency* — power
post	after	*postmortem* — after death
pre	before, earlier than	*prejudice* — judgment in advance
press	press	*impression* — a pressing into
prim	first	*primal* — first, original
pro	in favor of, in front of, forward	*proceed* — to go forward
		prowar — in favor of war
psych	mind	*psychiatry* — cure of the mind
quer, quir, quis, ques	ask, seek	*query* — to ask
		inquisitive — asking many questions
		quest — a search

WORD PART	MEANING	EXAMPLE
re	back, again	*rethink* — to think again *reimburse* — to pay back
rid, ris	laugh	*deride* — to make fun of *ridiculous* — laughable
rupt	break	*erupt* — to break out *rupture* — a breaking apart
sci, scio	know	*science* — knowledge *conscious* — having knowledge
scrib, script	write	*describe* — to write about *inscription* — a writing on
semi	half	*semiconscious* — half conscious
sent, sens	feel, think	*sensation* — feeling *sentient* — able to feel
sequ, secut	follow	*sequential* — following in order
sol	alone	*desolate* — lonely
solv, solu, solut	loosen	*dissolve* — to loosen the bonds of *solvent* — loosening agent
son	sound	*sonorous* — sounding
spec, spic, spect	look	*inspect* — to look into *spectacle* — something to be looked at
spir	breathe	*respiration* — breathing
stab, stat	stand	*establish* — to make stand, to found
string, strict	bind	*restrict* — to bind, to limit
stru, struct	build	*construct* — to build
super	over, greater	*superfluous* — overflowing, beyond what is needed
tang, ting, tact, tig	touch	*tactile* — of the sense of touch *contiguous* — touching
tele	far	*television* — machine for seeing far
ten, tain, tent	hold	*tenacity* — holding power *contain* — to hold together
term	end	*terminal* — last, ending
terr	earth	*terrain* — surface of the earth
test	witness	*attest* — to witness
therm	heat	*thermos* — container that retains heat
tort, tors	twist	*contort* — to twist out of shape
tract	pull, draw	*attract* — to pull toward
trans	across	*transport* — to carry across a distance
un	not	*uninformed* — not informed
uni	one	*unify* — to make one

WORD PART	MEANING	EXAMPLE
vac	empty	*evacuate* — to make empty
ven, vent	come	*convene* — to come together
ver	true	*verity* — truth
verb	word	*verbose* — wordy
vid, vis	see	*video* — means of seeing
		vision — sight
viv, vit	life	*vivid* — lively
voc, vok	call	*provocative* — calling for a response
		revoke — to call back
vol	wish, will	*involuntary* — not willed

Exercises

In each of the following exercises, the words in the left-hand column are built on roots given in the etymology chart. Match each word with its definition from the right-hand column. Refer to the chart if necessary. Can you identify the roots of each word? If there is any word you can't figure out, look it up in a dictionary.

EXERCISE 1

mutable	able to be touched
culpable	laughable
interminable	empty of meaning or interest
amiable	of the first age
vacuous	holding firmly
vital	necessary to life
primeval	unending
tenacious	stable, not able to be loosened or broken up
tangible	changeable
inoperable	friendly
risible	blameworthy
indissoluble	not working, out of order

EXERCISE 2

infinity	list of things to be done
duplicity	sum paid yearly
levity	a throwing out or from
brevity	shortness
ejection	endlessness
edict	body of teachings
infraction	killing of a race
genocide	lightness of spirit
agenda	a breaking
annuity	doubledealing
microcosm	official decree; literally, a speaking out
doctrine	world in miniature

EXERCISE 3

recede	state as the truth
abdicate	throw light on
homogenize	forswear, give up a power
illuminate	put into words
supervise	make freer
verbalize	go away
liberalize	bury
legislate	oversee
intervene	make laws
inter	draw out
aver	make the same throughout
protract	come between

EXERCISE 4

abduction	arrival, a coming to
fortitude	a pressing together
consequence	a flowing together
confluence	something added to
compression	a coming back to life
locus	place
status	truthfulness
disunity	that which follows as a result
veracity	strength
revival	lack of oneness
advent	a leading away, kidnapping
adjunct	standing, position

EXERCISE 5

nascent	being born
centennial	before the war
prospective	believing easily
circumspect	going against
multinational	in name only
clamorous	hard, unyielding
antebellum	looking forward
contrary	careful, looking in all directions
impassioned	hundred-year anniversary
credulous ·	having interests in many countries
obdurate	full of strong feeling
nominal	shouting

EXERCISE 6

dislocation	wrong name
misanthropy	withdrawal
misnomer	a knowing in advance
misconception	denial
negation	power of will
propulsion	forerunner
volition	a putting out of place
retraction	wrong idea
arbitration	a judging
inclination	a pushing forward
precognition	hatred of mankind
precursor	a leaning toward

EXERCISE 7

solitary	measuring time
altruistic	all-powerful
beneficial	badly shaped
benevolent	doing good, favorable
malefactor	alone, single
malformed	of the earth
malodorous	serving others
omniscient	bad-smelling
omnipotent	great-spirited, generous
magnanimous	all-knowing
chronometric	evil-doer
terrestrial	well-wishing

Word List

This list contains over 1300 words of the sort that may appear on the test you are planning to take. Many of them will be familiar to you, although you may not be sure precisely what they mean or how they are used. Many of them are words you will encounter later in the practice tests in this chapter.

Each word is briefly defined and then used in a sentence. For the roots of words and more extended definitions, check your dictionary.

The physical act of writing a word and its definition helps to reinforce this information so that you will remember new vocabulary words and their meanings. You may not have time to learn all the words listed here before your exam, but you can certainly learn many of them. Start by choosing 15 or 20 words whose meanings you don't know. Write them down, leaving enough space on your paper to write the correct definitions beside each word. Then copy the definition from the word list onto your paper. You now have a smaller word list that you can carry with you and review whenever you have some spare time. When you have mastered this list, prepare another in the same way.

Another useful study technique is to make flash cards. Write a word on one side of an index card or piece of paper and its meaning on the other side. Prepare 15 to 25 cards in this way. Study the words and definitions carefully, then test yourself. Arrange the cards in a deck so that the sides of the cards with the new vocabulary words are face up. Then, on a separate sheet of paper, write the meaning of the word shown on the top card. Turn this card over and write the meaning of the second word. Do this until you have gone through the entire deck. Check yourself by comparing your answers with the definitions on the backs of your flash cards. You can reverse the cards and, looking at the definitions, write the correct words. You can also have a friend hold the cards up for you as you tell her or him the correct meaning of each word.

abandon — to give up with the intent of never again claiming a right or interest in: *Abandoned* cars along the highway are unsightly.

abase — to cast down or make humble, to reduce in estimation: He refused to *abase* himself by admitting his mistake in front of the crowd.

abate — to lessen in intensity or number: After an hour the storm *abated* and the sky began to clear.

abdicate — to give up a power or function: The father *abdicated* his responsibility by not setting a good example for the boy.

aberrant — differing from what is right or normal: *Aberrant* behavior is frequently seen as a sign of emotional disturbance.

abet — to encourage or countenance the commission of an offense: Aiding and *abetting* a criminal makes one a party to crime.

abeyance — temporary suspension of an action: The strike motion was held in *abeyance* pending contract negotiations.

abhor — to regard with horror and loathing: The pacifist *abhors* war.

abject — miserable, wretched: Many people in underdeveloped and overpopulated countries live in *abject* poverty.

abide by — to live up to, submit to: We will *abide* by the decision of the court.

abolish — to do away with, as an institution: Slavery was *abolished* in Massachusetts shortly after the American Revolution.

abominate — to loathe: I *abominate* all laws that deprive people of their rights.

abort — to come to nothing, cut short: The mission was *aborted* when several of the helicopters broke down.

abrasive — scraping or rubbing, annoyingly harsh or jarring: The high-pitched whine of the machinery was *abrasive* to my nerves.

abridge — to shorten: The paperback book was an *abridged* edition.

absenteeism — condition of being habitually absent, as from work: *Absenteeism* at the plant becomes more of a problem around holidays.

abstain — to refrain voluntarily from some act: Alcoholics must *abstain* from any indulgence in alcoholic drinks.

abstract — not concrete; not material; not easy to understand; theoretical: His *abstract* ideas are difficult to apply to everyday situations.

absurd — clearly untrue, nonsensical: The parents dismissed the child's story of meeting men from outer space as *absurd*.

abundant — plentiful, more than enough: Rich soil and *abundant* rainfall make the region lush and fruitful.

abut — to touch, as bordering property: When estates *abut*, borders must be defined precisely.

accede — to consent: He *acceded* to their request.

accelerate — to increase in speed: Going downhill, a vehicle will naturally *accelerate*.

access — means of approach: Public libraries insure that the people have *access* to vast stores of information.

acclaim — to applaud, approve loudly: The crowd in the square *acclaimed* their hero as the new president.

accommodate — to make room for, adjust: The room can *accommodate* two more desks. We will *accommodate* ourselves to the special needs of those clients.

accumulate — to gather, pile up: Over the years she has *accumulated* a large collection of antique bric-a-brac.

accurate — careful and exact; conforming exactly to truth or to a standard; free from error: The witness gave an *accurate* account of the accident.

accustom — to get or be used to: The supervisor was not *accustomed* to having her instructions ignored.

acknowledge — to admit, recognize as true or legitimate: We do not *acknowledge* the state's authority to legislate people's beliefs.

acoustic — pertaining to hearing: The *acoustic* qualities of a room may be improved by insulation.

acquiesce — to comply or accept reluctantly: One must often *acquiesce* to the demands of a superior.

acquire — to come into possession or control of: Once *acquired*, the skill of swimming is not easily lost.

acquit — to set free from an accusation: The jury *acquitted* the defendant.

acuity — acuteness, sharpness: His unusual *acuity* of vision allowed him to spot the landmark before it was visible to the others.

acumen — sharpness of mind, keenness in business matters: The *acumen* of many early industrialists accounts for their success.

adamant — inflexible, hard: She was *adamant* in her determination to succeed.

adaptable — able to adjust to new circumstances: Thanks to the intelligence that has made technology possible, humans are more *adaptable* to a variety of climates than any other species.

adept — skilled, well-versed: A journalist is *adept* at writing quickly.

adequate — sufficient, enough: Without *adequate* sunlight, many tropical plants will not bloom.

adhere — to hold, stick to, cling: Many persons *adhere* to their beliefs despite all arguments.

adjacent — adjoining; having a common border: New York and Connecticut are *adjacent* states.

adjourn — to suspend proceedings, usually for the day: Since it is now five o'clock, I move that we *adjourn* until tomorrow morning.

adjunct — something joined to a thing but not necessarily a part of it: A rider is an *adjunct* to a legislative bill.

adroit — skillful in the use of the hands or mental faculties: The *adroit* juggler held the crowd spellbound.

adulterate — to corrupt or to make impure by addition of foreign substances: Many foods are *adulterated* by the addition of preservatives.

advantageous — useful, favorable: Our opponent's blunders have been *advantageous* to our campaign.

adverse — opposing, contrary: *Adverse* winds slowed the progress of the ship.

advocate — to plead for or urge: Socialists *advocate* public ownership of utilities.

affable — amiable, pleasant, easy to talk to: The smiling face and *affable* manner of the agent put the child at ease.

affect — to influence: The judge did not allow his personal feelings to *affect* his judgment of the case's legal merits.

affectation — artificial behavior or attitudes: Her upper-class manner of speaking was nothing but an *affectation*.

affidavit — sworn statement in writing: An *affidavit* may serve in place of a personal appearance.

affiliation — connection, as with an organization: His *affiliation* with the club has been of long standing; he has been a member for ten years at least.

affinity — relationship, kinship: There is a close *affinity* among many European languages, such as Spanish and Italian.

affirm — to avow solemnly, declare positively: Although as a Quaker he refused to take an oath, he *affirmed* in court that the defendant's account was true.

affix — to attach, fasten: A price tag was *affixed* to each item.

affluence — wealth: The new *affluence* of the family made them the object of curiosity and envy to their poorer neighbors.

agenda — list of things to be done: The *agenda* of the conference included the problem of tariffs.

agitate — to stir up or disturb: Rumors of change in the government *agitated* the population.

aggression — unprovoked attack: The invasion of Afghanistan was denounced in the Western press as *aggression*.

agoraphobia — unreasoning fear of open places: A person suffering from *agoraphobia* may be unable to go outdoors without experiencing panic.

agrarian — having to do with land: *Agrarian* reforms were one of the first measures adopted in the economic rehabilitation of the country.

alarm — disturb, excite: The parents were *alarmed* when their child's temperature rose suddenly.

alien — (*adj*) strange, foreign: Their customs are *alien* to us. (*n*) foreigner who has not become a citizen of the country where he lives: There are, it is estimated, over a million illegal *aliens* living in New York City.

allay — to pacify, calm: Therapy will often *allay* the fears of the neurotic.

allege — to declare without proof: The *alleged* attacker has yet to stand trial.

alleviate — to lessen, make easier: The morphine helped to *alleviate* the pain.

allocate — to distribute or assign: The new serum was *allocated* among the states in proportion to their population.

allude — to refer to indirectly or by suggestion: The book *alludes* to an earlier document that we have been unable to locate or even to identify.

aloof — distant, reserved or cold in manner: Her elegant appearance and formal politeness made her seem *aloof*, though in reality she was only shy.

altercation — angry dispute: The *altercation* stopped just short of physical violence.

altitude — height, especially above sea level or the earth's surface: The plane had reached an *altitude* of four miles.

altruist — person who acts unselfishly in the interests of others: She proved herself an *altruist* by volunteering to help the flood victims when there was no hope of recompense.

amalgamate — to make into a single unit: The new owners *amalgamated* several small companies into a single corporation.

amass — to collect, pile up: Through careful investment he had *amassed* a sizable fortune.

ambiguous — having more than one possible meaning: The *ambiguous* wording of some legislative acts requires clarification by the courts.

ambivalent — having conflicting feelings: I am *ambivalent* about the job; although the atmosphere is pleasant, the work itself is boring.

amenable — agreeable, open to suggestion: He was *amenable* to the proposed schedule change.

amicable — friendly: Courts often seek to settle civil suits in an *amicable* manner.

amnesty — pardon for a large group: The president granted *amnesty* to those who had resisted the draft.

amplify — to enlarge, expand: Congressmen may *amplify* their remarks for appearance in the Record.

analogous — having a similarity or partial likeness: Writers have often seen springtime as *analogous* to youth and winter to old age.

analyze — to study or determine the nature and relationship of the parts: After the chemist *analyzes* the powder, she will tell us what it is.

anemia — deficiency of red blood corpuscles or hemoglobin in the blood: Before it was treated, her *anemia* caused her to tire easily.

annals — chronological records: The *annals* of the scientific societies reflect the advances of our era.

annihilate — to destroy completely: If the government does not act to preserve the few remaining herds, the whole species will have been *annihilated* by the end of the century.

annotate — to provide explanatory notes: For the student edition, the editor *annotated* the more difficult passages in the essay.

annual — yearly, once a year: The company holds an *annual* picnic on the Fourth of July.

annuity — amount of money payable yearly: Investment in an *annuity* provides income for their retirement.

annul — to wipe out, make void: The Supreme Court can *annul* a law that it deems unconstitutional.

anomalous — out of place, inappropriate to the surroundings: An *anomalous* jukebox stood rusting in the square of the primitive village.

anonymous — bearing no name, unsigned: Little credence should be given to an *anonymous* accusation.

antecedent — a preceding event, condition or cause: The bombing of Pearl Harbor was *antecedent* to our entry into World War II.

anticipate — to foresee, give thought to in advance; expect: We *anticipate* that this movie will be a box office hit.

antipathy — dislike: His *antipathy* toward cats almost amounted to a phobia.

apathy — lack of interest or emotion: Voter *apathy* allows minorities of politically concerned people to influence elections out of proportion to their numbers.

apex — summit, peak: Some people reach the *apex* of their careers before forty.

apparition — phantom, anything that appears suddenly or unexpectedly: Dressed in the antique gown, the woman looked like an *apparition* from her grandmother's era.

appearance — outward aspect; act of coming in sight: The celebrity put in an *appearance* at the fund-raising activities.

appease — to give in to satisfy or make peace, to pacify: Only a heartfelt apology will *appease* his rage at having been slighted.

append — to attach as a supplement: Exhibits should be *appended* to the report.

appendix — extra material added at the end of a book: A chronology of the events described may be found in the *appendix*.

applicable — able to be applied, appropriate: Since you are single, the items on the form concerning your spouse are not *applicable*; leave those spaces blank.

appraise — to set a value on: The price at which authorities *appraise* a building determines its taxes.

appreciate — to think well of; to grasp the nature, worth, quality or significance of: I can *appreciate* the difficulties you are having with your invalid mother.

apprehend — to arrest; to understand: The police moved to *apprehend* the suspect. I could not *apprehend* what she was trying to say.

apprehensive — fearing some coming event: The students were *apprehensive* about the examination.

apprise — to give notice to: He was captured because none could *apprise* him of the enemy's advance.

approach — to come near to; taking of preliminary steps; a means of access: The *approach* to the haunted house was through a tangled arbor.

approbation — approval: The act was performed with the *approbation* of the onlookers.

apropos — pertinent, to the point: His remarks were *apropos* and well-founded.

arbitrary — despotic, arrived at through will or caprice: An *arbitrary* ruling of the civil commission is being challenged in the courts.

arbitration — settling a dispute by referring it to an outsider for decision: Under the agreement, disputes were to be settled by *arbitration*.

archaic — no longer in use: Some words like ''thou,'' once a common form of address, are now *archaic*.

archives — historic records: A separate building houses the United States *archives* in Washington.

ardent — passionately enthusiastic: His *ardent* patriotism led him to risk his life in the underground resistance movement.

aroma — fragrance: The *aroma* of good coffee stimulates the salivary glands.

arraign — to bring before a court of law, formally accuse: A person arrested must be *arraigned* within 24 hours.

arrogant — disposed to exaggerate one's own worth and importance: The *arrogant* waiter refused to acknowledge the disheveled man at the counter.

articulate — expressing oneself readily, clearly or effectively: The *articulate* child told of the day's events in school.

artifact — man-made object: *Artifacts* found by archeologists allow them to reconstruct the daily lives of ancient peoples.

ascertain — to find out with certainty: Because the woman's story was so confused, we have been unable to *ascertain* whether a crime was committed or not.

ascribe — to attribute, assign as a cause: His death was *ascribed* to poison.

asinine — stupid, silly: The argument was too *asinine* to deserve a serious answer.

asphyxiation — death or loss of consciousness caused by lack of oxygen: The flames never reached that part of the building, but several residents suffered *asphyxiation* from the smoke.

assailant — attacker: Faced with a line-up, the victim picked out his alleged *assailant*.

assemblage — a group or collection of things or persons: Out of the *assemblage* of spare parts in the garage, we found the pieces to repair the bicycle.

assent — to concur, comply, consent: All parties involved *assented* to the statement.

assert — to claim or state positively: She *asserted* her title to the property.

assess — to set a value on: The house has been *assessed* for taxes at far below its market value.

assiduous — performed with constant diligence: *Assiduous* attention to her assignment won her a promotion.

assign — to appoint, prescribe: The new reporters were *assigned* to cover local sports events.

assist — to give aid, help: The laboratory aide *assists* the chemist in researching the properties of chemical substances.

assumption — something taken for granted or supposed to be fact: I prepared dinner on the *assumption* that they would be home by seven.

assure — to make something certain, guarantee; to promise with confidence: The fact that they left their tickets *assures* that they will return. I *assured* her that someone would be there to meet her.

astute — difficult to deceive: An *astute* judge of character, he guessed that his opponent was bluffing.

attain — to get through effort, to achieve or reach: Thanks to their generous contribution, the campaign has *attained* its goal.

attest — to bear witness to by oath or signature: Disinterested witnesses must *attest* to the signing of a will.

attrition — a gradual wearing down: With the armies dug into the trenches, World War I became a war of *attrition*.

atypical — not normal or usual: The usually calm man's burst of temper was *atypical*.

auction — to sell to the highest bidder: Bidding started at five dollars, but the chair eventually was *auctioned* for thirty.

augment — to increase: He *augments* his wealth with every deal.

aural — of the ear or hearing: Since the sound system was not working properly, the *aural* aspect of the performance was a disappointment.

auspicious — predicting success: The first week's business was an *auspicious* start for the whole enterprise.

austerity — quality of being strict, rigorous, very simple, or unadorned: To save money they went on an *austerity* program, cutting down on driving and nonessential purchases.

authorize — to give official permission: The guard is *authorized* to demand identification from anyone entering the building.

autocratic — despotic, acting without regard for the rights or opinions of others: An *autocratic* attitude on the part of a supervisor is deeply resented by subordinates.

automatically — involuntarily, spontaneously, mechanically: The computer *automatically* records the amount of sales for each month.

autopsy — examination and partial dissection of a body to determine the cause of death: The *autopsy* revealed a brain tumor.

averse — having a dislike or reluctance: The local population disliked tourists and were *averse* to having their pictures taken.

avert — to turn aside or ward off: By acting quickly we *averted* disaster.

avow — to declare openly: She *avowed* her belief in the political system.

baffle — to perplex, frustrate: The intricacies of the game *baffle* description.

banal — commonplace, trite: There was nothing fresh or memorable in their *banal* exchange of opinions.

banter — light, good-natured teasing: The comments were mere *banter*, not intended to wound.

bar — to oppose, prevent, or forbid; to keep out: Conviction of committing a felony will *bar* you from voting.

barren — unfruitful, unproductive: Only a few scrubby trees clung to the rocky soil of that *barren* landscape.

barter — to trade by direct exchange of one commodity for another: At the Indian market I *bartered* my sleeping bag for a handwoven poncho.

basic — fundamental: The teacher explained the *basic* concepts of democracy.

belabor — to beat or attack, especially with words: The speech *belabored* at great length what everyone in the audience already knew.

belie — to lie about, to show to be false: Her laughing face *belied* her pretense of annoyance.

belittle — to make smaller or less important: He *belittled* the actress's talent by suggesting that her beauty, rather than her acting ability, was responsible for her success.

beneficiary — one who benefits, especially one who receives a payment or inheritance: The man named his wife as the *beneficiary* of the insurance policy.

benign — kindly: Her *benign* influence helped to alleviate their depression.

berate — to scold vehemently: The teacher who *berates* his class is rationalizing his own faults.

bespeak — to indicate, to speak for, especially in

advance: The success of the first novel *bespeaks* a promising career for the young author.

bestow — to grant or confer: The republic *bestowed* great honors upon its heroes.

bibliography — list of sources of information on a particular subject: She assembled a *bibliography* of major works on early American history published since 1960.

biennial — happening every two years: Many state legislatures convene on a *biennial* basis.

bigot — narrow-minded, intolerant person: A *bigot* is not swayed from his beliefs by rational argument.

blatant — too noisy or obtrusive, impossible to ignore: The children's *blatant* disregard for conventional manners appalled their older relatives.

bogus — false, counterfeit: Using a *bogus* driver's license, she had opened an account under an assumed name.

bolster — to prop up, support: The announcement that refreshments were being served *bolstered* the flagging spirits of the company.

bore — to weary by being dull, uninteresting, or monotonous: I get so *bored* when my aunt tells the same stories over and over.

bourgeois — middle-class: Thriftiness, respectability, and hard work are often thought of as *bourgeois* traits.

boycott — to refuse to do business with or use: Consumers *boycotted* the company's products to show support for the striking workers.

brazen — brassy, shameless: The delinquents demonstrated a *brazen* contempt for the law.

breach — opening or gap, failure to keep the terms, as of a promise or law: When they failed to deliver the goods, they were guilty of a *breach* of contract.

brevity — conciseness, terseness: *Brevity* is the essence of journalistic writing.

brochure — pamphlet: *Brochures* on many topics are available free of charge.

brunt — the principal force or shock, greater part: The *brunt* of the attack was borne by the infantry.

brusque — blunt, curt in manner: A *brusque* manner displeases many persons.

budget — plan for the spending of income during a certain period: The present *budget* allocates one fourth of our joint income for rent and utilities.

buoyant — rising or floating, cheerful: His *buoyant* nature would not allow him to remain glum for long.

cache — hiding place for loot or supplies: The *cache* left by the expedition was found many years later.

cadaver — dead body: The *cadaver* was dissected by the medical students.

cadence — rhythmic flow, modulation of speech, measured movement: The low and musical *cadence* of the actress's voice was a delight to hear.

cadre — framework, skeleton organization: A *cadre* of commissioned and noncommissioned officers was maintained.

cajole — to coax, persuade by artful flattery: He was *cajoled* into betting on the game.

caliber — capacity of mind, quality: It is a crucial and delicate job requiring personnel of the highest *caliber*. *Also:* **calibre**.

candid — honest, open: She was always *candid* about her feelings; if she liked you, you knew it.

candor — unreserved honest or sincere expression: *Candor* and innocence often go hand in hand.

cant — jargon, secret slang, empty talk: The conversational *cant* of the two social scientists was unintelligible to everyone else at the party.

capacious — roomy, spacious: The travelers had all their possessions in one *capacious* suitcase.

capitulate — surrender: The city *capitulated* to the victors.

capricious — changing suddenly, willfully erratic: He is so *capricious* in his moods that no one can predict how he will take the news.

captivate — to influence and dominate by some special charm; to hold prisoner: The Frenchman *captivated* the American dowager by kissing her hand.

carcinogenic — producing cancer: In tests on laboratory animals, the drug was shown to be *carcinogenic*.

caricature — distorted sketch: *Caricature* is the weapon of the political cartoonist.

carnage — destruction of life: The *carnage* of modern warfare is frightful to consider.

catalyst — substance that causes change in other substances without itself being affected: Platinum is a *catalyst* in many chemical processes.

cataract — a condition of the eye in which the lens becomes opaque: The old man's vision had been impaired by *cataracts*.

category — class or division in a system of classification: Patients are listed according to *categories* which designate the seriousness of their condition.

caustic — biting, burning, stinging: The surface of the wood had been marred by some *caustic* substance.

caveat — legal notice preventing some action; a warning: A *caveat* may be entered to stop the reading of a will.

cede — to yield, assign, transfer: A bill of sale will *cede* title of the property.

censure — to disapprove, blame, condemn as wrong: The unprofessional conduct of several of its members has been officially *censured* by the organization.

cerebral — pertaining to the brain: The stroke was the result of a *cerebral* hemorrhage.

chagrin — disappointment, vexation: The failure of the play filled the backers with *chagrin*.

chaos — complete confusion or disorder: By the time the children had finished playing with all the toys, the room was in *chaos*.

characteristic — typical trait, identifying feature: The curved yellow bill is a *characteristic* of this species.

charlatan — one who pretends to know more than he does: *Charlatans* who pretend they can cure cancer have been responsible for many deaths.

chastise — to punish or to censure severely: The disobedient boy was *chastised* by being sent to his room.

chauvinism — zealous, unreasoning patriotism: *Chauvinism* is the cause of many unnecessary wars.

chicanery — unethical methods, legal trickery: He accused the winning candidate of *chicanery* in manipulating the election.

chide — to rebuke, scold: The parents *chided* the disobedient child.

childish — immature; characteristic of a child: Constant quibbling about the details of past events is *childish*.

chronic — long-lasting, recurring: His *chronic* asthma flares up at certain times of the year.

chronology — arrangement by time, list of events by date: The book included a *chronology* of the poet's life against the background of the major political events of his age.

circumspect — watchful in all directions, wary: A public official must be *circumspect* in all his actions.

circumvent — to go around, frustrate: A technicality allowed people to *circumvent* the intention of the law.

citation — summons to appear in court; an official praise, as for bravery; reference to legal precedent or authority: Caught for speeding, I received a *citation*. Three firemen received *citations* for the heroic rescue effort. The attorney asked the clerk to check the *citations* to cases in the Supreme Court.

clamor — loud, continuous noise; uproar: The *clamor* of the protesting mob was unbearable.

clandestine — secret: The conspirators held a *clandestine* meeting.

clemency — leniency: The governor granted *clemency* to the prisoners.

cliché — trite, overworked expression: ''White as snow'' is a *cliché*.

climactic — of a climax: At the *climactic* moment of the film, the heroine walks out and slams the door.

climatic — of a climate: Over eons, *climatic* changes turned the swamp into a desert.

coalition — temporary union of groups for a specific purpose: Various environmentalist groups formed a *coalition* to work for the candidate most sympathetic to their cause.

coerce — to compel, force: He did not sign the confession freely but was *coerced*.

cogent — conclusive, convincing: A debater must present *cogent* arguments to win his point.

cognizant — having knowledge: She was *cognizant* of all the facts before she made a decision.

coherent — logically connected or organized: They were too distraught to give a *coherent* account of the crash.

cohesion — sticking together: The *cohesion* of molecules creates surface tension.

coincide — to be alike, to occur at the same time: This year Thanksgiving *coincides* with her birthday.

collaborate — to work together on a project: The friends decided to *collaborate* on a novel.

collate — to put the pages of a text in order: The photocopies have been *collated* and are ready to be stapled.

colleague — fellow worker in a profession: The biologist enjoyed shoptalk with her *colleagues* at the conference.

collusion — secret agreement, conspiracy: Higher prices were set by *collusion* among all the manufacturers.

colloquial — of speech and informal writing, conversational: Having studied only formal French, she was unable to understand many of her host's *colloquial* expressions.

comatose — lethargic, of or like a coma: The drug left him in a *comatose* state.

comical — causing amusement; funny; humorous: The poodle looked *comical* stepping out in its rubber boots.

commence — to make a beginning: High school *commencement* exercises mark the beginning of a new life.

commend — to praise: The supervisor *commended* them for their excellent work.

commensurate — equal, corresponding in measure: He asked for compensation *commensurate* with his work.

commiserate — to express sympathy for: It is natural to *commiserate* with the innocent victim of an accident.

commission — to authorize, especially to have someone perform a task or to act in one's place: I have *commissioned* a neighbor to collect the mail while I'm away.

commodious — comfortably or conveniently spacious: The *commodious* closets impressed the prospective tenants.

common — ordinary, widespread: Financial difficulty is a *common* problem for young married couples today.

commute — to substitute one thing for another; to travel regularly between home and work: The alchemists dreamed of *commuting* lead into gold. *Commuting* is a daily routine for most working people.

comparable — equivalent, able or worthy to be compared: His degree from a foreign university is *comparable* to our master's degree.

compatible — harmonious, in agreement: Since they have similar attitudes, interests, and habits, they are a *compatible* couple.

compel — to force: He was *compelled* by law to make restitution.

compensate — to be equal to, make up for: Money could not *compensate* him for his sufferings.

competent — fit, capable, qualified: I am not *competent* to judge the authenticity of this document; you should take it to an expert.

competition — rivalry for the same object: The theory of free enterprise assumes an unrestricted *competition* for customers among rival businesses.

complacent — self-satisfied: A *complacent* student seldom attains the heights of success achieved by those who demand more of themselves.

complete — to finish, bring an end to: The project was *completed* in time for the fall science fair.

complex — a whole made up of interconnected or related parts: The school grew from a few classrooms to a whole *complex* of buildings organized around the computer center.

complicity — partnership in wrongdoing: By withholding evidence she became guilty of *complicity* in the crime.

comply — to go along with, obey: The crowd *complied* with the order to disperse.

compose — to put together, create: I spent an hour *composing* a formal letter of protest.

comprehensible — able to be understood: The episode was only *comprehensible* to those who knew the story thus far.

comprise — to include, be made up of, consist of: The test will *comprise* the subject matter of the previous lessons.

compromise — a settlement of a difference in which both sides give up something: We are willing to make some concessions in order to reach a *compromise*.

compulsory — required, forced: Attendance is *compulsory* unless one has a medical excuse.

compunction — remorse, uneasiness: He showed no *compunction* over his carelessness.

compute — to figure, calculate: He *computed* the total on his pocket calculator.

concede — to yield, as to what is just or true: When the candidate realized she could not win, she *conceded* gracefully.

concentration — strength or density as of a solution; close or fixed attention: The high *concentration* of frozen orange juice requires that it be diluted with three cans of water before drinking.

concept — idea, general notion: The *concept* that all individuals have inherent and inalienable rights is basic to our political philosophy.

conception — beginning, original idea, mental image: The *conception* of the plan was originally his.

conciliatory — tending to placate or to gain goodwill: After the quarrel he sent flowers as a *conciliatory* gesture.

concise — brief, to the point: A précis must be *concise* yet cover the topic.

concoction — combination of various ingredients: The drink was a *concoction* of syrup, soda and three flavors of ice cream.

concurrent — running together, happening at the same time: *Concurrent* action by the police and welfare authorities reduced juvenile crime.

condole — to express sympathy: His friends gathered to *condole* with him over his loss.

condone — to pardon, overlook an offense: The law will not *condone* an act on the plea that the culprit was intoxicated.

conducive — leading to, helping: Mother found the waterbed *conducive* to a restful sleep.

conduct — behavior: Her cool *conduct* in the emergency inspired confidence in those around her.

confidence — a relationship of trust, intimacy or certitude: To ski with abandon, you must have *confidence* in your bindings.

confidential — private, secret: Respondents were assured that the census was *confidential* and would be used for statistical purposes only.

confiscate — to seize, appropriate: The government has no right to *confiscate* private property without just compensation.

conflagration — large fire: New York City was almost destroyed in the 1835 *conflagration*.

conformity — harmony, agreement: In *conformity* with the rule, the meeting was adjourned.

congruent — in agreement or harmony: *Congruent* figures coincide entirely throughout.

conjecture — to make a guess based on information at hand: He *conjectured* that vocabulary questions would appear on the test.

connive — to pretend ignorance of or assist in wrongdoing: The builder and the agent *connived* in selling overpriced homes.

conscientious — honest, faithful to duty or to what is right: He is *conscientious* in his work, and so has won the trust of his employers.

consensus — general agreement: The *consensus* of the jury was that the defendant was not guilty.

consequence — result of an action or process; outcome; effect: The *consequence* of eating too much chocolate may be a bad complexion.

conservative — tending to preserve what is, cautious: At the annual conference, they presented their *conservative* views on the future of education.

considerable — important, large, much: The director has *considerable* clout among the members of the board; they value her recommendations highly.

consign — to entrust, hand over: The child was *consigned* to the care of her older sister until the court could appoint a guardian.

consolidate — to combine into a single whole: Let us *consolidate* our forces before we fight the gang from the next neighborhood.

consonance — harmony, pleasant agreement: Their *consonance* of opinion in all matters made for a peaceful household.

constant — unchanging, fixed, continual: It is difficult to listen to his *constant* complaining.

constituency — body of voters: The congressman went home to discuss the issue with his *constituency*.

constitute — to set up, make up, compose: In industrialized countries farmers *constitute* only a small percentage of the population.

constrain — to force, compel: She felt *constrained* to make a full confession.

construe — to interpret, analyze: His attitude was *construed* as one of opposition to the proposal.

consume — to do away with completely; to use up: The fire totally *consumed* the frame dwelling.

contagious — transmittable by direct or indirect contact: Hepatitis is a *contagious* disease.

contaminate — to pollute, make unclean or unfit: The pesticide seeped into the water table, *contaminating* the wells.

contentious — quarrelsome: One *contentious* student can ruin a debate.

contiguous — next to, adjoining: Alaska is not *contiguous* to other states of the United States.

contingent — depending upon something's happening: Our plans were *contingent* on the check's arriving on time.

contort — to twist out of shape: Rage *contorted* her features into a frightening mask.

contract — a formal agreement, usually written: The company signed a *contract* to operate a bookstore on campus.

contrition — sincere remorse: They were overwhelmed by *contrition* when they realized the damage they had caused.

contrive — to devise, plan: They *contrived* a way to fix the unit using old parts.

controversy — highly charged debate, conflict of opinion: A *controversy* arose over whether to use the funds for highway improvement or for mass transit.

contusion — bruise: He suffered severe *contusions* in the accident.

convene — to gather together, as an assembly: The graduates will *convene* on the campus.

conventional — ordinary; traditional; not unusual: The architecture of the new building is unimaginative and very *conventional*.

converge — to move nearer together, head for one point: The flock *converged* on the seeded field.

convulsion — violent, involuntary contracting and relaxing of the muscles: Epilepsy is accompanied by *convulsions*.

cooperate — to work together for a common goal: If everyone *cooperates* on decorations, entertainment and refreshments, the party is sure to be a success.

coordinate — to bring different elements into order

or harmony: In a well-run office, schedules are *coordinated* so that business is uninterrupted.

corpulent — very fat: The *corpulent* individual must choose clothing with great care.

correlate — to bring into or show relation between two things: Studies have *correlated* smoking and heart disease.

correspondence — letters, communication by letter: A copy of the book order will be found in the *correspondence* file under the name of the publisher.

corroborate — to provide added proof: Laws of evidence require that testimony on a crime be *corroborated* by other circumstances.

counter — to offset or to nullify; to oppose or check: Their fine offense is *countered* by our exceptional defensive line.

countermand — to revoke an order or command: The wise executive will not hesitate to *countermand* an unwise order.

courteous — polite, considerate: A *courteous* manner — friendly but not personal — is essential for anyone who deals with the public.

covert — hidden, secret: To avoid public outcry, the President ordered a *covert* military action and publicly denied that he was sending out combat troops.

cranium — skull of a vertebrate: The *cranium* affords great protection to the brain.

crass — stupid, unrefined: The *crass* behavior of some tourists casts discredit on their nation.

craven — cowardly: His *craven* conduct under stress made him the butt of many jests.

credence — faith, belief: One could have little *credence* in the word of a known swindler.

credible — worthy or able to be believed: The tale, though unusual, was entirely *credible*, considering the physical evidence.

credulous — inclined to believe on slight evidence: The *credulous* man followed every instruction of the fortune-teller.

creed — statement of faith: The Apostles' *creed* is the basis of their belief.

cremate — burn a dead body: He left instructions that he was to be *cremated* upon his death.

criterion — standard of judging: Logical organization was a *criterion* for grading the essays. *Plural:* **criteria**.

crucial — of utmost importance: The discovery of the letter was *crucial* to the unraveling of the whole mystery.

cryptic — having a hidden meaning, mysterious: The *cryptic* message was deciphered by the code expert.

cull — to select, pick out from a group: From the pile we *culled* all the mail for local delivery.

culmination — acme, highest attainment: Graduation with highest honors was the *culmination* of her academic efforts.

culpable — faulty, blameworthy, guilty: The *culpable* parties should not escape punishment.

curative — concerning or causing the cure of disease: The grandmother had faith in the *curative* powers of certain herbs.

cursory — superficial, hurried: *Cursory* examination of the scene revealed nothing amiss, but later we discovered that some jewelry was missing.

curtail — to reduce, shorten: Classes were shortened in the winter to *curtail* heating costs.

customary — usual, according to habit or custom: Today, because of the traffic jam, he did not take his *customary* route.

dearth — scarcity: A *dearth* of water can create a desert in a few years.

debase — to reduce in dignity or value: Inflation has *debased* the currency so that a dollar now buys very little.

debate — to argue formally for and against: The candidate challenged the incumbent to *debate* the issues on television.

debilitate — to enfeeble, weaken: Constant excesses will *debilitate* even the strongest constitution.

deceive — to trick, be false to: They *deceived* us by telling us that our donations would be used to provide food to the needy; in reality, the money was used to supply guns to the rebels.

decelerate — to slow down or to reduce the rate of progress: When you see a stop sign in the distance, it is wise to *decelerate*.

decimate — to destroy a large part (literally, one tenth) of a population: The Black Death had *decimated* the town.

decompose — to break up or separate into basic parts or elements: After some time, vegetation that lies on damp ground tends to *decompose*.

decorum — that which is suitable or proper: He had gentlemanly notions of *decorum;* he always held doors open for ladies and held their chairs when they sat down to dinner.

decrepit — broken down by old age: With its crumbling brick and peeling paint, the building had a *decrepit* appearance.

decry — to clamor against: Critics *decry* the lack of emotion on the stage.

deduct — to subtract, take away: Because the package was damaged, the seller *deducted* two dollars from the price.

deem — to judge, think: The newspaper did not *deem* the event worthy of coverage.

de facto — actual as opposed to legal: Although he holds no official position, he is the *de facto* head of the government.

default — failure to do what is required: In *default* of the payment, the property was seized by the creditor.

deference — act of respect, respect for another's wishes: Out of *deference* to her age, we rose when she entered.

deficient — not up to standard, inadequate: The child is *deficient* in reading but excels in arithmetic.

definition — description, explanation or meaning: Look in the dictionary for *definitions* of words you do not know.

defoliate — to strip of leaves: All the trees in the yard had been *defoliated* by an infestation of moths.

defray — to pay (costs): The company *defrayed* the costs of a vacation trip for the winner of the essay contest.

defunct — dead, no longer functioning: The business has been *defunct* since the big fire.

defy — to oppose or resist openly or boldly; to challenge; to dare: Some people *defy* No Smoking rules and light up.

degenerate — to decline from a higher or normal form: The discussion eventually *degenerated* into a shouting match.

degrade — to lower in status, value or esteem: The celebrity refused interviews, feeling that it was *degrading* to have her personal life publicly discussed.

de jure — according to law: Although it had successfully seized power, the junta was not recognized as the *de jure* government by the neighboring countries.

delegate — to authorize or assign to act in one's place: Since I will be unable to attend the conference, I have *delegated* my assistant to represent me.

delegation — group of persons officially authorized to act for others: Our *delegation* to the United Nations is headed by the ambassador.

delete — to strike out, erase: Names of those who fail to pay their dues for over a year are *deleted* from the membership rolls.

deleterious — injurious, harmful: DDT, when taken internally, has a *deleterious* effect on the body.

deliberate — intended, meant: It was no accident but a *deliberate* act.

delineate — to mark off the boundary of: They asked him to *delineate* the areas where play was permitted.

delinquent — delaying or failing to do what rules or law require: Since she was *delinquent* in paying her taxes, she had to pay a fine.

demagogue — leader who uses mob passions to gain power: Hitler was a *demagogue* who played on the irrational fears and hatred of the mob.

demean — to degrade, debase: He would not *demean* himself by making personal attacks.

demolish — to destroy, especially a building: The wrecking crew arrived and within a few hours the structure was *demolished*.

demonstrable — able to be shown: The tests showed that the consumers' preference was justified by that brand's *demonstrable* superiority.

demote — to lower in rank: He was stripped of his rank and *demoted* to private.

denigrate — to blacken, defame: The lawyer tried to *denigrate* the character of the witness by implying that he was a liar.

dense — thick, compact or crowded; difficult to penetrate: The underbrush was so *dense* that it was impossible to follow the path.

deny — to declare untrue, refuse to recognize: I categorically *deny* the accusation.

deplete — to empty, use up: At the present rates of consumption, the known reserves will be *depleted* before the end of the century.

deplore — to lament, disapprove strongly: Pacifists *deplore* violence even on behalf of a just cause.

deposition — written testimony taken outside court: A *deposition* had to be taken from the hospitalized witness.

deprecate — to plead against; to express disapproval: Do not *deprecate* what you cannot understand.

depreciate — to lessen in value: Property will *depreciate* rapidly unless kept in good repair.

deprive — to take away, often by force: No person may be *deprived* of his liberty without due process of law.

deride — to mock, laugh at: Many passersby *derided* the comical figure of the street-corner orator.

derelict — (*adj*) abandoning duty, remiss: The policeman was *derelict* in his duty. (*n*) thing or person abandoned as worthless: In winter the city's *derelicts* frequent the bus station to stay warm.

derogatory — disparaging, disdainful: Her *derogatory* remarks hid feelings of envy.

designate — to name, appoint: We will rendezvous at

the time and place *designated* on the sheet.

despicable — contemptible: The villain in melodramas is always a *despicable* character.

despotic — ruling with absolute authority: The *despotic* king is out of place in a constitutional monarchy.

destitute — in extreme want: Three successive years of crop failures had left the peasants *destitute*.

desuetude — lack of use: The law, which had never been repealed, had fallen into *desuetude* and was never enforced.

deteriorate — to get worse: Storing it in a cedar chest will keep the antique fabric from *deteriorating* further.

determine — to find out; to be the cause of, decide: The doctor interviewed the mother to *determine* whether there was a family history of diabetes. The result of this test will *determine* our next step.

deterrent — a thing that discourages: The absolute certainty of apprehension is a powerful *deterrent* to some types of crime.

detonate — to explode: An electrical charge can be used to *detonate* certain explosives.

detract — to take away a part, lessen: The old-fashioned engraving *detracted* from the value of the piece of jewelry.

detrimental — causing damage or harm: The support of fringe groups can be *detrimental* to the campaign of an office-seeker.

deviate — to stray, turn aside from: The honest man never *deviated* from telling the truth as he saw it.

device — a piece of equipment designed to serve a certain purpose: The *device* at the end of the clothes brush serves as a shoe horn.

devious — roundabout, indirect, underhanded: When no one would tell her anything, she resorted to *devious* means to uncover the truth.

devise — to contrive, invent: I will *devise* a plan of escape.

devoid of — completely without: The landscape was flat and barren, *devoid of* interest or beauty.

dexterity — quickness, skill and ease in some act: The art of juggling is one that calls for the highest degree of *dexterity*.

dichotomy — division into two parts, often opposed: The *dichotomy* of her position, half instructor, half administrator, made efficient work in either field impossible.

diffidence — shyness, lack of assertiveness: His *diffidence* before such a distinguished visitor prevented him from expressing his own views.

diffuse — to spread out, scatter widely: When the bottle broke, the fragrance *diffused* throughout the room.

digress — to wander from the subject: To *digress* from the main topic may lend interest to a theme, but at the cost of its unity.

dilate — to expand: Some drugs will cause the pupil of the eye to *dilate*.

dilemma — choice of two unpleasant alternatives, a problem: Even a wrong decision may be preferable to remaining in a *dilemma*.

diminish — to make less: Inflation *diminishes* the value of the dollar.

diplomatic — employing tact and conciliation in dealing with people, especially in stressful situations: It would be *diplomatic* not to constantly refer to that woman's previous husbands.

disability — loss of ability: The accident resulted in a temporary *disability*; the employee was out for two weeks.

discard — to throw away: Dead files more than ten years old may be *discarded*.

discern — to perceive, identify: The fog was so thick we could barely *discern* the other cars.

disciple — follower of a teacher: The renowned economist won over many *disciples* with her startling theories.

disclaim — to renounce, give up claim to: To obtain United States citizenship, one must *disclaim* any title or rank of nobility from another nation.

disclose — to reveal: The caller did not *disclose* the source of her information.

disconcert — to throw into confusion: An apathetic audience may *disconcert* even the most experienced performer.

discordant — harsh, not harmonious: The *discordant* cries of the gulls made me long for the familiar sounds of the city.

discount — to underestimate the importance of; to minimize; to disregard: Do not *discount* the value of experience.

discretion — power of decision, individual judgment: The penalty to be imposed in many cases is left to the *discretion* of the judge.

disdain — to reject as unworthy: Many beginners *disdain* a lowly job that might in time lead to the position they desire.

disengage — to loosen or break a connection: Depressing the clutch *disengages* the driving force from the wheels.

disinterested — not involved in, unprejudiced: A *disinterested* witness is one who has no personal stake in the outcome of the case.

dismantle — to take apart: The machine must be *dismantled*, cleaned, repaired and reassembled.

disparage — to speak slightingly of, belittle: A teacher who *disparages* the efforts of beginners is not helping the students.

disparity — inequality, difference in degree: A *disparity* in age need not mean an incompatible marriage.

dispatch — to send on an errand: The bank *dispatched* a courier to deliver the documents by hand.

dispel — to drive away, make disappear: The good-humored joke *dispelled* the tension in the room.

dispense with — to get rid of, do without: Let's *dispense with* the formalities and get right down to business.

disquieting — disturbing, tending to make uneasy: There have been *disquieting* reports of a buildup of forces along the border.

dissemble — to conceal or misrepresent the true nature of something: He *dissembled* his real motives under a pretence of unselfish concern.

disseminate — to spread, broadcast: With missionary zeal, they *disseminated* the literature about the new religion.

dissension — lack of harmony or agreement: There was *dissension* among the delegates about which candidate to support.

dissipate — to scatter aimlessly, spend foolishly: He soon *dissipated* his inheritance.

dissuade — to advise against, divert by persuasion: Her friends *dissuaded* her from the unwise plan.

distend — to stretch: If you *distend* a balloon beyond a certain point, it breaks.

distinct — clear, notable: There is a *distinct* difference between these two musical compositions.

distortion — a twisting out of shape, misstatement of facts: The *distortions* of the historians left little of the man's true character for posterity.

distract — to divert, turn aside: The loud crash *distracted* the attention of the students.

distraught — crazed, distracted: The young woman was *distraught* over the tragedy of her husband's death.

diverge — to extend in different directions from a common point: The map showed a main lode with thin veins *diverging* in all directions.

diverse — varied, unlike: A realistic cross-section must include citizens of *diverse* backgrounds and opinions.

divert — to amuse or entertain; to distract attention: A visit from her sister *diverted* the hospital patient.

divest — to deprive, strip: After the court martial, he was *divested* of his rank and decorations.

divisive — tending to divide, causing disagreement: The issue of abortion, on which people hold deep and morally-based convictions, was *divisive* to the movement.

divulge — to reveal, make public: Newspaper reporters have long fought the courts for the right not to *divulge* their sources of information.

docile — easily led: The child was *docile* until he discovered his mother was gone.

document — an original or official paper serving as proof: The *document* appears to be the legal deed to the property.

dogmatic — arbitrary, believing or believed without proof: The politician, *dogmatic* in his opposition, refused to consider alternative solutions.

domicile — residence: Some people have one *domicile* in winter, another in summer.

dormant — sleeping, inactive: Perennial flowers such as irises remain *dormant* every winter and burgeon in the spring.

dossier — file on a subject or person: The French police kept a *dossier* on every person with a criminal record.

dross — waste matter, scum: The process of separating the valuable metal from the *dross* may be so expensive that a mining claim is worthless.

dubious — doubtful: He had the *dubious* distinction of being absent more than any other student.

duplicity — hypocrisy, double-dealing: The *duplicity* of the marketplace may shock the naive.

durable — long-lasting, tough: Canvas, unlike lighter materials, is a *durable* fabric.

dwindle — to become steadily less; to shrink: As we consume more oil, our supply *dwindles*.

dynamic — in motion, forceful, energetic: A *dynamic* leader can inspire followers with enthusiasm and confidence.

eclectic — drawing from diverse sources or systems: His *eclectic* record collection included everything from Bach cantatas to "punk" rock.

ecology — science of the relation of life to its environment: Persons concerned about *ecology* are worried about the effects of pollution on the environment.

ecstasy — extreme happiness: The lovers were in *ecstasy*, oblivious to their surroundings.

edict — public notice issued by authority: The *edict* issued by the junta dissolved the government.

efface — to obliterate, wipe out: The tablets honoring Perón were *effaced* after his fall from power.

effect — (*v*) to bring about: New regulations have *effected* a shift in policy on applications. (*n*) a result: The headache was an *effect* of sinus congestion.

effective — producing a decided, decisive or desired result: The soft lighting in the restaurant was very *effective*.

efficacious — able to produce a desired effect: The drug is *efficacious* in the treatment of malaria.

effigy — image of a person, especially of one who is hated: They burned his *effigy* in the public square.

effrontery — audacity, rude boldness: He had the *effrontery* to go up to the distinguished guest and call him by his first name.

egocentric — self-centered: The *egocentric* individual has little regard for the feelings of others.

egress — a going out, exit: The building code requires that the apartment have at least two means of *egress*.

elate — to make joyful, elevate in spirit: A grade of 100 will *elate* any student.

electorate — body of persons entitled to vote in an election: Less than 50 percent of the *electorate* actually voted in the last election.

elective — filled or chosen by election: Although she had served on several commissions by appointment, she had never held *elective* office.

elicit — to draw out, evoke: Her direct questions only *elicited* further evasions.

eligible — fit to be chosen, qualified: Veterans are *eligible* for many government benefits, including low-cost loans.

eliminate — to remove, do away with: By consolidating forms, the new procedures have *eliminated* some needless paperwork.

elocution — style of speaking, especially in public: His *elocution* was so clear that everyone in the assembly could hear every word.

elusive — hard to find or grasp: Because the problem is so complex, a definitive solution seems *elusive*.

emaciated — very thin, wasted away: He had a tall, bony figure, as *emaciated* as a skeleton.

emanate — to derive from, issue forth: American law *emanates* largely from English common law.

embargo — governmental restriction or prohibition of trade: In retaliation for the invasion, the government imposed an *embargo* on grain shipments to the Soviet Union.

embellish — to decorate, adorn: She would *embellish* her narratives with fanciful events.

embody — to render concrete, give form to: He tried to *embody* his ideas in his novel's characters.

emendation — change or correction in a text: The author corrected typographical errors and made a few other *emendations* in his manuscripts.

emigrate — to leave a country permanently to settle in another: Many people applied for visas, wishing to *emigrate* and escape persecution at home.

emissary — one sent to influence opponents politically: The rebels sent an *emissary* to negotiate a truce.

empathy — sense of identification with another person: Her *empathy* with her brother was very strong; she generally knew what her sibling was feeling without his having to explain.

employ — to use: The artist *employed* charcoal in many of her sketches.

enable — to make able or possible: A summer job will *enable* you to pay for the course you need to take.

enact — to put into law, do or act out: A bill was *enacted* lowering the voting age to eighteen.

encounter — to come upon face to face unexpectedly; to meet in conflict: The burglar ran around the corner and suddenly *encountered* a policeman.

encroach — to infringe or invade: Property values fall when industries *encroach* upon residential areas.

encyclopedic — covering a wide range of subjects: The knowledge of a good instructor must be *encyclopedic*, ranging far beyond his specialized field.

endearment — a word or an act expressing affection: Have you seen the movie *Terms of Endearment*?

endeavor — to attempt by effort, try hard: I *endeavored* to contact them several times but they never returned my calls.

endemic — peculiar to or prevalent in an area or group: Severe lung disease is *endemic* in coal-mining regions.

endocrine — of a system of glands and their secretions that regulate body functions: The thyroid gland is part of the *endocrine* system.

endorse — to declare support or approval for: Community leaders were quick to *endorse* a project that would bring new jobs to the neighborhood.

enervate — to weaken, enfeeble: A poor diet will *enervate* a person.

enforce — to make forceful, to impose by force:

Because of the holiday, parking restrictions are not being *enforced* today.

engender — to produce, cause, beget: Angry words may *engender* strife.

engross — to absorb fully, monopolize: He was so *engrossed* in his hobbies that he neglected his studies.

engulf — to swallow up: The rising waters *engulfed* the village.

enhance — to improve, augment, add to: The neat cover *enhanced* the report.

enigma — riddle, anything that defies explanation: The origin of the statues on Easter Island is an *enigma*.

enlightened — free from prejudice or ignorance, socially or intellectually advanced: No *enlightened* society could condone the exploitation of children as it was once practiced in American industry.

enmity — state of being an enemy, hostility: The *enmity* between China and Vietnam is traditional and unabated.

enormity — state of being enormous or outrageous: The age of the victim added to the *enormity* of the crime.

ensue — to follow immediately or as a result: One person raised an objection and a long argument *ensued*.

entail — to involve or make necessary: Getting the report out on time will *entail* working all weekend.

entitle — to give a right or claim to: This pass *entitles* the bearer to two free admissions.

entrenched — firmly established: Protestant fundamentalism is deeply *entrenched* in the lives of those people.

enumerate — to count, specify in a list: In her essay she *enumerated* her reasons for wanting to attend the school.

enunciate — to pronounce clearly: He could not *enunciate* certain sounds because of a speech impediment.

envenom — to make poisonous, embitter: Out of jealousy he tried to *envenom* the relationship between his friend and his rival.

environs — surroundings, suburbs: We searched the campus and its *environs*.

ephemeral — short-lived, temporary: *Ephemeral* pleasures may leave lasting memories.

epitome — an abstract, part that typically represents the whole: He prepared an *epitome* of his work to show to the editor.

epoch — distinctive period of time: Hemingway's writings marked an *epoch* in American literature.

equanimity — calm temper, evenness of mind: Adversity could not ruffle her *equanimity*.

equivocal — having more than one possible meaning, deliberately misleading while not literally untrue: His *equivocal* statements left us in doubt as to his real intentions.

eradicate — to pluck up by the roots, wipe out: They tried to *eradicate* the hordes of rabbits by introducing a deadly epidemic.

erode — to eat into, wear away: The glaciers *eroded* the land, leaving deep valleys.

ersatz — substitute, imitation: The burger consisted of *ersatz* beef made from soybeans.

escapade — an adventurous prank, reckless adventure for amusement: Relieved from duty at last, the soldiers went on a three-day *escapade*.

esoteric — limited to a few, secret: The *esoteric* rites of the fraternity were held sacred by the members.

espouse — to take up and support, as a cause: Our congressman *espouses* government funding for housing for the homeless.

essential — necessary, basic: A person must eat a variety of foods to obtain all the *essential* vitamins and minerals.

estimate — rough calculation: The contractor submitted a written *estimate* of the cost of a new roof.

estranged — alienated, separated: Her *estranged* husband had moved out six months previously.

etymology — origin and history of a word, study of the changes in words: The *etymology* of "bedlam" has been traced back to "Bethlehem," the name of a London hospital for the mentally ill.

eulogize — to praise highly in speech or writing: The deceased was *eulogized* at his funeral.

euphemism — substitution of an inoffensive or mild expression for a more straightfoward one: Like many other people, he used "gone" and "passed away" as *euphemisms* for "dead."

euphoria — extreme sense of well-being: Their *euphoria* at their ascent of the mountain was heightened by their narrow escape from death.

evacuate — to empty, clear out: The authorities ordered the town *evacuated* when the waters rose.

evaluate — to determine the value of: The purpose of the survey is to *evaluate* the effect of the new teaching methods on the students' progress.

evasive — avoiding direct confrontation: She admitted that she had been there but was *evasive* about her reasons.

evince — to make evident, display: His curt reply *evinced* his short temper.

evolution — gradual change: Through the discovery of ancient bones and artifacts, anthropologists hope to chart the *evolution* of the human species.

exacerbate — to make worse, aggravate: A generous portion of french fries is sure to *exacerbate* an upset stomach.

exacting — severe in making demands: She was an *exacting* tutor, never content with less than perfection from her pupils.

examine — to investigate, to test: The doctor *examined* the patient for symptoms of pneumonia.

exceed — to go beyond, surpass: The business's profits for this year *exceeded* last year's profits by $16,000.

excess — amount beyond what is necessary or desired: When the pieces are in place, wipe away the *excess* glue.

excise — to remove by cutting out: The surgeon will have to *excise* the tumor.

exclude — to shut out, not permit to enter or participate: The children made a pact that all adults were to be *excluded* from the clubhouse.

execrable — extremely bad: Although her acting was *execrable*, she looked so good on stage that the audience applauded.

execute — to put into effect, perform: He *executed* the duties of his office conscientiously.

exemplary — serving as a pattern, deserving imitation: The leader's *exemplary* behavior in both her private and public life made her a model for all to follow.

exempt — excused: Having broken his leg, the child was *exempt* from gym for the rest of the term.

exhibit — to show, display: The paintings were *exhibited* in the municipal museum.

exhort — to incite by words or advice: The demagogue *exhorted* the crowd to attack the station.

exonerate — to free from blame: The confession of one prisoner *exonerated* the other suspects.

exorbitant — too much, beyond what is reasonable or acceptable: Though they had won the battle, there was no celebrating; the cost in lives had been *exorbitant*.

expansion — enlargement in scope or size: The company's *expansion* into foreign markets has increased its profits.

expedient — advantageous, appropriate to the circumstances, immediately useful though not necessarily right or just: Under pressure to reduce the deficit, the mayor found it *expedient* to cut funds for social services.

expedite — to speed, facilitate: In order to *expedite* delivery of the letter, he sent it special delivery.

expel — to push or force out: When a balloon bursts, the air is *expelled* in a rush.

expenditure — a spending: The finished mural more than justified the *expenditure* of time and money necessary for its completion.

experiment — test undertaken to demonstrate or discover something: *Experiments* were devised to test how motor skills were affected by emotional states.

expertise — skill or technical knowledge of an expert: The *expertise* with which she handled the animal delighted the spectators.

explicate — to explain, develop a principle: He *explicated* the parts of the text that the students had found confusing.

explicitly — openly, without disguise: When the annoying visitor refused to take a hint, the host told him *explicitly* that it was time he left.

exploit — to use, especially unfairly or selfishly: Some employers *exploit* the labor of illegal immigrants, who are afraid to complain about long hours and substandard wages.

exposé — exposure of a scandal: Following the newspaper's *exposé* of corruption in the state capitol, the two assemblymen were indicted for influence-peddling.

expressly — specifically, especially: I wrote it *expressly* for you.

exquisite — perfect, especially in a lovely, finely tuned or delicate way: The handmade lace was *exquisite* in every detail.

extend — to stretch out or to prolong in time; to broaden; to present for acceptance: A snow day *extended* Christmas vacation this year.

extensive — broad, of wide scope, thorough: Several hundred persons were interviewed as part of an *extensive* survey.

extenuate — to partially excuse, seem to lessen: His abrupt rudeness was *extenuated* by his distraught state of mind; no one could blame him for it.

extinct — no longer existing; no longer active, having died down or burnt out: Prehistoric animals are now all *extinct*.

extraneous — not forming an essential or vital part; having no relevance; coming from outside; foreign: The lengthy report was filled with *extraneous* information.

extricate — to free from an entanglement: Carefully removing each prickly branch, she *extricated* herself from the briars.

facade — front of a building: People come from miles around to admire the *facade* of St. Marks' Church.

facetious — amusing, joking at an inappropriate time: His *facetious* criticisms were out of place at that moment; we were too upset to see the humor of the situation.

facilitate — to make easy or less difficult, free from impediment, lessen the labor of: This piece of machinery will *facilitate* production.

facility — ease: Her *facility* in reading several languages made her ideal for the cataloguing job.

facsimile — exact copy: A *facsimile* edition of a book is a photographic reproduction of an original manuscript or printed version.

factotum — employee with miscellaneous duties: He was the chief *factotum* of the plant.

faint — feeble, languid, exhausted: The children were *faint* with fatigue and hunger.

fallacious — untrue, misleading, containing a mistake in logic: Her arguments were transparently *fallacious*.

fallible — capable of erring or being deceived in judgment: It is a shock for children to discover that their parents are *fallible*.

falter — to hesitate, stammer, flinch: He speaks with a *faltering* tongue.

fanatic — person with an unreasoning enthusiasm: The *fanatics* were eager to die for the glory of their religion.

fantastic — fanciful, produced or existing only in the imagination: Her story was so *fantastic* that no one could believe it.

fastidious — disdainful, squeamish, delicate to a fault: The homeowner was so *fastidious* that she had the exterminator come every week.

fatigue — mental or physical weariness: After a full day's work, their *fatigue* was understandable.

favoritism — unfair favoring of one person over others: *Favoritism* in the office based on personal friendship is resented.

feasible — able to be performed or executed by human means or agency, practicable: It is *feasible* to plan to complete the project by July.

fickle — likely to change: None is so *fickle* as a neglected lover.

finalize — to make final: A notarized signature will *finalize* the agreement.

finesse — artifice, subtlety of contrivance to gain a point: She directed the conversation with such *finesse* that in the end he not only agreed to the plan but thought it was his own idea.

finite — having a limit, bounded: There was only a *finite* number of applicants to be considered.

fiscal — financial, having to do with funds: The administration's *fiscal* policy entailed tighter controls on credit.

fissure — crack: The earthquake created a *fissure* two feet wide down the center of the street.

flammable — capable of being kindled into flame: They were careful to keep the material away from sparks because it was *flammable*.

flaunt — to display freely, defiantly, or ostentatiously; *Flaunting* expensive jewelry in public may be an invitation to robbery.

flourish — to achieve success; to prosper or thrive: A baby needs much loving attention in order to truly *flourish*.

flout — to mock, show contempt for: He *flouted* public opinion by appearing with his lover in public.

fluctuate — to change continually from one direction to another: Stock market prices *fluctuate* unpredictably when the economy is unstable.

foray — plundering raid: The bandits made a *foray* into town to steal supplies.

foreclose — to rescind a mortgage for failure to keep up payments: The bank *foreclosed* the mortgage and repossessed the house, putting it up for sale.

foresight — a looking ahead: She had the *foresight* to realize that the restaurant would be busy, so she called ahead for reservations.

forfeit — to lose because of a fault: The team made a couple of decisive errors and so *forfeited* their lead.

formality — fixed or conventional procedure, act or custom; quality of being formal: Skipping the *formality* of a greeting, she got straight to the point. The *formality* of his attire was entirely appropriate to the ceremonious occasion.

formidable — causing fear or awe: He had a *formidable* enemy.

fortuitous — occurring by chance; bringing or happening by luck: It was *fortuitous* that I chose the winning lottery numbers.

forum — place for public business or discussion: A television interview would be the best *forum* for bringing our views to the attention of the public.

forward — to promote, send, especially to a new address: The secretary promised to *forward* the request to the person in charge.

fracture — a break, split: He sustained a compound *fracture* of the left leg.

frail — physically weak: Very old people are often *frail*.

fraud — intentional deceit for the purpose of cheating: The land development scheme was a *fraud* in which gullible investors lost tens of thousands of dollars.

fraudulent — false, deceiving for gain: His claim to be the true heir was exposed as *fraudulent*.

frenetic — frenzied: *Frenetic* activity is evident in the dormitory just before exam time.

frivolous — not serious: The atmosphere of the gathering was entirely *frivolous* as everyone got dressed up in costumes and played children's games.

fundamental — basic: Education is *fundamental* to your future security.

furious — full of madness, raging, transported with passion: The animal was so *furious* that it had to be confined.

furnish — to provide with what is needed: The army recruit will be *furnished* with uniforms and other equipment.

fuse — to blend thoroughly by melting together: The plastic parts were heated and *fused* to make a unit.

futile — trifling, useless, pointless: The entire matter was dropped because the arguments were *futile*.

gall — to chafe, rub sore, annoy, vex: That saddle will *gall* the horse's back.

gamut — complete range: The singer demonstrated the entire *gamut* of her vocal skills in her performance of the operetta.

garish — excessively vivid, flashy, glaring or gaudy: The streetwalker wore a *garish* outfit and makeup.

gauche — without social grace, tactless: It is considered *gauche* to ask acquaintances how much they earn or how much they paid for something.

genealogy — history of family descent, a family tree: They were able to trace their *genealogy* back four generations to a small village in Sicily.

generalization — induction, a general conclusion drawn from specific cases: From his experience with his own pets he made the *generalization* that all kittens love paper bags.

generate — to beget, procreate, produce: Every animal *generates* its own species.

generic — pertaining to a race or kind: The *generic* characteristics of each animal allow us to identify its species.

genial — pleasant, friendly: The president's rotund and *genial* face made him the perfect Santa Claus.

genus — kind or class: Biologists classify these related species as members of the same *genus*.

geriatrics — science of care for the aged: Our longer life span has made the study of *geriatrics* increasingly important.

germane — pertinent, on the subject at hand: The point, though true, was not *germane* to the argument.

gist — essential part, core: That all men are not equal was the *gist* of his speech.

glutton — person habitually greedy for food and drink: The man was too much of a *glutton* to stick to any diet.

goad — to drive with a stick, urge on: Although she was naturally lax, she was *goaded* on by her parents' ambitions.

gorge — to eat greedily: The neighborhood children *gorged* themselves on Halloween candy.

gracious — socially graceful, courteous, kind: A *gracious* host puts his guests at ease and is concerned only that they enjoy themselves.

graft — illegal use of position of power for gain: He was charged with *graft* in selling contracts for public works projects.

grandeur — splendor, magnificence, stateliness: The *grandeur* of the lofty mountains was admired by all.

grant — to bestow or transfer formally: The bright student was *granted* a scholarship.

graphic — described in realistic or vivid detail: The soldiers returned with *graphic* descriptions of the battle.

grapple — to seize, lay hold of, either with the hands or with mechanical devices: He *grappled* with the man who had attacked him.

gratuitous — free, voluntary, unasked for or unnecessary: Her spiteful temper expressed itself in *gratuitous* insults.

gratuity — tip: He left a *gratuity* for the chambermaid.

gregarious — fond of company: They are a *gregarious* couple who cultivate many friendships among diverse people.

grueling — exhausting: The labor was so *grueling* that two workers fainted.

gruff — rough: His manner was so *gruff* that most of the children feared him.

guile — deceit, cunning: His brief success was due to flattery and *guile* rather than to genuine talent.

gullible — easily deceived: Naive people are often *gullible*.

gyrate — to revolve around a point, whirl: The tornado *gyrates* around a moving center.

gyroscope — rotating wheel apparatus that maintains

direction regardless of position of surrounding parts: The automatic *gyroscope* holds an airplane on its course even when the machine is upside down.

habitable — capable of being inhabited or lived in, capable of sustaining human beings: The climate of the North Pole makes it scarcely *habitable*.

haggard — gaunt, careworn, wasted by hardship or terror: After three days of being lost on the mountain, the *haggard* campers staggered into the village.

hallucination — apparent perceiving of things not present: In her *hallucinations* she saw bizarre faces and heard voices calling to her.

haphazard — random, without order: He studied in such a *haphazard* manner that he learned nothing.

harass — to annoy with repeated attacks: The students perpetually *harassed* the teacher with unnecessary questions.

harp — to dwell constantly on a particular subject: The employee *harped* so continually on the difficulty of his job that he was eventually fired.

harrowing — severely hurtful or trying, emotionally or physically: The survivors of the crash went through a *harrowing* ordeal before their rescue.

hazardous — dangerous: Trucks carrying *hazardous* materials such as explosives are not permitted on the bridge.

hectic — fevered, hurried and confused: The tour turned out to be somewhat *hectic*, covering three cities in as many days.

hegemony — predominance: Hitler's aim was German *hegemony* over the world.

heinous — hateful, atrocious: The deed was so *heinous* that the perpetrator was despised for it.

herald — to announce the arrival of, usher in: Crocuses *herald* the advent of spring.

heterodox — not orthodox, not conforming, especially in religious belief: Her *heterodox* opinions and outlandish behavior earned her a reputation as an eccentric.

heterogeneous — composed of unlike elements: Since the school favored *heterogeneous* groupings, there was a wide range of ability and achievement in every class.

hinder — to retard, slow down, prevent from moving forward: Cold weather has *hindered* the growth of the plants.

hindsight — a looking backward: With *hindsight* I realize that everything she said to me was true, though I couldn't accept it at the time.

historical — famous in history: We can gain insights into the present by relating current to *historical* events.

holocaust — great destruction of living beings, especially by fire: As the fire raged out of control, thousands of lives were lost in the *holocaust*.

homage — respect, expression of veneration or extreme admiration: She paid *homage* to her mentor by dedicating her book to him.

homicide — killing of one person by another: Killing in self-defense is considered justifiable *homicide*.

homogeneous — same, uniform throughout: The entering class was fairly *homogeneous*; nearly all the students were the same age and from similar middle-class homes.

homily — discourse on a moral problem, sermon: The judge read the boy a *homily* on his conduct before sentencing him.

horizontal — flat, parallel to the horizon: *Horizontal* stripes are frequently unflattering because they make the figure appear wider.

hospitable — welcoming, generous to guests: It was a *hospitable* room, with a soothing color scheme and deep, comfortable chairs.

hostile — conflicting, antagonistic, expressing enmity: Many tribes were *hostile* to the white settlers, just as the settlers viewed the Indians as enemies and rivals for the land.

huddle — to crowd together, press together without order or regularity: The crowd *huddled* under the shelter to get out of the rain.

humanities — branch of learning concerned with philosophy, literature, the arts, etc., as distinguished from the sciences and sometimes the social sciences: The essence of the *humanities* is a concern with human nature, experience and relationships.

humble — insignificant or unpretentious; not proud or self-assertive; conscious of one's shortcomings: I feel very *humble* in the presence of wealthy or powerful people.

humility — humbleness of spirit: The minister spoke to the members of his congregation with sincere *humility* about his own failings.

hyperbole — obvious exaggeration as a figure of speech: "He was as big as a house" is a common *hyperbole*.

hypertension — high blood pressure: *Hypertension* is often linked with serious diseases.

hypochondria — abnormal anxiety about health: Although no one could find anything wrong with him, his *hypochondria* drove him to get frequent checkups.

hypothesis — theory, tentative explanation yet to be proved: The *hypothesis* that life is common throughout the universe cannot as yet be supported by direct evidence.

identify — to establish the distinguishing characteristics of a group of individuals or activities: Can you *identify* the special skills needed by an athlete?

ideologue — one who believes in and propagates a social doctrine: The communist *ideologue* argued that the state was more important than any individual.

idiosyncrasy — peculiar tendency of an individual: Her *idiosyncrasy* of dropping in on her friends without warning has proved embarrassing on more than one occasion.

idolatry — worship of idols or false gods: His awestruck respect for the older boy amounted to *idolatry*.

ignominy — a discrediting, disgrace: After the *ignominy* of the impeachment he retired from public life.

illicit — not licensed: *Illicit* love is the root of many divorce actions.

illuminate — to throw light on, explain: The editor's notes *illuminated* the more obscure passages in the text.

illusion — false appearance, vision that is misleading: The optical *illusion* made the lines of equal length appear to be unequal.

illusory — unreal, only apparent: The money we hope to make is *illusory* until we have it in hand.

immolate — to kill as a sacrifice: The Buddhist monk *immolated* himself in a public square as a gesture of protest against the war.

immune — not susceptible, protected, as from disease: An inoculation for smallpox makes one *immune* to the disease.

impalpable — not able to be felt: The seismograph can measure tremors in the earth's crust *impalpable* to humans.

impartial — not favoring one side or another: The squabbling children appealed to the babysitter for an *impartial* judgment.

impassioned — animated, excited, expressive of passion or ardor: The *impassioned* performance of the actor was moving and convincing.

impeach — to challenge one's honesty or reputation, to call before a tribunal on a charge of wrongdoing: President Nixon resigned before he could be *impeached* by the Senate for high crimes and misdemeanors.

impeccable — faultless: Successful comedy depends on *impeccable* timing.

impediment — hindrance, something that delays or stops progress: Lack of training may be an *impediment* to advancement.

impel — to drive forward, push, incite: Although she was not personally involved, her sense of justice *impelled* her to speak out.

imperative — of greatest necessity or importance: This is an emergency; it is *imperative* that I reach them at once.

imperil — to put in danger: The incompetence of the pilot *imperiled* the safety of all on board.

imperceptible — not easily seen or observed: The daily growth of the plant was *imperceptible*.

imperishable — not subject to decay, indestructible: Through thousands of years his fame as a philosopher has been *imperishable*.

imperturbable — not easily excited: His *imperturbable* expression was a great aid in poker.

impervious — not to be penetrated or passed through: Heavy cardboard is *impervious* to light.

impetuous — impulsive, acting suddenly and without forethought: The *impetuous* boy leaped before he looked.

implement — to put into effect, to realize in practice: When they *implemented* the program, they realized that some of the planned procedures were not practicable and would have to be modified.

imply — to suggest, say without stating directly: Although they said nothing about it, their cool manner *implied* strong disapproval of the scheme.

imponderable — not capable of being weighed or measured: The results of the negotiations constitute an *imponderable* at this time.

importune — to beg: Do not *importune* me for another loan; you never paid back the last one.

impotent — lacking power, helpless: The disease left him *impotent* to walk across the room.

impoverish — to make poor: She was an exceptionally effective administrator; the company has been *impoverished* by her loss.

imprecise — not precise, vague, inaccurate: The description was *imprecise* because the witness had had only a fleeting glimpse of the man.

impressive — having the power of affecting or of exciting attention and feeling: The view was so *impressive* that we'll never forget it.

imprimatur — license to publish: He was given an *imprimatur* for all the works of Benjamin Franklin.

impromptu — spontaneous, not planned or prepared in advance: *Impromptu* remarks, spoken on the spur of the moment, often tell voters more about a candidate's real opinions than his carefully edited speeches do.

improvise — to make, invent or arrange offhand, using what is conveniently available: If we do not have all of the proper ingredients, we shall have to *improvise*.

impudence — shamelessness, want of modesty, assurance accompanied by a disregard for the opinions of others: His *impudence* in denying having made the promise left us flabbergasted.

impugn — to cast doubt on someone's motives or veracity: Do not *impugn* his testimony unless you can substantiate your charges.

impunity — exemption from punishment, penalty, injury or loss: No person should be permitted to violate the laws with *impunity*.

impute — to attribute, ascribe: The difficulties were *imputed* to the manufacturer's negligence.

inadequate — not equal to the purpose, insufficient to effect the object: He could not maintain his car because of *inadequate* funds.

inalienable — not transferable, not able to be taken away: As humans we are endowed with certain *inalienable* rights.

inarticulate — not able to speak or speak clearly, not distinct as words: The *inarticulate* noises of the infant soon give way to recognizable words.

incarcerate — to imprison: The sheriff ordered the prisoner *incarcerated*.

inception — beginning: The scheme was harebrained from its *inception*; it was no surprise when it was abandoned.

incessant — unceasing, uninterrupted, continual: The *incessant* rain kept the children indoors all day.

incidence — range of occurrence or effect: The *incidence* of reported alcoholism among teenagers is increasing.

incision — cut: The surgeon made an *incision* above the navel.

incognito — with identity concealed: The prince was traveling *incognito*.

incompatibility — inconsistency, lack of agreement, inability to get along: The *incompatibility* of their tastes made for endless disagreement.

inconsiderable — not worthy of consideration or notice, unimportant, small, trivial: The distance between Minneapolis and St. Paul is *inconsiderable*.

incontestable — not able to be disputed or denied: With the development of the atomic bomb, U.S. military superiority became *incontestable*.

inconvenience — to cause trouble or bother to; to incommode: We were *inconvenienced* but not injured by the accident directly in front of us on the highway.

incorrigible — beyond reform, not capable of being corrected: The dog's viciousness was *incorrigible*, so it had to be destroyed.

increase — to enlarge, become greater, multiply: Class attendance *increased* by 30 percent after the flu epidemic ended.

incriminate — to accuse or implicate in a crime or fault: Picked up by the police, the boy *incriminated* his companions by naming them as accomplices in the theft.

inculcate — to instill, to impress on the mind by repetition: From earliest childhood they had been *inculcated* with the tenets of the community's belief.

incur — to acquire or meet with through one's own actions: The debts *incurred* in the legal proceedings were to be paid off in monthly installments.

indelible — not able to be erased, blotted out or washed away: The form must be signed in *indelible* ink; pencil is not acceptable.

indict — to accuse formally: The grand jury *indicted* two of the company's executives.

indigenous — native to a country: The *indigenous* trees of the Rockies are mostly evergreens.

indigent — poor, penniless: The home is for the *indigent* aged who depend on the state for support.

indiscriminate — not selective: The police made *indiscriminate* arrests, taking into custody scores of people who had broken no law.

indoctrinate — to instruct in a set of principles or beliefs: Children are sent to Sunday school to be *indoctrinated* in the basic tenets of a particular religion.

indolent — lazy: An *indolent* student never learns much.

induct — to bring in, initiate a person: A volunteer must pass a physical before being *inducted* into the army.

inept — incompetent, clumsy, inefficient: The basketball team's center is tall and powerful but so physically *inept* that he frequently loses the ball.

inexhaustible — unfailing: The city has an *inexhaustible* supply of water.

inexorable — relentless: The *inexorable* logic of history points to an eventual decline of every great power.

infer — to conclude from reasoning or implication: From hints that the student dropped, the instructor *inferred* that she was having problems at home.

infiltrate — to pass through or into, especially secretly or as an enemy: The radical organization had been *infiltrated* by federal agents who monitored its membership and activities.

inflammable — easily set on fire, excited or provoked: The tanker truck bore a warning: "Caution: *inflammable* substance!"

inflammatory — tending to arouse to anger or violence: An *inflammatory* speech incited the crowd to riot.

inflate — to blow up or swell, to increase beyond what is right or reasonable: The store is able to get away with charging *inflated* prices because of its convenient location and long hours.

infraction — violation, breaking of a law or regulation: The building inspector noted several *infractions* of the health and safety codes.

ingenuous — free from pretence or trickery: An *ingenuous* approach is often better than guile.

ingratiate — to establish in favor: He tried to *ingratiate* himself with his teacher by bringing her apples.

inherent — inborn, existing as a basic or natural characteristic: A love of hunting is *inherent* in cats.

initiate — to begin, introduce: The fraternity *initiates* new members every semester.

innocuous — harmless: His words were *innocuous*, but his look could have killed.

innovation — something new, a change, as in custom or method: The celebration of the Mass in languages other than Latin is a major twentieth-century *innovation* in the Roman Catholic Church.

innuendo — indirect intimation, hint, especially of something negative: There was an *innuendo* of threat in the phrases she chose.

inopportune — inconvenient or unseasonable; not appropriate: To suggest a change of itinerary at this time would be *inopportune*.

input — anything put in, such as power into a machine or information into a discussion: The *input* of time and money in market analysis paid off in a profitable investment.

inquest — judicial investigation: The state held an *inquest* to examine the cause of the disaster and determine whether charges should be brought against any parties.

inquisitive — curious, asking questions: Private eyes in detective fiction often get into trouble for being too *inquisitive*.

insatiable — never satisfied, always greedy: His appetite for wealth was *insatiable*; no matter how rich he became, he always craved more.

insert — to put into something else: The nurse *inserted* the needle into the patient's arm.

insidious — secretly dangerous, tending to entrap: The casino games were *insidious*; before he realized it, he had gambled away all of his savings.

insignificant — not important, too small to matter: The difference in scores between the two groups was statistically *insignificant*.

insinuate — to suggest subtly, especially something negative: When you say I remind you of Lincoln, are you *insinuating* that I'm dead?

inspect — to examine, view closely: She *inspected* the cloth for rips or tears.

instigate — to urge a bad action: The propaganda was designed to *instigate* a pogrom against the oppressed minority.

instill — to impart gradually: A skillful teacher can *instill* in children a love of learning.

instruct — to teach, direct: The employees were *instructed* in the use of the computers during the training session.

insubordinate — failing to obey: Ignoring a direct order is an *insubordinate* act with grave consequences.

insufficient — inadequate to a need, use or purpose: The provisions are *insufficient* in quantity.

insular — pertaining to an island: Puerto Rico is an *insular* commonwealth.

insure — to make certain, guarantee: Bail is set to *insure* the defendant's appearance in court. *Also:* **ensure**.

intact — entire or uninjured: Though he is past eighty, my father's mental faculties are *intact*.

intangible — not able to be touched or easily defined: The company's goodwill among its customers is a genuine but *intangible* asset.

integral — necessary to the whole: The woodwind section is an *integral* part of an orchestra.

integrate — to absorb into an organization or group: Company orientation programs help to *integrate* new employees into an existing organization.

intelligible — capable of being understood or comprehended; clear: The baby's gibberish is *intelligible* only to its parents.

intend — to mean, signify, plan: They *intend* to make repairs on their old car.

intensive — concentrated, intense: *Intensive* private tutoring is needed to take care of this student's reading problem.

intent — firmly directed or fixed; concentrated attention on something or some purpose; act or instance of intending: The tailor was very *intent* at his sewing machine.

intercede — to interpose in behalf of: She asked the minister to *intercede* with her family.

intercept — to cut off, meet something before it reaches its destination: The missile was *intercepted* and destroyed before it reached its target.

interpolate — to change a text by inserting new material: The editor *interpolated* the latest news into the proofs.

intractable — stubborn, unruly: An *intractable* person is slow to learn a new way of life.

intransigent — uncompromising: Their *intransigent* attitude antagonized the opposition and made negotiations difficult.

intravenous — through a vein: The patient was given *intravenous* feedings of glucose because he could not swallow.

intrepid — brave, fearless: The *intrepid* explorers stepped out onto the lunar surface.

intrinsic — belonging naturally: The *intrinsic* value of diamonds lies in their hardness.

inundate — to flood: When the craze was at its height, the police were *inundated* daily with reports of UFO sightings.

inventory — the stock or goods of a business, list of stock or property: The annual *inventory* check showed that several cartons of paper had been damaged by water.

investigation — close examination and observation, inquiry: The *investigation* showed that arson was the cause of the blaze.

inveterate — firmly established over a long period; habitual: The early alarm was a rude shock, for he was an *inveterate* late-sleeper.

invoice — a bill, itemized list of goods sent to a buyer: The book was packed with the *invoice* charging $24.00, including shipping.

irate — intensely angry: The *irate* farmer shot the fox in his barnyard.

irradiate — to spread out, expose to radiant energy, heat by radiant energy: The heat from the fireplace *irradiated* the room, warming the company.

irreconcilable — unable to be harmonized: His statements about liking school were *irreconcilable* with the distaste he expressed for books in general.

irreplaceable — not able to be replaced: The painting is priceless in the sense that it is *irreplaceable*; it is the only one of its kind.

irritate — to annoy, inflame: The harsh cleansers used in the job can *irritate* the skin.

isotopes — chemical elements differing in atomic weight but having the same atomic number: Radioactive *isotopes* are used to follow the flow of oil in a pipeline.

itinerary — plan or schedule of travel: Our *itinerary* includes three days in Florence and a week in Rome.

jaded — wearied, sated with overuse: *Jaded* with the pleasures of the idle rich, she decided to find a useful occupation.

jargon — confusing unintelligible talk, usually a specialized language used by experts: The computer programmers spoke a *jargon* rife with undecipherable acronyms.

jeopardy — risk, danger, especially the legal situation of a person on trial: Do not put your health in *jeopardy* by exposing yourself to infection needlessly. A person shall not be put in *jeopardy* twice for the same offense.

jettison — to cast overboard: They had to *jettison* the cargo to lighten the plane.

judicial — having to do with courts or judges: Chief Justice of the Supreme Court is the highest *judicial* position in the United States.

judicious — prudent: His policy was *judicious*; he got results without taking great risks.

jurisprudence — philosophy or theory of law: The courses for the most part emphasize the practical application of the law rather than *jurisprudence* or legal history.

justify — to prove by evidence, verify, absolve: The defendant was able to *justify* her statement with evidence.

juxtaposition — a placing close together: The *juxtaposition* of the Capitol and White House was avoided by the planners to emphasize the separation of the branches of government.

kindred — alike, related: Though from diverse backgrounds, they were *kindred* spirits, alike in intellect and ambition.

kinetic — of or caused by motion: *Kinetic* energy is produced by a stream turning a water wheel.

knead — to mix, squeeze and press with the hands: She *kneaded* the dough before shaping it into four loaves for baking.

labyrinth — maze, complex and confusing arrangement: The ancient town within the city walls was a *labyrinth* of narrow, winding streets.

lacerate — to tear tissue roughly: The baby swallowed the safety pin, which *lacerated* his intestine.

laconic — terse, pithy: Her *laconic* replies conveyed much in few words.

lament — to bewail, mourn for: The boy *lamented* the death of his father.

languid — listless, slow, lacking energy: His *languid* walk irritated his companions, who were in a hurry.

lapse — slip, minor or temporary fault or error: I was embarrassed by a momentary *lapse* of memory when I couldn't recall her name.

larceny — legal term for theft: The shoplifter was apprehended and charged with petty *larceny*.

lassitude — feeling of weariness, languor: The *lassitude* caused by the intense heat led them to postpone their sightseeing.

laudable — praiseworthy: The girl listened to the old man's endless and repetitive stories with *laudable* patience.

lavish — to give generously or extravagantly: The doting grandfather *lavished* his grandchild with gifts.

laxity — looseness, lack of strictness: In summer, when business was slow, the manager allowed the employees some *laxity* in their hours.

legacy — something inherited: He acquired the house as a *legacy* from his grandmother.

legible — written clearly, able to be read: Please print or type if your handwriting is not easily *legible*.

legislature — lawmaking body: The federal *legislature* of the United States, the Congress, has two houses.

legitimate — lawful, genuine: The government is a *legitimate* one, duly elected by the people in free elections.

leisurely — without haste; slow: The lovers took a *leisurely* stroll around the pond.

leniency — mercy, gentleness, lack of strictness: The *leniency* of the court in suspending the sentence was well repaid by the convicted man's later contribution to the community.

lethargic — drowsy, slothful, sleepy: The convalescent moved in a *lethargic* manner.

levity — lightness of spirit, frivolity, playfulness: The party toys and silly costumes epitomized the *levity* of the occasion.

liability — debt, something disadvantageous: An older person returning to the job market may find his or her age a *liability*.

liable — legally responsible; likely, in a negative sense: If you trip and hurt yourself on the stairs because the light is out, the landlord is *liable*. He is *liable* to lose his temper when he hears the news.

liaison — connection, linking: He had served as a *liaison* between the Allied command and the local government.

libel — written defamation, anything tending to lower reputation: The report was a *libel* on the man's professional standing.

limitation — restriction, finitude: There is a *limitation* on time in which you can redeem the ticket.

lithe — gracefully flexible: Her *lithe* figure suggested that she was a dancer.

litigation — lawsuit, process of carrying on a lawsuit: As long as the estate is tied up in *litigation* by the would-be heirs, no one has use of the property.

loathe — to dislike with disgust; to detest; to hate: I *loathe* people who mistreat animals.

logical — according to reason or logic: Using the data from the experiment, he made a *logical* conclusion about the eating habits of white mice.

longevity — life span, long life: The Bible credits the first generations of men with a *longevity* unheard of today.

longitudinal — pertaining to length: They measured the *longitudinal* distance carefully.

loyal — faithful and unswerving in allegiance: His *loyal* friend stood by the disgraced politician.

lucid — clear, transparent: The directions were written in a style so *lucid* that a child could follow them.

lucrative — profitable: A *lucrative* enterprise is attractive to investors.

ludicrous — apt to raise laughter, ridiculous: The scene was so *ludicrous* that the audience roared with laughter.

lugubrious — excessively mournful in a way that seems exaggerated or ridiculous: The bloodhound had an endearingly *lugubrious* look.

luminous — emitting or reflecting a glowing light: In total darkness a cat appears to have *luminous* eyes.

lurid — shocking, sensational, tastelessly violent or passionate: The cheap novel told a *lurid* tale of murder and lust.

machination — scheme or secret plot, especially an evil one: The *machinations* of his influential uncle landed him a well-paid sinecure in a prestigious company.

maelstrom — whirlpool: The ship was twisted in the *maelstrom*.

magisterial — authoritative, arrogant, dogmatic: In front of a class the normally humble man assumed a *magisterial* air.

magma — molten rock within the earth: Far beneath the solid crust, the *magma* flows.

magnanimous — noble-minded, extremely generous, especially in overlooking injury: The painter was *magnanimous* enough to praise the work of a man he detested.

magnate — important business person: The steel *magnate* refused to approve the consolidation.

magnitude — size: The apparent *magnitude* of the moon is greater near the horizon than at the zenith.

maladroit — tactless in personal relations: His *maladroit* remarks embarrassed the hostess.

malaise — general bodily weakness: She complained of a *malaise* that caused her to sleep ten hours a day.

malfeasance — wrongdoing, especially in public office: The governor was accused of acts of *malfeasance*, including taking graft.

malign — (*adj*) evil, malicious, very harmful: *Malign* comments are often motivated by jealousy. (*v*) to speak ill of: The students often *maligned* the strict professor.

malleable — able to be shaped, adaptable: Children are more *malleable* than adults and adapt to new environments more readily.

malpractice — improper professional conduct: The surgeon was sued for *malpractice* after a sponge was found in his patient's abdomen.

mandate — specific order: Some islands are still ruled by United Nations *mandate*.

manifest — to appear, make clear, show: He claims a greater devotion to that cause than his actions *manifest*.

manual — (*adj.*) involving the hands: In my neighborhood bank, the tellers still use *manual* typewriters. (*n.*) a handbook: Each new employee is issued a *manual* of office procedures.

mar — to damage: The floor has been *marred* by scratches and scuff marks.

margin — edge, border: Cattails grow in the swampy area at the *margin* of the pond.

marquee — roof projecting from a building over the sidewalk: The theater's *marquee* protects patrons from the rain.

martinet — rigid, petty disciplinarian: The captain was a *martinet* who considered an unpolished button criminal negligence.

matriarch — mother who rules a family or clan: All important decisions were referred to the *matriarch* of the tribe.

matrix — something that gives form, as a mold: The linotype machine is equipped with a brass *matrix* for each letter so that a line can be assembled and cast in lead.

mawkish — slightly nauseating, insipidly sentimental: Her constant display of fawning affection was *mawkish*.

maximum — most: In this course the *maximum* number of cuts allowed is six.

meager — deficient in quality and quantity; inadequate: It is impossible to feed a family of four on that *meager* salary.

median — middle, middle item in a series: In a series of seven items the fourth is the *median*.

medicinal — having the property of healing: The plants had a high *medicinal* value.

mediocre — of average or middle quality: A *mediocre* student in high school will rank low among candidates for college.

memorandum — written reminder, informal written interoffice communication. The office manager circulated a *memorandum* outlining the procedures to be followed in the fire drill. *Plural:* **memoranda.**

menace — to threaten, express an intention to inflict injury: The periodic floods *menaced* the city with destruction.

mend — to improve; to correct; to reform: A frugal housewife *mends* small holes in socks.

merchant — shopkeeper, one who buys and sells goods for a profit: The *merchants* who operate businesses in the mall have formed an association.

meritorious — deserving reward: Medals were awarded for *meritorious* service.

metamorphose — to transform: Two months abroad *metamorphosed* him into a man of the world.

meticulous — showing careful attention to detail, very precise: The sewing in the jacket was so *meticulous* that one could hardly see the stitches.

militant — defiant, ready to fight, especially for a cause: *Militant* in their political beliefs, they considered any compromise a sellout.

militate — to operate against, work against: A poor appearance at the interview will *militate* against your being hired.

mingle — to mix, join a group: The mayor *mingled* with the crowd at the reception, shaking hands and thanking her supporters.

miniature — very small, done on a scale smaller than usual: The *miniature* microphone could be concealed in a piece of jewelry.

minuscule — tiny, minute (after a small cursive script): Such *minuscule* particles cannot be viewed with the usual classroom microscope.

minute — tiny; very precise: The device records the presence of even *minute* amounts of radiation. The writer's *minute* attention to the refinements of style resulted in an elegantly worded essay.

misanthropy — dislike or distrust of mankind: The *misanthropy* of the hermit was known to all.

misappropriation — act of using for a wrong or illegal purpose: The *misappropriation* of the funds was uncovered and those responsible were formally charged.

miscalculate — to calculate erroneously: *Miscalculating* the distance, he fell short.

miscellany — collection of various or unlike things: The old steamer trunk contained a *miscellany* of papers, clothes and assorted junk.

misconstruction — wrong interpretation of words or things, a mistaking of the true meaning: His *misconstruction* of the situation caused him to act unjustly.

misdemeanor — a misbehaving, a minor legal offense: The *misdemeanor* resulted in a $50 fine.

misnomer — wrong or inaccurate name: At this season Muddy River is a *misnomer*; the waters are sweet and crystalline.

mitigate — to lessen, make milder: He sought to *mitigate* their grief with soothing words.

modicum — a little, a small quantity: The girl had only a *modicum* of learning, having never finished the fifth grade.

modulation — adaptation or variance in pitch, intensity, volume, musical key: The distinctive *modulations* of the Jamaican accent sound lilting to other English ears.

molest — to disturb, annoy, bother: The children were warned not to *molest* the bulldog.

mollify — to soothe, placate: The irate customer was *mollified* by the manager's prompt action and apology.

monetary — pertaining to money, consisting of money, financial: A penny is the smallest *monetary* unit in this country.

monitor — to watch over, check on: An office was set up to *monitor* all radio broadcasts originating within the country.

monolith — large piece of stone: The obelisk in New York's Central Park is a *monolith* brought here from Egypt.

montage — picture made up of pictures or material from several sources: The illustration was a *montage* of various European scenes.

morale — level of spirits, mental or emotional condition: After a landslide victory at the polls, *morale* in the party was at a peak.

morass — swamp, bog, messy or troublesome state: The application became mired in a *morass* of paperwork; there was no response for several weeks.

mores — customs, principles of conduct of a culture: The *mores* of any group are enforced by indoctrination and social pressure to conform.

moribund — dying: The *moribund* tree put out fewer and fewer leaves each spring.

morose — gloomy, sulking, unreasonably unhappy: The boy was *morose* for days over his failure to get tickets for the concert.

mortgage — to pledge property as security for a loan: Few people can afford to buy a house without taking a *mortgage* on it.

motivation — reason for doing something: The *motivation* for her questions was not mere curiosity but a genuine desire to help.

motley — variegated, composed of clashing elements: A *motley* crowd attacked the consulate.

muddle — to confuse or stupefy: The liquor had gotten him badly *muddled*.

mundane — worldly, humdrum, unexciting: The film was undistinguished, a *mundane* exercise in horror movie clichés.

mutation — change, especially a sudden one: He deplored the *mutations* of fortune that had altered his position so drastically.

mutilate — to cut up, damage severely: The computer cannot read a *mutilated* card.

mutiny — forceable resistance to lawful authority; revolt; rebellion: The sailors threatened their overbearing captain with *mutiny*.

myopia — nearsightedness: The optometrist prescribed glasses for the patient's *myopia*.

nadir — lowest point: Enrollment hit its *nadir* last year and has been rising slowly since.

naive — unsophisticated, artless: He was *naive* as the result of a sheltered life in the country.

nape — back of neck: The collar chafed his *nape*.

nascent — coming into being; being born; beginning to develop: I am afraid that I have a *nascent* cold.

nauseate — to cause disgust and nausea to: Food *nauseates* the patient.

navigation — the science of locating position and plotting course of ships; ship traffic or commerce: The invention of radar was a terrific *navigational* aid.

nebulous — hazy, indistinct: He had a *nebulous* theory about memorizing key words as an aid to study.

necessitate — to render unavoidable, compel: Sickness *necessitated* a long hospital stay.

negate — to make nothing, undo or make ineffective: The witness's full confession *negated* the need for further questions.

neglectful — careless, heedless: Because he was *neglectful* of his duties as principal, he was asked to resign by the board of education.

negligible — too small or insignificant to be worthy of consideration: The difference in their ages is *negligible*.

negotiate — to bargain, confer with the intent of reaching an agreement: As long as both sides are willing to *negotiate* in good faith, a strike can be avoided.

nepotism — favoritism shown to a relative: The civil service system helped to do away with *nepotism* in hiring.

neuralgia — pain along the course of a nerve: *Neuralgia* is often confused with rheumatism.

neurotic — suffering from or typical of neurosis, a range of mental disorders less severe than psychosis: Hysterical pain — physical discomfort without organic cause — is a common *neurotic* symptom.

neutralize — to reduce to a state of indifference between different parties, make inactive: The marine territory was *neutralized* by the nations through a treaty.

node — knot or protuberance: A *node* appears at the joints of a plant.

nomadic — roaming from place to place without a fixed home: Gypsy life is typically *nomadic*.

nomenclature — names of things in any art or science, whole vocabulary of technical terms appropriate to any particular branch of science: The *nomenclature* used in medical science is almost entirely derived from Latin.

nonchalant — indifferent, cool, unconcerned: The woman acted in a *nonchalant* manner, pretending not to notice the celebrities.

noncompliance — failure to comply: His *noncompliance* with the terms of the contract forced them to sue.

nonsensical — meaningless, characterized by nonsense: Until analyzed and interpreted, dreams often seem *nonsensical*.

normal — regular, average, usual: The doctor found that her blood pressure and temperature were *normal*.

notary — person empowered to attest signatures, certify documents, etc.: The signature is valid if witnessed by a *notary* public.

notify — to let know, inform: Applicants will be *notified* of the results by mail.

notorious — famous in an unfavorable way: The official was *notorious* among his associates for failing to keep appointments.

novice — person new to a job or activity, someone inexperienced: A *novice* in the job, she needed more time than an experienced worker to complete the same tasks.

noxious — harmful, injurious, unwholesome: The *noxious* fumes from the refinery poisoned the air.

null and void — legal expression for not valid, without legal force: If it is not properly signed, the will may be declared *null and void*.

nullify — to make void or without effect: The new contract *nullifies* their previous agreement.

numerical — expressed in or involving numbers or a number system: Please arrange all of your test papers in *numerical* order.

numerous — consisting of great numbers of units or individuals: The grains of sand on the beach are too *numerous* to count.

obesity — excessive fatness: Her *obesity* was due to her love of rich foods.

obituary — account of the decease of a person: Newspapers keep files on famous people in case they have to run an *obituary*.

objective — (*adj*) unbiased, not influenced by personal involvement, detached: It is extremely difficult to be *objective* about one's own weaknesses. (*n*) aim, goal: Our *objective* is greater efficiency; we must study the possible means to that goal.

obligatory — required, morally or legally binding: He feels nothing in common with his family, yet he makes an *obligatory* visit to them once or twice a year.

obliterate — to demolish, destroy all trace of: The building had been *obliterated*; we could not even be sure exactly where it had stood.

oblivious — so preoccupied as not to notice: The patron, absorbed in her reading, was *oblivious* to the librarian's question.

obnoxious — odious, hateful, offensive, repugnant: They left because of the *obnoxious* odors.

obscure — dim, murky, not easily seen or understood: Despite attempts at interpretation, the meaning of the passage remains *obscure*.

obsequious — servile, overly willing to obey: His *obsequious* obedience to the conquerors turned our stomachs.

observable — able to be seen, noticeable: There has been no *observable* change in the patient's condition.

obsess — to beset, haunt the mind: He was *obsessed* with the idea that he was being followed.

obsolete — outmoded, no longer in use or appropriate: Since several offices have been relocated, the old directory is *obsolete*.

obstacle — hindrance, something that bars a path or prevents progress: She refused to think of her handicap as an *obstacle* to a fulfilling career.

obtain — to gain; to attain or get by means of effort or planned action; to reach or achieve: *Obtain* a social security number by applying for one at your local social security office.

obtrude — to enter when not invited or welcome: It was unfair to *obtrude* upon their privacy.

obtuse — not sharp or pointed; slow to understand or perceive; dull or insensitive: The bigot feels no sympathy for the homeless because of his *obtuse* perception.

obvious — self-evident: The truth was *obvious* to the well-informed.

occasionally — sometimes, from time to time: We go to the theater *occasionally*.

occidental — pertaining to the western hemisphere: The finest gems come from *occidental* countries, according to some experts.

ocular — of the eye: An *ocular* injury impaired her vision temporarily.

odious — deserving hatred or repugnance: Publicly comparing the talents of children is an *odious* habit.

omit — to leave out, pass by, neglect: He *omitted* an important passage when he read his speech.

omnipotent — all-powerful: By the end of the third match the champion felt *omnipotent*.

onerous — difficult and unpleasant, burdensome: The work was so *onerous* she often thought of quitting.

onus — burden, responsibility: The *onus* of proof is on the accuser; the defendant is presumed innocent until proved guilty.

operational — in working order, able to be operated: The elevator will not be *operational* until tomorrow; it is being repaired.

opportune — suitable or appropriate: Since you have just made an impressive sale, this is an *opportune* time to ask for a raise.

opprobrium — reproach for disgraceful conduct, infamy: He deserved all the *opprobrium* he received for turning his back on his friend.

optimum — best for a purpose, most favorable: Under *optimum* conditions of light and moisture, the plant will grow to over three feet.

option — (*v*) to purchase the right to buy or sell something within a specified time: For a thousand dollars she *optioned* the novel for one year, wrote a script, and sold it to the movies. (*n*) choice, power of right to choose: Before acting, consider your *options*.

optional — not required, open to choice: Air conditioning is *optional*; its cost is not included in the sticker price.

ordinance — city statute: The city council passed an *ordinance* requiring all dogs to be leashed.

ordure — filth, excrement: As soon as he entered the hovel, he could smell the stench of *ordure*.

orifice — opening into a cavity: The surgeon worked through an *orifice* below the ribs.

origin — beginning, source: The *origin* of the irrational fear was in a childhood trauma.

oscillate — to swing in regular motion: The pendulum continued to *oscillate*, but the clock hands did not move.

ostensible — avowed, apparent: The *ostensible* purpose of the withdrawal was to pay a debt, but actually the money was used for entertainment.

ostentatious — pretentious, showy: Some people abhor large diamonds as too *ostentatious*.

ostracize — to exclude from a group by common consent: The family of the gangster was *ostracized* by the community.

outcome — result, end: The *outcome* of the race was never in doubt.

pacific — calm, tranquil, placid: The explorer who named the ocean *pacific* found it free from storms and tempests.

pacify — to make peaceful; to soothe; to settle; to subdue: Perhaps a bottle of juice will *pacify* the crying baby.

painstaking — very careful or diligent: The search for the lost ring was long and *painstaking*.

palliate — to make an offense seem less grave: She attempted to *palliate* her error by explaining the extenuating circumstances.

palpitate — to beat rapidly, flutter, or move with slight throbs: Her heart *palpitated* with fright.

pamper — to gratify to the full, coddle, spoil: She *pampered* her pet dog in every possible manner.

pamphlet — very brief, paperbound book or treatise: The planes dropped *pamphlets* urging the population to surrender and promising fair treatment.

panacea — remedy for all ills: Complete honesty was the *panacea* she recommended for all personal conflicts.

panic — sudden and overwhelming terror, exaggerated alarm: When the children heard the noise, they fled the old house in *panic*.

panoramic — offering a broad or unlimited view: From the summit of the mountain one has a *panoramic* view of the whole range.

paradigm — model, pattern to be copied: The teacher handed out a sample letter as a *paradigm* of the correct form.

paradox — internal contradiction, statement that appears to contradict itself: "If he doesn't watch his health, he's going to wake up dead" is an example of a *paradox*.

paralyze — to unnerve or render ineffective: The catastrophe *paralyzed* the community.

paraphrase — a rewording, repetition of the meaning of something in different words: To *paraphrase* someone's work without acknowledging the source of one's information is a form of plagiarism.

parasite — person or creature who lives at the expense of another without giving anything in return: The members of the ruling class were for the most part *parasites* who enjoyed wealth produced by others and contributed nothing of value to the economic or cultural life of their nation.

parity — comparative equality: Municipal employees demanded wage *parity* with workers in the private sector.

parochial — narrow in viewpoint: Having no acquaintance with other cultures and ways of life made his outlook *parochial*.

parole — conditional release, release from prison before full sentence is served: Freed on *parole*, the convict was required to report periodically to an officer assigned to his case.

parsimonious — frugal or stingy: The extravagant person may consider the average man to be *parsimonious*.

partially — in part: Bald tires were *partially* responsible for the skid; however, slick road conditions also contributed.

participate — to take part in: At the meet all contestants will *participate* in the opening festivities.

partisan — devoted or committed to a party or cause, especially blindly or unreasonably so: *Partisan* loyalty can no longer be taken for granted; voters are now attracted to individuals more than to parties.

partition — division into parts: The present *partition* of Germany followed from the occupation of the country by the Allied forces in World War II.

passionate — expressing intense feeling; enthusiastic; intense; easily angered: The *passionate* young musician practiced five hours each day.

patent — (*adj*) obvious, easily seen: The promise of tax relief was a *patent* attempt to win last-minute support from the farmers. (*n*) exclusive right, as to a product or invention: The company's *patent* on the formula expires after a certain number of years.

peculiar — odd, special, unique, not ordinary: The fragrance is *peculiar* to violets; no other flower smells the same.

pecuniary — financial: She had no *pecuniary* interest in the project.

pedantic — making a needless display of learning: The *pedantic* lecturer made several allusions to literary works that his audience had no acquaintance with.

peevish — fretful, hard to please: The girl was unpopular because she was so *peevish*; she was always complaining about something.

pejorative — disparaging: Calling a man a skunk is *pejorative*.

penal — concerning legal punishment: The *penal* code defines crimes and their legal penalties.

penchant — strong inclination: He has a *penchant* for making friends.

pending — waiting to be decided: Our petition is still *pending*; we don't know what will be decided.

penetrate — to pierce or to pass into or through: With luck, our night attack will *penetrate* the enemy lines.

pension — regular payments to someone who has fulfilled certain requirements: After twenty years of service she retired on a full *pension*.

per capita — for each person: The country has a *per capita* income of under $800.

perceive — to feel, comprehend, note, understand: I *perceived* that the beast was harmless.

perception — act of receiving impressions by the senses: *Perception* is that act of the mind whereby the mind becomes aware of anything, such as hunger or heat.

peremptory — imperative, dictatorial: He announced his opinions in a *peremptory* tone extremely rankling to his listeners.

perforate — to make holes in: The top of the box had been *perforated* to allow the air to circulate.

peripheral — of an edge or boundary: The person who notices people almost behind him has excellent *peripheral* vision.

permeable — capable of having fluids pass through: Most clay dishes are *permeable* unless glazed.

permutation — rearrangement of the order of a group of items: The sequences CBA and BCA are *permutations* of ABC.

pernicious — causing much harm: Excessive drinking is a *pernicious* habit.

perpendicular — in an up-and-down direction, vertical, upright, at a right angle: The lamp post, having been grazed by the truck, was no longer *perpendicular*.

perpetrate — to do something evil, to commit, as a crime: The committee *perpetrated* the hoax in an attempt to defame the rival candidate.

perquisite — incidental compensation: A chauffeured car is one of the *perquisites* of a commissioner's position.

persist — to continue, especially against opposition: Despite the rebuffs, he *persisted* in his efforts to befriend the disturbed youngster.

pertinent — relevant, concerning the matter at hand: Since those circumstances were vastly different, that example is not *pertinent* to this case.

perturb — to disturb greatly; to disquiet: We were greatly *perturbed* by strange noises in the night.

peruse — to read carefully, study: She *perused* the text, absorbing as much information as she could.

petty — small, trivial, unimportant, small-minded: Don't bother the supervisor with *petty* problems but try to handle them yourself.

petulance — petty fretfulness, peevishness: Her *petulance* in demanding her own way reminded me of a two-year-old's behavior in demanding parental attention.

picturesque — having a rough, unfamiliar, or quaint natural beauty: The mountains with their rugged crags and steep ravines present a *picturesque* landscape.

pinnacle — peak, acme: She had reached such a *pinnacle* of fame that everywhere in the country her name was a household word.

piquant — stimulating the sense of taste, agreeably pungent: Mustard and chutney are both *piquant* in different ways.

pique — fit of resentment: His *pique* at being scolded lasted all day.

placate — to soothe the anger of, pacify: A quick temper is often easily *placated*.

placid — peaceful, undisturbed: The drug had relieved her anxiety, leaving her in a *placid* and jovial mood.

platitude — trite remark: He spouts *platitudes* all day but can't solve a practical problem.

plausible — seeming credible, likely, trustworthy: Since his clothes were soaked, his story of falling into the creek seemed *plausible*.

plenary — full, fully attended: The issue was so serious that the committee called a *plenary* meeting of the board to decide on a course of action.

plethora — oversupply: There is a *plethora* of bad news and a paucity of good news.

pliable — flexible, able to bend, readily influenced, yielding: Having no preconceived opinion on the matter, we were *pliable*, ready to be swayed by a forceful speech.

podium — raised platform, as for use by speakers or musical conductors: The poet stepped to the *podium* to address the audience.

pogrom — organized massacre of a certain class of people: The Russian *pogroms* in the 1880s forced a huge exodus of Jews.

poignant — having sharp emotional appeal, moving: Reading the *poignant* story, he began to cry.

polemics — art of disputing: She is an expert at *polemics* and is studying for a career in law.

polymer — compound of high molecular weight: *Polymers* are basic to the creation of plastics.

pontificate — to speak pompously: He would rise slowly, *pontificate* for half an hour, and sit down without having said a thing we didn't know before.

portable — able to be carried easily: The *portable* typewriter was equipped with a carrying case.

portentous — foreshadowing future events, especially somber ones: The thunderstorm that broke as we were leaving seemed *portentous* but in fact the weather was lovely for the rest of the trip.

posterity — succeeding generations: Many things we build today are for *posterity*.

postmark — official mark on a piece of mail showing the post office from which it was delivered and the date: Although the letter had been written in Tulsa, the *postmark* showed that it had been mailed from Omaha.

potent — powerful or effective in action: Penicillin is a *potent* medicine that should be administered with care.

potential — possible, not yet realized: If she qualifies for the promotion, her *potential* earnings for the next year might be close to $20,000.

pragmatic — concerned with practical values: He has a *pragmatic* mind, willing to try whatever promises to get results.

precarious — insecure: The animal had found a *precarious* perch on the window ledge.

precedent — similar earlier event, especially one used as a model or justification for present action: The lawyer's brief argued that the legal *precedents* cited by the opposition were not relevant because of subsequent changes in the law.

precipitous — steep like a precipice: The road had a *precipitous* drop on the south side.

precise — exact: The coroner determined the *precise* time of murder by examining the victim.

preclude — to make impossible: Obeying the speed limit would *preclude* my getting home in five minutes.

precocious — advanced in development: *Precocious* children should be given enriched programs of study to develop their talents.

precursor — predecessor, forerunner: The Continental Congress was the *precursor* of our bicameral legislature.

predatory — plundering, hunting: The hawk is a *predatory* bird.

predecessor — one who has preceded or gone before another in a position or office: In his inaugural address the new president of the association praised the work done by his *predecessor.*

predicament — troublesome or perplexing situation from which escape seems difficult: Having promised to balance the budget, to cut taxes and to increase defense spending, the newly-elected president found himself in a hopeless *predicament.*

predilection — preference, liking: She had a *predilection* for gourmet food at any price.

predominantly — for the most part: Although there are a few older students, the class is *predominantly* made up of eighteen-year-olds.

preeminent — most outstanding: She is the *preeminent* authority in her field.

preempt — to exclude others by taking first: Regularly scheduled programs were *preempted* by convention coverage.

prejudiced — biased, judging in advance without adequate evidence: Since I have never liked Westerns, I was *prejudiced* against the film before I ever saw it.

preliminary — going before the main event or business, introductory: A few easy *preliminary* questions put the applicant at ease.

premature — not yet mature or ripe, happening too soon: As she got to know him better, she decided that her initial judgment of him had been *premature.*

premeditation — act of meditating beforehand, previous deliberation: The *premeditation* of the crime was what made it so heinous.

premise — proposition or idea on which an argument or action is based: I waited to call on the *premise* that they wouldn't be home until evening.

preoccupied — having one's thoughts elsewhere, inattentive: *Preoccupied* by her dilemma, she missed her stop on the train.

preposterous — very absurd: The idea of the president's visiting our class was *preposterous.*

prerogative — exclusive privilege or right: As the child's guardian, she had the *prerogative* of deciding whether he would attend private or public school.

prescribe — to recommend, especially in a professional capacity: For the headache the physician *prescribed* aspirin.

presume — to accept as true without proof; to anticipate or take for granted, overstep bounds: An accused person is *presumed* innocent until proved guilty. I was furious that she had *presumed* to take the car without permission.

prevalent — current, widely found, common: Feelings of anger and helplessness are *prevalent* among the voters in that district.

preventive — aiming to prevent or keep from happening: *Preventive* measures must be taken to guard against malaria.

previous — occurring before in time or order: The *previous* month's electric bill included only 29 days.

primary — first, most important: Our *primary* goal is to train people for jobs that are actually available; other aspects of the program are secondary.

prime — of highest quality, value or importance: The *prime* reason for donating to charity should be to help others.

principal — main, most important: The *principal* city economically is also the most populous in the state.

prior — earlier, and therefore usually taking precedence: The director will not be able to meet with you today due to a *prior* engagement.

privileged — exempt from usual conditions, receiving special benefit; not to be made known, confidential: Only a few *privileged* outsiders have been permitted to observe the ceremony. Since communications between spouses are *privileged*, a man cannot be compelled to testify against his wife.

probability — likelihood: The *probability* that your plane will crash is practically nil.

probation — period of testing or evaluation: After a week's *probation* the employee was hired permanently.

proceed — to go forward, continue: Because of numerous interruptions, the work *proceeded* slowly.

proclaim — to announce loudly, publicly and with conviction: When the victory was announced, a holiday was *proclaimed* and all work ground to a halt.

proclivity — tendency: The child has a *proclivity* for getting into trouble.

procrastinate — to delay doing something, put off without reason: Since you'll have to get it done eventually, you might as well stop *procrastinating* and get started.

procure — to get, obtain, cause to occur: At the last minute the convict's attorney *procured* a stay of execution.

prodigal — extravagant, spending freely: She is more *prodigal* with her advice than with more concrete assistance.

prodigious — very large: He had a *prodigious* nose and a tiny mouth.

profit — valuable return, income or gain: We hope to realize a nice *profit* from the sale of our home.

profligate — utterly immoral: The *profligate* son was a regular source of income for his father's attorney.

prohibit — to prevent by authority; to forbid: During the era of *Prohibition*, the sale of alcoholic beverages was forbidden.

prolong — to draw out to greater length: The treatment *prolongs* life but cannot cure the disease, which is terminal.

promote — to help bring about; to raise or advance to a higher position: The object of many service clubs is to *promote* athletic events for the handicapped.

prompt — quick, following immediately: Correspondents appreciate *prompt* replies to their inquiries.

promulgate — to announce publicly as a law or doctrine: The revolutionary government *promulgated* some of the promised reforms.

proofread — to read and mark corrections: Always *proofread* and correct your work before you turn it in.

proper — suitable, appropriate: It is *proper* to write a letter of thanks to someone who has given you a present.

proportionate — in correct proportion or relation of amount, fairly distributed: An area's representation in the House of Representatives is *proportionate* to its population.

proscribe — to outlaw, forbid by law: Theft is *proscribed* mostly by state law.

prosecute — to carry on legal proceedings against: In return for information, the attorney general has agreed not to *prosecute* your client.

prospectus — booklet describing a business enterprise, investment or forthcoming publication distributed to prospective buyers: The *prospectus* for the real estate development was mailed to potential investors.

prosper — to thrive, do well, grow richer: An expensive suit and a new car suggested that the man's business was *prospering*.

protagonist — leading character: Mike Hammer is the *protagonist* of a whole series of detective stories.

protocol — rigid code of correct procedure, especially in diplomacy: *Protocol* demands that we introduce the ambassador before the special envoy; to fail to do so would be interpreted as an effront.

prototype — original model, first example: Homer's *Iliad* became the *prototype* for much of the later epic poetry of Europe.

protract — to draw out in time or space, lengthen: The jury's deliberations were *protracted* by confusion over a point of law.

provisional — temporary, for the time being only: The *provisional* government stepped down after the general elections.

provocation — a provoking, a cause for resentment or attack: The attack, coming without *provocation*, took them by surprise.

proximity — nearness: The *proximity* of the shopping mall is a great advantage to those residents who don't drive.

psychic — of the mind, acting outside of known physical laws: He claimed special *psychic* powers, including the ability to foresee the future.

punctuality — being on time: The train had an excellent record for *punctuality*; it almost always arrived precisely at 8:15.

purchase — to buy: We need to *purchase* or borrow a tent before we can go camping.

putative — supposed, reputed: His *putative* wealth was exaggerated by his ostentation.

quadrennial — lasting four years, occurring once in four years: The *quadrennial* games were anticipated eagerly.

qualification — that which makes one qualified or eligible: The applicant's *qualifications* for the position are a degree in library science and two years' experience in a small branch library.

quandary — doubt, uncertainty, state of difficulty or perplexity: She was in a *quandary* because the problem was so complex.

quantity — amount: Speeding up the process would result in an increased *quantity* but a poorer quality.

queasy — causing or affected by nausea; squeamish: The thought of riding the subway at night makes me *queasy*.

quench — to extinguish, put out: She *quenched* the flames with water.

query — to question: He *queried* the witness about his alibi.

quirk — turn, twist, caprice: A sudden *quirk* of fancy caused her to change her mind.

quiver — to shake, tremble, shudder: The dog *quivered* with excitement.

quorum — minimum number of members that must be present for an assembly to conduct business: No votes may be taken until there are enough representatives present to constitute a *quorum*.

quota — proportional share: The school had an unwritten *quota* system that set limits on the proportion of applicants accepted from different geographical areas.

quote — to cite word for word, as a passage from some author, to name or repeat: He *quoted* the words of Woodrow Wilson in his acceptance speech.

quotient — in arithmetic, the number resulting from the division of one number by another: The *quotient* of ten divided by five is two.

rabid — furious, raging; suffering from rabies: The *rabid* animal was destroyed before it could bite anyone.

radiation — divergence in all directions from a point, especially of energy: Solar *radiation* is the *radiation* of the sun as estimated from the amount of energy that reaches the earth.

rambunctious — wild; marked by uncontrollable exuberance; unruly: When the children get together with all of their cousins, the group tends to get *rambunctious*.

ramification — breakdown into subdivisions, a branching out: The *ramifications* of the subject were complex.

rampant — springing or climbing unchecked, rank in growth: The *rampant* growth of the weeds made the lawn look extremely unsightly.

ramshackle — tumbling down, shaky, out of repair: It was impossible to be comfortable in such a *ramshackle* house.

rancor — malice, ill will, anger: In spite of the insults of his opponent, the man remained calm and spoke without *rancor*.

randomly — in an unplanned or haphazard way, without order or pattern: The papers had been strewn *randomly* about the room.

rapidity — speed: The *rapidity* with which her hands flew over the piano keys was too great to follow with the eye.

ratify — to give formal approval to: The proposed amendment must be *ratified* by the states before it can become law.

ratio — proportion, fixed relation of number or amount between two things: The *ratio* of women to men in middle-level positions in the firm is only one to seven.

rationale — rational basis, explanation or justification supposedly based on reason: They defended their discrimination with the *rationale* that women were incompetent physically to handle the job.

raze — to destroy down to the ground, as a building: Buildings in the path of the highway construction will be *razed*.

reactionary — extremely conservative, marked by opposition to present tendencies and advocating a return to some previous or simpler condition: The pamphlet expressed a *reactionary* hatred of innovation and a nostalgia for "the good old days."

rebuff — a snub, repulse, blunt or impolite refusal: When overtures of friendship are met with *rebuff*, they are not likely to be renewed.

rebuke — to reprimand, criticize sharply: He *rebuked* the puppy in stern tones for chewing up the chair.

rebuttal — contradiction, reply to a charge or argument: Each side was allowed five minutes for *rebuttal* of the other side's arguments.

recalcitrant — stubborn, refusing to obey: A *recalcitrant* child is difficult to teach.

recapitulate — to mention or relate in brief, summarize: The abstract *recapitulated* the main points of the argument.

recede — to go back or away: The waters *receded* and left the beach covered with seaweed.

receptive — able and tending to receive and accept, open to influence: The manager, unsatisfied with the store's appearance, was *receptive* to the idea of a major remodeling.

recessive — tending to recede or not make itself felt: The characteristic encoded in a *recessive* gene may be passed on to an individual's offspring even though it is not apparent in the individual.

recipient — one who receives: The *recipient* of the award had been chosen from among 200 candidates.

reciprocal — done in return, affecting both sides, mutual: The United States has *reciprocal* trade agreements with many nations.

reckless — not thinking of consequences, heedless, causing danger: People who feel they have nothing to lose often become *reckless*.

recondite — profound; obscure; concealed: Anthropologists try to discover the *recondite* facts about human origins.

reconsider — to think over again: When he refused the appointment, the committee asked whether he would *reconsider* his decision if more money were offered.

recourse — seeking of aid or remedy in response to some action or situation: Unless you correct this error immediately, I will have no *recourse* but to complain to the manager.

recreation — relaxation, play: Physical *recreation* often relieves tension and improves the emotional outlook.

recriminate — to return accusation for accusation: They *recriminated* constantly over the most trivial setbacks, each blaming the other whenever anything went wrong.

rectitude — honesty, integrity, strict observance of what is right: Her unfailing *rectitude* in business dealings made her well trusted among her associates.

recumbent — lying down: The painting depicted the goddess *recumbent* on a sumptuous couch.

recuperate — to become well, get better: It is best to stay home from work until you have *recuperated* completely.

recur — to happen again: Unless social conditions are improved, the riots are bound to *recur*.

redeem — to save, ransom, free by buying back: Though the film is boring in parts, it is *redeemed* by a gripping finale.

redress — compensation for a wrong done: The petitioners asked the state for a *redress* of grievances for which they had no legal recourse.

reduce — diminish in size, amount, extent or number: We all hope that the arms agreement will include a *reduction* of atomic warheads.

redundant — wordy, repeating unnecessarily: The expressions "more preferably" and "continue to remain" are *redundant*.

referral — a being referred from one person or agency to another, as for employment: A *referral* service arranged appointments for women who wished to obtain abortions.

refinement — act of clearing from extraneous matter, purification: The *refinement* of the metals freed them from the impurities that made them unfit for commercial use.

refrain — to keep from doing something, to not do: Considerate parents *refrain* from criticizing their children in front of others.

refuse — to decline to accept; reject: If you *refuse* this assignment, you may not be offered another.

regal — pertaining to a monarch, royal: He had a *regal* air that impressed even those who knew him for an imposter.

regimen — regular manner of living: The *regimen* of army life bored him.

rehabilitate — to restore to a former state or capacity: The stated object of the program is to *rehabilitate* ex-offenders.

reimburse — to refund, pay back: The company found it difficult to *reimburse* the salesman for all his expenses.

reiterate — to repeat: The instructions were *reiterated* before each new section of the test.

related — connected by some common relationship: The circumstances of this fire are closely *related* to those of the hotel fire that occurred last week.

relegate — to transfer to get rid of, assign to an inferior position: He *relegated* the policeman to a suburban beat.

relevant — concerning the matter at hand, to the point, related: Her experience in government is *relevant* to her candidacy; her devotion to her family is not.

relinquish — to give up, hand over: The aunt *relinquished* custody of the child to its mother.

reluctant — opposing or unwilling: The bank is *reluctant* in giving further loans until all obligations are paid.

reminisce — to remember, talk about the past: When old friends get together, they love to *reminisce*.

remit — to pay, to send payment: The invoice was *remitted* by check; you should be receiving it shortly.

remuneration — reward, payment, as for work done: Health benefits are part of the *remuneration* that goes with the position.

renege — to go back on a promise or agreement: Their assurances of good faith were hollow; they *reneged* on the agreement almost at once.

renounce — to give up or disown, usually by formal statement: The nation was urged to *renounce* its dependence on imports and to buy more American cars.

replenish — to supply again, to make full or complete again something that has been depleted: Some natural resources, such as lumber, can be *replenished*.

reprehensible — deserving rebuke or blame: Conduct that selfishly endangers the safety of others is *reprehensible*.

repress — to subdue, hold back, keep down, keep from expression or consciousness: We could not *repress* a certain nervousness as the plane bumped along the runway.

reprieve — postponement of some evil, such as punishment: You have a *reprieve*; the test has been put off for a week.

reprimand — severe criticism, especially a formal rebuke by someone in authority: Since it was a first offense, the judge let the teenager off with a *reprimand*.

reprisal — injury in return for injury: The Israelis launched a raid in *reprisal* for the night attack.

reprove — to censure, rebuke, find fault with: The instructor *reproved* the student for failing to hand in the assignments on time.

repudiate — to refuse to accept, reject: The candidate *repudiated* the endorsement of the extremist group.

request — to ask for: The students *requested* a meeting with the college president to discuss the new policy.

requisite — required, necessary: No matter when he starts work, an employee may take vacation time as soon as he has worked the *requisite* number of weeks.

requisition — formal written order or request: The office manager sent in a *requisition* for another desk and chair.

requite — to give in, return, to repay: The man's sympathy and good humor were *requited* by the enthusiastic affection of his nephews.

rescind — to cancel formally or take back: They *rescinded* their offer of aid when they became disillusioned with the project.

reserve — to keep back or save for use at a later time; to set aside for the use of a particular person: The runner had *reserved* energy for a burst of speed in the final lap. Call the restaurant to *reserve* a table for four.

residence — place where a person lives, fact of living in a place: According to the phone company, that number is a *residence*, not a business.

residue — something left over, remainder: A *residue* of coffee grounds was left at the bottom of the cup.

resilient — able to spring back: The spring was still *resilient* after years of use.

resplendent — very bright, shining: She was *resplendent* in the jewelry and sequined dress.

respondent — person who responds or answers: Several *respondents* refused to answer most of the questions in the survey.

response — a reply or reaction: His *response* to my question was another question.

restitution — restoration to a rightful owner, reparation for an injury: He agreed to make *restitution* for the money he had stolen.

restrict — to confine, keep within limits: Use of the computer room is *restricted* to authorized personnel.

resume — to begin again after an interruption: The courtroom proceedings *resumed* after an hour's recess for lunch.

resurgent — rising again: The *resurgent* spirit of nationalism caused riots in Cyprus.

resuscitate — to bring back to life: Artificial respiration was used to *resuscitate* the swimmer.

retain — to keep: Throughout the grueling day she had managed somehow to *retain* her sense of humor.

retaliate — to give injury for injury: The boxer *retaliated* for the punch with a stunning blow to the head.

retard — to slow: Drugs were successfully used to *retard* the progress of the disease.

reticent — restrained in speech, unwilling to talk: People are *reticent* to confess such anxieties for fear of appearing weak.

retroactive — applying to what is past: A law cannot be made *retroactive*; it can only apply to future actions.

retrogress — to go backward, lose ground: Because of the devastation of the recent earthquakes, living conditions in the region have *retrogressed*.

reveal — to make known, display: His dishonesty was *revealed* during the trial.

reverence — feeling of deep respect or awe, as for something sacred: The great novelist was disconcerted by the *reverence* with which her students greeted her most casual remark.

revive — to come or bring back to life: A cool drink and a bath *revived* her spirits.

revision — a correction or a change; revised form or version: A playwright makes many *revisions* to the script before the play is produced.

revolutionize — to change fundamentally or completely: Understanding of its addictive properties has *revolutionized* the public's attitude toward crack.

rheostat — electrical resistor with changeable resistance: A *rheostat* is used to make lights dimmer.

rife — widespread, prevalent, filled with: The city was *rife* with rumors that a coup was imminent.

robust — hardy, strong, healthy: Her *robust* health was apparent in her springy walk and glowing skin.

rouse — to stir up; to excite or to awaken: Let us *rouse* the citizenry to a new era of patriotism.

rubicund — ruddy; reddish: The heavy drinker has a bulbous, *rubicund* nose.

rue — to be sorry for, regret: He *rued* the day he made that mistake.

ruminate — to chew the cud, to think over at leisure: A cow *ruminates* after it eats. I will *ruminate* on your proposal and let you know my decision later this week.

rupture — a breaking off, breach: The bungling of the rescue operation, which resulted in the death of the ambassador, led to a *rupture* of diplomatic relations between the two nations.

saccharine — pertaining to sugar, having the qualities of sugar, overly sweet: The *saccharine* sentimentality of the film is cloying to any audience over the age of twelve.

sacrament — sacred rite: The Roman Catholic Church recognizes seven *sacraments*, including baptism, matrimony and extreme unction.

sagacious — wise, discerning: Teachers are more *sagacious* than students give them credit for.

salient — conspicuous, noticeable, prominent: The *salient* points of the speech could not be forgotten by the audience.

salutary — promoting health, conducive to good: The preacher's anecdotes provided a *salutary* lesson.

salvage — to save or recover from disaster, such as shipwreck or fire: Divers *salvaged* gold coins and precious artifacts from the sunken Spanish galleon.

salvation — act of preserving from danger, destruction or great calamity: The governor's strategy of delay proved to be the *salvation* of the province.

sanction — to authorize, approve, support: The parent organization refused to *sanction* the illegal demonstration staged by the splinter group.

sanguine — ardent, confident, optimistic: The leader was *sanguine* about the movement's chances for success.

satiate — to gratify completely, surfeit: Employees at candy factories soon get so *satiated* that they never eat the sweets.

saturate — to fill fully, soak, cause to become completely penetrated: The cloth was thoroughly *saturated* with the soapy water.

scant — barely or scarcely sufficient; inadequate: They made do with the *scant* rations in the lifeboat for two days.

schematic — in the form of an outline or diagram: A *schematic* drawing of the circuitry illustrated how the radio worked.

schism — a split, breakup: The Great *Schism* created two rival Christian churches, the Eastern and the Western.

scintillating — sparkling, brilliant, witty: Absorbed in the *scintillating* conversation, the guests lost track of the time.

scrupulous — having scruples, conscientiously honest and upright: That attorney is too *scrupulous* to get involved in racketeering.

secular — not religious, not concerned with religion: The *secular* authorities often have differences with the church in Italy.

secure — (*adj*) safe, reliable, free from fear or danger: Her *secure* job assured her of a steady income for as long as she chose to work. (*v*) to make safe, to obtain: I have *secured* two tickets for tonight's performance.

sedition — incitement to rebel against the government: *Sedition* is an offense punishable under state laws.

seismic — caused by earthquake: A seismograph measures the strength of *seismic* tremors in the earth.

semantics — study of the meanings of words: In

English *semantics*, many synonyms have quite different connotations.

semester — a period of six months; either of the two approximately 18-week periods into which the academic year is commonly divided: Each *semester* every student must produce at least two lengthy research papers.

seminar — a group of graduate students working under the direction of a professor, each doing original research and sharing results through group discussions: The advanced psychology class enjoyed its *seminars* concerning personality disorders.

sentimental — having or showing tender feelings; such feelings in excess; influenced more by emotion than reason or thought: The *sentimental* actress wept when she watched herself in her old movies.

sequester — to seize by authority, set apart in seclusion: The jury was *sequestered* until the members could reach a verdict.

severe — harsh, extreme, serious: The tough drug laws required *severe* penalties for repeat offenders.

shibboleth — password, identifying phrase of a group or attitude: ''Power to the people'' was a popular *shibboleth* of the 1960s.

silicosis — lung disease resulting from inhaling dust: Miners are often victims of *silicosis* and resultant tuberculosis.

simulate — to pretend, feign, give a false appearance of: Although she had guessed what the gift would be, she *simulated* surprise when she unwrapped the package.

simultaneous — happening or existing at the same time: There were *simultaneous* broadcasts of the game on local television and radio stations.

sinecure — job requiring little work: The person who is looking for a *sinecure* should avoid working here; this job is very demanding.

site — piece of land considered as a location for something, such as a city: The archeologists began excavations at the *site* of the ancient city.

skepticism — doubt, partial disbelief: He listened to the fantastic story with patent *skepticism*. *Also:* **scepticism**.

slander — spoken false statement damaging to a person's reputation: The witness was guilty of *slander* when he falsely testified that his partner had connived in the tax fraud scheme.

slate — to put on a list, to schedule: The meeting is *slated* for next Tuesday.

slipshod — shabby; careless in appearance or work-manship: *Slipshod* work habits tend to lead to faulty products.

slovenly — untidy in personal and work habits: The *slovenly* housekeeper was of very little use.

smirk — annoyingly smug or conceited smile: His arrogant behavior and *smirk* of satisfaction whenever he won made him unpopular with the fans.

smother — to destroy life by depriving of air; to suppress expression or knowledge: One way to extinguish a small fire is to *smother* it with thick foam.

solicitude — concern, anxiety, uneasiness of mind occasioned by the fear of evil or the desire for good: The teacher had great *solicitude* for the welfare of her students.

solution — in chemistry, a homogeneous molecular mixture in which a substance is dissolved in a liquid: To relieve her sore throat she gargled with a saline *solution*.

somatic — bodily, physical: Psychological disturbances often manifest themselves indirectly as *somatic* symptoms.

sonorous — resonant: His *sonorous* voice helped make him a success as a stage actor.

soporific — causing sleep: Because of the drug's *soporific* effect, you should not try to drive after taking it.

sparkling—glittering or shining; brilliant or lively. The sailors polished the brass until it was *sparkling*.

spartan — very simple, frugal, hardy, disciplined, or self-denying: In addition to the usual classes, the military school imposed a *spartan* regimen of physical training.

specialize — to adapt to a special condition, concentrate on only one part of a field or endeavor: The assembly line caused labor to become more *specialized* as each worker performed only a small part of the whole manufacturing process.

specific — precise, well-defined, not general: The patron was not looking for any *specific* book but had just come in to browse.

specious — deceptively plausible: He advanced his cause with *specious* arguments and misinformation.

spontaneous — coming from natural impulse, having no external cause, unplanned: Oily rags improperly disposed of may cause a fire by *spontaneous* combustion.

sporadic — occasional, happening at random

intervals: He made *sporadic* attempts to see his estranged wife.

spurious — false, counterfeit, phony: The junta's promise of free elections was *spurious*, a mere sop to world opinion.

spurt — a sudden brief burst of activity; a squirt or shooting forth: There is a *spurt* of extra retail business right before Christmas.

squalid — wretched, filthy, miserable: The *squalid* shantytown was infested with vermin and rife with disease.

stalemate — deadlock, situation in which neither side in a game or contest can make a move: Talks have reached a *stalemate*; neither side is authorized to make the necessary concessions.

stamina — power of endurance, physical resistance to fatigue or stress: While younger swimmers tend to be faster over short distances, older swimmers often have more *stamina*.

stature — height, elevation (often used figuratively): His work in physics was widely admired in the profession and his *stature* as an expert in his field unquestioned.

status — position, rank, present condition: Her *status* as vice president allows her to take such action without prior approval by the board of directors.

stealthy — furtive, secret: While their grandfather was distracted by the phone, the children made a *stealthy* raid on the refrigerator.

sterile — free from germs, barren, infertile, unproductive, lacking in liveliness or interest: The room was depressingly *sterile* with its drab colors, bare walls, and institutional furniture.

stigma — distinguishing blemish inflicted by others: The *stigma* caused by gossip lasted long after the accusation had been disproved.

stipulate — to make an express demand or condition: The lease *stipulated* that the rent could be raised by a certain percentage every year.

stoical — showing calm fortitude: She was *stoical* in the face of great misfortunes.

stratagem — scheme that outwits by cleverness or trickery: His *stratagem* created confusion among the other team and allowed his side to take the lead.

strenuous — rugged, vigorous, marked by great energy or effort: Climbing the volcano was *strenuous* exercise even for the physically fit.

strident — harsh-sounding: She had a *strident* voice that sent shivers down my back.

stringent — severe, strict, compelling: The buying and selling of securities is governed by *stringent* SEC rules.

sturdy — strongly built; hardy: The *sturdy* oak tree has withstood many hurricanes.

suave — smoothly polite: His *suave* manners reflected great confidence and poise.

subdue — to overcome, calm, render less harsh or less intense: The understanding actions of the nurse helped to *subdue* the stubborn and unruly child.

submit — to give in, surrender, yield; to give, hand in: Although the doctors were dubious of his full recovery, the patient refused to *submit* to despair. The couple *submitted* their application to the loan officer.

subordinate — under the power or authority of another; one occupying a lower rank: The private is *subordinate* to his sergeant.

subpoena — writ summoning a witness: They issued *subpoenas* to all necessary witnesses.

subsequent — following in time, order or place: *Subsequent* to his arrest, the suspect was arraigned before the judge.

subsidy — financial aid granted by the government: Ship operators and airlines receive federal *subsidies* in the form of mail delivery contracts.

substantial — real or actual; of considerable wealth or value; significantly large; ample: Wise parents put aside *substantial* sums of money toward their children's education.

substitute — person or thing put in place of something else: A temporary worker filled in as a *substitute* for personnel on vacation.

subterfuge — deceitful means of escaping something unpleasant: The lie about a previous engagement was a *subterfuge* by which they avoided a distasteful duty.

subversive — tending to undermine or destroy secretly: The editor was accused of disseminating propaganda *subversive* to the national security.

successive — following one after another without interruption: Last week it rained on four *successive* days.

successor — one who follows another, as in an office or job: Retiring from office, the mayor left a budget crisis and a transit strike to his *successor*.

succinct — to the point, terse: A *succinct* communiqué summed up the situation in four words.

succor — aid, help in distress: Despite the threat of harsh reprisals, many townspeople gave *succor* to the refugees.

succumb — to yield to superior strength or force; to give in; to die: The reluctant novice *succumbed* to the pleading of the swimming counselor and plunged into the water.

sufficient — ample, adequate, enough: Our supplies are *sufficient* to feed an army for a week.

suffuse — to overspread: The floor was *suffused* with a disinfectant wax.

summarize — to cover the main points: The newscaster *summarized* the content of the president's speech.

sundry — miscellaneous, various: *Sundry* errands can be consolidated into a single trip in order to save gas.

supercilious — proud and haughty: The *supercilious* attitude of the old and wealthy families has contributed to many social upheavals.

superficial — on or concerned with the surface only, shallow: The *superficial* review merely gave a synopsis of the movie's plot.

superfluous — extra, beyond what is necessary: It was clear from the scene what had happened; his lengthy explanations were *superfluous*.

superior — one who is above another in rank, station or office; placed higher up; good or excellent in quality: The manager of the typing pool is *superior* to the other typists.

supersede — to take the place of: The administration appointed new department heads to *supersede* the old.

supervise — to oversee, direct work, superintend: A new employee must be carefully *supervised* to insure that he learns the routine correctly and thoroughly.

supine — lying on the back, passive, inactive: The girls were *supine* on the beach, roasting in the sun.

supplant — to take the place of, especially unfairly: The mother claimed that her sister had deliberately tried to *supplant* her in the daughter's affections.

supple — capable of being bent or folded without creases, cracks or breaks; limber; easily changed or influenced: The *supple* gymnast danced on the exercise bar.

supplement — to add to, especially in order to make up for a lack: The dietician recommended that she *supplement* her regular meals with iron pills.

supplicate — to beg: He *supplicated* the emperor for a pardon.

support — to uphold, assist: I *support* our country's policy of aid to underdeveloped nations.

surcharge — an additional tax or cost above the usual: On Amtrak there is a *surcharge* for use of a sleeping car.

surfeit — excess: There was a *surfeit* of food at the table, and no one could finish the meal.

surpass — to excel, go beyond: The success of our program *surpassed* even our high expectations.

surreptitious — secret, unauthorized, clandestine: A *surreptitious* meeting in the basement of one of the conspirators was arranged for midnight.

surrogate — acting in place of another, substituting: The housekeeper acted as *surrogate* mother for the children after their own mother died.

surveillance — a watching: The suspect was kept under *surveillance*.

susceptible — easily affected, liable: She is *susceptible* to colds because of her recent illness.

suspend — to stop or cause to be inactive temporarily; to hang: Service on the line was *suspended* while the tracks were being repaired. The light fixture was *suspended* from the beam by a chain.

suture — stitch on a wound: The surgeon made several *sutures* to close the wound.

syllabus — a summary or outline containing main points; a course of study or examination requirements: The course *syllabus* should give an overview of what is to be taught in that course.

symposium — meeting for discussion of a subject: They listened to a television *symposium* on the subject of better schools.

synthesis — combination of parts into a whole: The decor was an artful *synthesis* of traditional and contemporary styles.

systematic — orderly, following a system: A *systematic* review of hiring in the past two years revealed discrepancies between official policy and actual practice.

tabulate — to arrange data in some order: The election results were *tabulated* by township.

tachometer — instrument for measuring rotational speed: They watched the *tachometer* closely, keeping an eye on the engine's rpm's.

tacitly — silently, without words, by implication: He *tacitly* assented to his friend's arguments but wouldn't admit to being convinced.

tangential — digressing, off the point, not central: Facts about the author's life, while they may be fascinating, are *tangential* to an evaluation of her works.

tangible — capable of being touched, having objective reality and value: The new position offered an opportunity for creativity as well as the more *tangible* reward of a higher salary.

taper — progressively narrowed toward one end; to lessen; diminish: The *tapered* shape of a funnel makes it suitable for transferring liquids into narrow-neck containers.

tardiness — lateness: His *tardiness* was habitual; he was late getting to class most mornings.

taxonomy — science of classification: The *taxonomy* of the law first separates the civil from the criminal.

tedious — boring, long and tiresome: The film was so *tedious* that we walked out in disgust before it was half over.

temerity — contempt for danger or opposition; recklessness; nerve; audacity: The arsonist had the *temerity* to offer to help fight the fire.

temporize — to evade immediate action, to stall for time: He sought to *temporize* while the sun was in his eyes.

tenable — capable of being held or defended: The club had no *tenable* reasons for the exclusion; it was purely a case of prejudice.

tenacity — persistence, quality of holding firmly: His *tenacity* as an investigator earned him the nickname ''Bulldog.''

tenancy — state or time of being a tenant: We observed nothing unusual during the first few weeks of our *tenancy* at the cottage.

tendon — in anatomy, a hard, insensible cord or bundle of connective tissue by which a muscle is attached to a bone: The Achilles *tendon* connects the heel with the calf of the leg.

tenement — dwelling place, apartment, especially a building that is run-down, dirty, etc.: Rows of dilapidated *tenements* lined the streets of the impoverished neighborhood.

tension — tautness or stress; mental or nervous strain: The intense hostility between the opposing sides created a great deal of *tension* in the room.

tentative — done as a test, experiment, or trial: The negotiators have reached a *tentative* agreement, the details of which have yet to be worked out.

tenuous — held by a thread, flimsy: The business survived on a *tenuous* relationship with a few customers.

terminate — to end: She *terminated* the interview by standing up and thanking us for coming.

terminology — special vocabulary used in a field of study: Use proper *terminology* in technical writing so that your meaning will not be ambiguous.

terse — to the point, using few words: The official's *terse* replies to our questions indicated that he did not welcome being interrupted.

textile — cloth, woven material: New England in the nineteenth century was dotted with *textile* mills operated by water power.

theory — speculative truth, proposition to be proved by evidence or chain of reasoning: The professor emphasized that the explanation was only a *theory* subject to verification, not an established fact.

thesaurus — dictionary of synonyms: A good *thesaurus* distinguishes the shades of meaning among words with similar definitions.

thesis — essay, proposition to be debated: She completed her doctoral *thesis*.

thorough — done to the end; omitting nothing; complete; very exact: The search for the missing airplane was *thorough* and painstaking.

timorous — fearful; timid: The abused child was a *timorous* little waif.

tirade — vehement speech: He shouted a long *tirade* at the driver who had hit his car from behind.

tolerate — to permit, put up with: We *tolerate* ignorance in ourselves because we are too indolent to study.

torrent — a swift, violent stream of liquid; rush of words or mail; heavy rain: When a dam breaks, it releases a *torrent* of water.

torsion — twisting or wrenching: Too much bodily *torsion* may lead to backaches.

total — complete, entire, whole: The *total* cost of our European vacation will be more than $4,000.

toxic — poisonous: Alcohol consumed in very large quantity may prove highly *toxic*.

tractable — easily led: A *tractable* worker is a boon to a supervisor but is not always a good leader.

tradition — handing down of beliefs and customs from generation to generation; long established customs or practice: Our family *tradition* is to sing at the table at Sunday dinner.

tranquil — quiet, calm, peaceful: The *tranquil* morning was disturbed by the appearance on the lake of a motorboat.

transcribe — to make a written copy of: These almost illegible notes must be *transcribed* before anyone else will be able to use them.

transcript — written copy: The court reporter read from the *transcript* of the witness's testimony.

transfusion — a pouring from one container into another: They gave the victim a blood *transfusion*.

transgression — a breaking of a law or commandment: We ask God to forgive our *transgressions*.

transition — change, passage from one place or state to another: The weather made a quick *transition* from sweltering to freezing.

transitory — fleeting, passing, not permanent: It is normal to feel a *transitory* depression over life's setbacks.

translate — to change from one medium to another, especially from one language or code to another: The flight attendant *translated* the announcement into Spanish for the benefit of two of the passengers.

transverse — lying across: They placed the ties *transversely* on the tracks and waited for the train to crash.

trauma — wound: Many emotional ailments in adults are related to psychic *traumas* in childhood.

travesty — imitation of a serious work so as to make it seem ridiculous: His production of Shakespeare in modern language was a *travesty*.

treaty — formal agreement between nations: An economic alliance between the governments was established by *treaty*.

trenchant — sharp, penetrating, forceful: His *trenchant* remarks cut to the heart of the matter.

trepidation — involuntary trembling, as from fear or terror: The ghost story caused them to feel a certain *trepidation* walking home late at night.

tribulation — great trouble or hardship: The Pilgrims faced many *tribulations* before the colony was firmly established.

tribunal — court of justice: The decision was left to an international *tribunal*.

truculent — ferocious, savage, harsh in manner: The champion affected a *truculent* manner to intimidate the young challenger.

truncate — to shorten by cutting: The shrubs were uniformly *truncated* to form a neat hedge.

truncheon — club: British police are armed with *truncheons*.

truss — to support, tie up in a bundle: The chicken should be *trussed* with string before roasting.

turbulent — violent, in wild motion, agitated: The *turbulent* stream claimed many lives.

turgid — swollen: The river was *turgid* from the incessant rains.

tutelage — guardianship: She grew up under her cousin's *tutelage*.

tyro — novice, beginner: He is a *tyro* in finance.

ubiquitous — existing everywhere: Papaya trees, *ubiquitous* in the region, bear large yellow fruits.

ultimate — final, last: After hours of soul-searching, her *ultimate* decision was no different from her original one.

unaccountable — mysterious, not able to be explained: The *unaccountable* disappearance of the family led to wild stories of flying saucers.

unavoidable — not preventable: Because he is a republican and she is a democrat, their disagreement about economic policy was *unavoidable*.

uncanny — weird, so acute as to appear mysterious: After a lifetime of fishing those waters, the old man was able to predict weather changes with *uncanny* precision.

uncouth — unrefined, awkward: The girl was so *uncouth* she could hardly handle a knife and fork and had no notion of table manners.

unethical — without or not according to moral principles: Although he did not break any law, the man's conduct in taking advantage of credulous clients was certainly *unethical*.

ungainly — not expert or dexterous, clumsy, physically awkward: She walked in an *ungainly* way, as if her shoes were two sizes too large.

unilateral — one-sided, coming from or affecting one side only: The decision to separate was *unilateral*; one spouse moved out against the other's wishes.

uniformity — sameness, lack of variation: Although the temperature is pleasant, the *uniformity* in weather from season to season can become boring.

unique — without a like or equal, unmatched, single in its kind: The statue was valuable because of its *unique* beauty.

unkempt — uncombed, not cared for, disorderly: He was recognized by his *unkempt* beard.

unmitigated — not lessened, not softened in severity or harshness: According to President Eliot of Harvard, inherited wealth is an *unmitigated* curse when divorced from culture.

unprecedented — never before done, without precedent: Sputnik I accomplished *unprecedented* feats.

unravel — to untangle, explain, clear from complication: The detective was able to *unravel* the mystery.

unreliable — not dependable: Because of his *unreliable* attendance at conferences, the professor was not asked to prepare a speech.

unscrupulous — unprincipled, not constrained by moral feelings: The *unscrupulous* landlord refused to return the security deposit, claiming falsely that the tenant had damaged the apartment.

unwieldy — ponderous, too bulky and clumsy to be moved easily: I need help moving this *unwieldy* mattress.

upbraid — to charge with something disgraceful, reproach, reprove with severity: The husband *upbraided* his wife for her extravagance.

urbane — smoothly polite, socially poised and sophisticated: He travels in *urbane* circles and is as suave as any of his friends.

urgent — pressing, having the nature of an emergency: We received an *urgent* message to call the hospital.

utensil — implement, tool: Forks and other *utensils* are in the silverware drawer.

utilize — to use, put to use: We will *utilize* all the resources of the department in the search for the missing child.

vacant — empty, unoccupied: The *vacant* lot was overgrown with weeds.

vacate — to leave empty: The court ordered the demonstrators to *vacate* the premises.

vacillate — to fluctuate, change back and forth, be inconsistent: The employer's manner *vacillated* between oppressive friendliness and peremptory command.

vacuity — dullness of comprehension, lack of intelligence, stupidity: The *vacuity* of her mind was apparent to all who knew her.

vacuous — empty, without substance: His *vacuous* promises were forgotten as soon as they were uttered.

vain — unsuccessful, useless: A *vain* rescue attempt only made the situation worse.

valid — well-grounded or justifiable on principle or evidence; correctly derived; sound: On the basis of the evidence, the verdict appears to be *valid*.

validity — strength, force, being supported by fact, proof or law: The bill was never paid because its *validity* could not be substantiated.

valor — worthiness, courage, strength of mind in regard to danger: His *valor* enabled him to encounter the enemy bravely.

vandal — one who deliberately disfigures or destroys property: *Vandals* broke all the windows in the vacant building.

vanguard — troops who march in front of any army, advance guard: The uniforms of the *vanguard* were the most colorful of all.

vanquish — to conquer, overcome, overpower: Napoleon *vanquished* the Austrian army.

vapid — tasteless, dull, lifeless, flat: Their conversation was so *vapid* and predictable that I lost interest in talking to them.

variable — changing, fluctuating: The weather report stated that winds would be *variable*.

varicose — swollen, said of veins: *Varicose* veins sometimes cause large bulges on the legs.

variegate — to diversify in external appearance, mark with different colors: The builder created a *variegated* facade with marble of different hues.

various — dissimilar; characterized by variety; numerous; separate: We visited *various* national parks on our trip across the United States.

venal — able to be corrupted or bribed: The *venal* judge privately offered to hand down the desired verdict for a price.

vendetta — blood feud: The two families carried on a *vendetta* through three generations.

venial — forgivable; pardonable; excusable: A *venial* sin stands in marked contrast to a mortal sin, which is extreme, grave and totally unpardonable.

venerate — to respect: She was a great philanthropist whose memory deserves to be *venerated*.

venous — pertaining to a vein or veins: *Venous* blood is carried by the veins to the right side of the heart.

verbatim — word for word, in the same words: The lawyer requested the defendant to repeat the speech *verbatim*.

verbose — using more words than are necessary, tedious because of wordiness: The paper is well-organized but *verbose*; it should be cut to half its present length.

verdant — green, fresh: The *verdant* lawn made the old house look beautiful.

verdict — decision, especially a legal judgment of guilt or innocence: In our legal system, the *verdict* of a jury in convicting a defendant must be unanimous.

verge — to be on the border or edge: Their behavior *verged* on hysteria.

verify — to prove to be true, establish the proof of: You should *verify* the rumor before acting on it; it may not be true at all.

verisimilitude — appearance of truth: The movie set reproduced the ancient city with great *verisimilitude*; every detail seemed correct.

verity — truthfulness, honesty, quality of being real or actual: The *verity* of the document could not be questioned.

vernacular — native language: He spoke in the *vernacular* of southern Germany.

vernal — pertaining to spring: The *vernal* influence was everywhere, as trees blossomed and lovers strolled hand in hand.

versatile — competent in many things, subjects,

fields; flexible: *Versatility* is the hallmark of the good handyman.

vertical — upright, in an up-and-down position: A graph is constructed around a *vertical* and a horizontal axis.

vertex — top, highest point, apex: The view from the *vertex* of the hill was breathtaking.

vertigo — dizziness, giddiness, sense of apparent rotary movement of the body: The physician explained that the *vertigo* was due to some chronic disease of the heart.

verve — enthusiasm; energy; vitality: The optimist greets each new day with *verve* and a big smile.

vestige — remnant, remainder, trace: The appendix is a useless *vestige* of an earlier human form.

veterinary — concerning the medical treatment of animals: Reliable *veterinary* services are indispensable in areas where people raise animals for their livelihood.

vex — to irritate, distress, cause disquiet: She was periodically *vexed* by anonymous phone calls.

viability — capacity to live, survival ability: The chart compared the *viability* of male and female infants.

vibrate — to swing or oscillate rapidly: The strings of the instrument produce sound waves by *vibrating*.

vicarious — experienced secondhand through imagining another's experience: She took *vicarious* pleasure in the achievements of her daughter.

vigilant — watchful, on guard: As a Supreme Court justice he has always been *vigilant* against any attempt to encroach on the freedoms guaranteed by the Bill of Rights.

vilify — to defame, attempt to degrade by slander: She was sued for attempting to *vilify* her neighbor.

vindicate — to uphold, confirm: The judgment of the editor was *vindicated* by the book's success.

vindictive — unforgiving, showing a desire for revenge: Stung by the negative reviews of his film, in the interview the director made *vindictive* personal remarks about critics.

viscosity — in physics, the resistance to flow of a fluid: Motor oil has a greater *viscosity* than water.

vitality — life, energy, liveliness, power to survive: She had been physically active all her life and at the age of eighty still possessed great *vitality*.

vocation — regular occupation or work: My son's *vocation* is carpentry; his avocation (hobby) is fishing.

volatile — changing to vapor, quickly changeable, fickle: She had a *volatile* temper — easily angered and easily appeased.

volition — deliberate will: He performed the act of his own *volition*.

voracious — ravenous, very hungry, eager to devour: The *voracious* appetite of the man startled the other guests.

vulnerable — open to attack or danger; easily wounded or physically hurt; sensitive: A person undergoing chemotherapy is very *vulnerable* to infection.

waive — to forego, give up voluntarily something to which one is entitled: In cases of unusual hardship, the normal fee may be *waived*.

warp — to bend slightly throughout: The board had *warped* in the sun.

warrant — to deserve, justify: The infraction was too minor to *warrant* a formal reprimand.

wayward — perverse, capricious, willful, erratic: The *wayward* flight of the bat was difficult to follow.

weaken — to lose strength or effectiveness: His argument was *weakened* by the evidence.

weld — to join pieces of metal by compression and great heat: Steel bars were *welded* to make a frame.

welter — a wallowing, wavelike rolling, commotion: She took the *welter* of the crowd in stride, slipping down the street as quickly as she could.

wield — to use with full command or power: The soldier was skilled at *wielding* his sword.

wily — artful, cunning: He was *wily* enough to avoid detection.

wince — to shrink, as from a blow or from pain, flinch: She *winced* when the dentist touched the tooth.

winch — crank with a handle that is turned, usually for hoisting or hauling: They couldn't hoist the cargo onto the deck because the *winch* was too rusty to turn.

winnow — to examine, sift for the purpose of separating the bad from the good: Her statement was so garbled that it was impossible to *winnow* the falsehoods from the truth.

wooden — stiff; lifeless; dull; insensitive; made of wood: His *wooden* facial expression led us to believe that he was not at all interested in what we were telling him.

worthless — having no value or use: His suggestions are *worthless* because he has not studied the problems thoroughly.

wrest — to take by violence: It was impossible for the child to *wrest* the toy from the hands of the bigger boy.

xenophobia — fear and hatred of strangers or foreigners: The *xenophobia* of the candidate expressed itself in his extreme and unrealistic isolationism.

yearn — to feel longing or desire: The parents *yearned* for their recently deceased child.

zeal — ardor, fervor, enthusiasm, earnestness: She left a record for *zeal* that cannot fail to be an inspiration.

zenith — point directly overhead, highest point: The sun reaches its *zenith* at noon.

SYNONYMS

Two words are **synonyms** if they mean the same thing. In a synonym question, you must pick the word or phrase closest in meaning to the given word. This is the simplest kind of vocabulary question.

Synonym questions on civil service exams may take one of two forms. One form of synonym question offers you a key word followed by a number of choices. You are to choose the word that means the same or most nearly the same as the key word. The other form of synonym question provides a sentence in which one word is italicized. The sentence is then followed by a list of words, one of which is to be chosen as the best synonym for the indicated word. At the very least, finding the word in a sentence gives you an idea of how the word is used—as a noun, verb, or modifier. At its best, the use of the word in the sentence could give you a clue to its meaning even if you did not know the word. If your synonym questions are in the form of a sentence, begin by substituting each choice for the indicated word. You can probably eliminate some of the choices simply because they are the wrong part of speech or because they do not make sense in the sentence. Then, unless you already know the answer, try to figure out which word makes the *most* sense in the context of the sentence.

Take a look at some examples of synonyms questions presented as sentences:

1. If you have a question, please raise your hand to *summon* the test proctor.
    (A) ticket
    (B) fine
    (C) give
    (D) call

First eliminate *give*, (C). Its use in the sentence makes no sense at all. Your experience with the word *summons* may be with relation to *tickets*, (A), and *fines*, (B), but *tickets* and *fines* have nothing to do with having questions to ask while taking a test. Even if you are unfamiliar with the word *summon*, you should be able to choose (D), *call*, as its best synonym in this context.

2. The increased use of dictation machines has severely *reduced* the need for office stenographers.
    (A) enlarged
    (B) cut out
    (C) lessened
    (D) expanded

The meaning of the sentence should cause you to immediately rule out (A), *enlarged*, and (D), *expanded*, as being exactly contrary to the sentence. (B), *cut out*, implies that there is no need at all for office stenographers. If this were so, then the modifier *severely* would be unnecessary. Therefore, (C), *lessened*, is your best choice as a synonym of *reduced*.

3. We had to *terminate* the meeting because a fire broke out in the hall.
    (A) continue
    (B) postpone
    (C) end
    (D) extinguish

The correct answer is (C), *end*. Even if you don't know what *terminate* means, you can eliminate choice (A), *continue*, because it doesn't make much sense to say, "We had to continue the meeting because a fire broke out in the hall." Choice (B), *postpone*, means "to put off until another time." It makes sense in the given sentence, but it also changes the meaning of the sentence. Choice (D), *extinguish*, is similar in meaning to *terminate* but not as close as *end*. One can *extinguish* (put an end to) a fire but not a meeting.

4. The surface of the *placid* lake was as smooth as glass.
    (A) cold   (C) deep
    (B) muddy   (D) calm

Any one of the choices might be substituted for the word *placid*, and the sentence would still make sense. However, if the surface of the lake was as smooth as glass, the water would have had to be very *calm*. Thus, while a *cold*, *muddy*, or *deep* lake could have a smooth surface, it is most reasonable to assume, on the basis of the sentence, that *placid* means *calm* and that (D) is the answer.

5. The camel is sometimes called the ship of the *desert*.
    (A) abandon   (C) sandy wasteland
    (B) ice cream   (D) leave

Here the sentence is absolutely necessary to the definition of the word. Without the sentence, you would not know whether the word *desert* is the verb *de•sert'*, which means "to leave" or *to abandon*, or the noun *des'•ert*, which means a *sandy wasteland*, (C). If you are not sure of your spelling, the sentence can also spare you the confusion of *desert* with "dessert," which is the last course of a meal.

On the other hand, the sentence may be of little or no use at all in helping you to choose the synonym. The sentence may help you to determine the part of speech of the indicated word but not its meaning, as in:

6. The robbery suspect had a *sallow* complexion.
    (A) ruddy   (C) pockmarked
    (B) pale   (D) freckled

The sentence shows you a use of the word *sallow*, that it is used to describe a complexion, but it gives no clue that *sallow* means *pale*, (B). You either know the meaning of the word or you must guess.

If the given word is not part of a sentence, or if the sentence is of no use in defining the word, you must rely on other clues. Perhaps you have seen the word used but were never sure what it meant. Look carefully. Can you see any part of a word whose meaning you know?

7. remedial
    (A) reading   (C) corrective
    (B) slow   (D) special

Your association is probably "remedial reading." Be careful. *Remedial* does not mean *reading*. *Remedial* is an adjective; *reading* is the noun it modifies. Slow readers may receive remedial reading instruction in special classes. The *remedial* reading classes are intended to *correct* bad reading practices. Do you see the word *remedy* in *remedial*? You know that a *remedy* is a "cure or a correction for an ailment." If you combine all of the information you now have, you can choose (C), *corrective*, as the word that most nearly means *remedial*.

8. infamous
    (A) well known   (C) disgraceful
    (B) poor   (D) young

The first word you see when you look at *infamous* is "famous." "Famous," of course, means *well known*. Since the prefix *in* often means "not," you will eliminate (A) as the answer. A person who is not *well known* might be *poor*, but not necessarily. *Poor* should not be eliminated as a possible answer, but you should carefully consider the other choices before choosing *poor*. Since *in*, meaning "not," is a negative prefix, you should be looking for a negative word as the meaning of *infamous*. There is no choice meaning "not famous," so you must look for negative fame. *Disgrace*, (C), is a negative kind of fame. A person who behaves *disgracefully* is "well known for his bad behavior"; he is *infamous*.

Sometimes it helps to make up your own sentence using the given word and then to try substituting the answer choices in your sentence.

9. pertinent
    (A) relevant   (C) true
    (B) prudent   (D) respectful

The correct answer is (A), *relevant*. *Pertinent* means "having some bearing on or relevance to." In the sentence, "Her testimony was pertinent to the investigation," you could put *relevant* in the place of *pertinent* without changing the meaning. Choice (B), *prudent*, means "careful" or "wise." Although it sounds somewhat like *pertinent*, its meaning is different. Choice (C) may seem possible, since something that is *pertinent* should also be *true*. However, not everything that is *true* is *pertinent*. Choice (D), *respectful*, is misleading. Its opposite, "disrespectful," is a synonym for the word "impertinent." You might logically guess, then, that *respectful* is a synonym for *pertinent*. The best way to avoid a trap like this is to remember how you've heard or seen the word used. You never see *pertinent* used to mean *respectful*.

10. facsimile
    (A) summary   (C) list
    (B) exact copy   (D) artist's sketch

The correct answer is (B), *exact copy*. A *facsimile* is "a copy that looks exactly like the original, for instance, a photocopy." The word contains the root *simile*, meaning "like." Choice (C), *list*, has no connection with *facsimile*. Both a *summary*, (A), and an *artist's sketch*, (D), are in a sense copies of something else, but neither one is an *exact copy*.

11. severe
    (A) cut off   (C) serious
    (B) surprising   (D) unusual

The correct answer is (C), *serious*. *Severe* means "harsh," "extreme," or *serious*. You cannot use *serious* in every place you can use *severe*, or vice versa, but it is the best of the choices offered. For instance, in the sentence, "A severe case of the flu kept him in bed for three weeks," you could use *serious* in place of *severe* without changing the meaning. A synonym for choice (A), *cut off*, is "sever," which looks like *severe* but means something different. You can eliminate (A). Choices (B) and (D) seem more likely. Something that is *severe*—a snowstorm, for example—may also be *surprising* or *unusual*; it doesn't have to be, though. *Severe* snowstorms are not unusual in cold, wet climates. *Severe* hardship is not unusual in poverty-stricken areas.

12. feasible
    (A) simple   (C) visible
    (B) practical   (D) lenient

Something that is *feasible* is "possible or is capable of being done"; hence, (B), *practical*, is the correct answer.

How to Answer Synonym Questions

1. Read each question carefully.

2. If you know that some of the answer choices are wrong, eliminate them.

3. Use all of the clues. Try to figure out the meanings of words you do not know.

4. From the answer choices that seem possible, select the one that *most nearly* means the same as the given word, even if it is a word you yourself don't normally use. The correct answer may not be a perfect synonym, but of the choices offered it is the *closest* in meaning to the given word.

5. Make up a sentence using the given word. Then test your answer by putting it in the place of the given word in your sentence. The meaning of the sentence should be unchanged.

6. First answer the questions you know. You can come back to the others later.

7. When all else fails, make an educated guess.

Synonyms—Practice Tests

Circle the letter before each answer you choose. Answer Key on page 227.

Test 1

Directions: Select the word or phrase closest in meaning to the given word.

1. retain
   - (A) pay out
   - (B) play
   - (C) keep
   - (D) inquire

2. endorse
   - (A) sign up for
   - (B) announce support for
   - (C) lobby for
   - (D) renounce

3. intractable
   - (A) confused
   - (B) misleading
   - (C) instinctive
   - (D) unruly

4. correspondence
   - (A) letters
   - (B) files
   - (C) testimony
   - (D) response

5. obliterate
   - (A) praise
   - (B) doubt
   - (C) erase
   - (D) reprove

6. legitimate
   - (A) democratic
   - (B) legal
   - (C) genealogical
   - (D) underworld

7. deduct
   - (A) conceal
   - (B) withstand
   - (C) subtract
   - (D) terminate

8. mutilate
   - (A) paint
   - (B) damage
   - (C) alter
   - (D) rebel

9. egress
   - (A) extreme
   - (B) extra supply
   - (C) exit
   - (D) high price

10. horizontal
    - (A) marginal
    - (B) in a circle
    - (C) left and right
    - (D) up and down

11. controversy
    - (A) publicity
    - (B) debate
    - (C) revolution
    - (D) revocation

12. preempt
    - (A) steal
    - (B) empty
    - (C) preview
    - (D) appropriate

13. category
    - (A) class
    - (B) adherence
    - (C) simplicity
    - (D) cataract

14. apathy
    - (A) sorrow
    - (B) indifference
    - (C) aptness
    - (D) sickness

15. tentative
    - (A) persistent
    - (B) permanent
    - (C) thoughtful
    - (D) provisional

16. per capita
    - (A) for an entire population
    - (B) by income
    - (C) for each person
    - (D) for every adult

17. deficient
    - (A) sufficient
    - (B) outstanding
    - (C) inadequate
    - (D) bizarre

18. inspect
    - (A) disregard
    - (B) look at
    - (C) annoy
    - (D) criticize

19. optional
    - (A) not required
    - (B) infrequent
    - (C) choosy
    - (D) for sale

20. implied
    (A) acknowledged   (C) predicted
    (B) stated   (D) hinted

21. presumably
    (A) positively   (C) recklessly
    (B) helplessly   (D) supposedly

22. textile
    (A) linen   (C) page
    (B) cloth   (D) garment

23. fiscal
    (A) critical   (C) personal
    (B) basic   (D) financial

24. stringent
    (A) demanding   (C) flexible
    (B) loud   (D) clear

25. proceed
    (A) go forward   (C) refrain
    (B) parade   (D) resume

Test 2

Directions: Select the word or phrase closest in meaning to the given word.

1. brochure
   (A) ornament   (C) breakage
   (B) flowery statement (D) pamphlet

2. permeable
   (A) penetrable   (C) unending
   (B) durable   (D) allowable

3. limit
   (A) budget   (C) point
   (B) sky   (D) boundary

4. scrupulous
   (A) conscientious   (C) intricate
   (B) unprincipled   (D) neurotic

5. stalemate
   (A) pillar   (C) maneuver
   (B) deadlock   (D) work slowdown

6. competent
   (A) inept   (C) capable
   (B) informed   (D) caring

7. somatic
   (A) painful   (C) indefinite
   (B) drowsy   (D) physical

8. obstacle
   (A) imprisonment   (C) retaining wall
   (B) hindrance   (D) leap

9. redundant
   (A) concise   (C) superfluous
   (B) reappearing   (D) lying down

10. supplant
    (A) prune   (C) uproot
    (B) conquer   (D) replace

11. haphazard
    (A) devious   (C) aberrant
    (B) without order   (D) risky

12. commensurate
    (A) identical   (C) proportionate
    (B) of the same age (D) measurable

13. accelerate
    (A) drive fast   (C) decline rapidly
    (B) reroute   (D) speed up

14. purchased
    (A) charged   (C) ordered
    (B) bought   (D) supplied

15. zenith
    (A) depths
    (B) astronomical system
    (C) peak
    (D) solar system

16. succor
    (A) assistance   (C) vitality
    (B) Mayday   (D) distress

17. restrict
    (A) limit   (C) watch
    (B) replace   (D) record

18. strident
    (A) booming
    (B) austere
    (C) swaggering
    (D) shrill

19. dispatch
    (A) omit mention of
    (B) send out on an errand
    (C) hurry up
    (D) do without

20. inventory
    (A) catalog of possessions
    (B) statement of purposes
    (C) patent office
    (D) back order

21. assiduous
    (A) untrained
    (B) unrestricted
    (C) diligent
    (D) negligent

22. portable
    (A) drinkable
    (B) convenient
    (C) having wheels
    (D) able to be carried

23. annual
    (A) yearly
    (B) seasonal
    (C) occasional
    (D) infrequent

24. endeavored
    (A) managed
    (B) expected
    (C) attempted
    (D) promised

25. acumen
    (A) caution
    (B) strictness
    (C) inability
    (D) keenness

Test 3

Directions: Select the word or phrase closest in meaning to the given word.

1. excess
   (A) surplus
   (B) exit
   (C) inflation
   (D) luxury

2. verbose
   (A) vague
   (B) brief
   (C) wordy
   (D) verbal

3. collusion
   (A) decision
   (B) connivance
   (C) insinuation
   (D) conflict

4. subversive
   (A) secret
   (B) foreign
   (C) evasive
   (D) destructive

5. vacillating
   (A) changeable
   (B) equalizing
   (C) decisive
   (D) progressing

6. coincide
   (A) agree
   (B) disregard
   (C) collect
   (D) conflict

7. petty
   (A) lengthy
   (B) communal
   (C) small
   (D) miscellaneous

8. concede
   (A) confess
   (B) ebb
   (C) enact
   (D) give in

9. intrepid
   (A) willing
   (B) fanciful
   (C) cowardly
   (D) fearless

10. prolonged
    (A) refined
    (B) drawn out
    (C) tiresome
    (D) ardent

11. transcribe
    (A) write a copy
    (B) invent
    (C) interpret
    (D) dictate

12. negotiate
    (A) suffer
    (B) think
    (C) speak
    (D) bargain

13. credible
    (A) believable
    (B) correct
    (C) intelligent
    (D) gullible

14. objective
    (A) strict
    (B) courteous
    (C) fair
    (D) pleasant

15. examine
    (A) file
    (B) collect
    (C) distribute
    (D) inspect

16. quantity
    (A) flow
    (B) type
    (C) amount
    (D) difficulty

17. expedite
    (A) obstruct
    (B) advise
    (C) accelerate
    (D) demolish

18. coordinator
    (A) enumerator
    (B) organizer
    (C) spokesman
    (D) advertiser

19. reprisal
    (A) retaliation
    (B) warning
    (C) advantage
    (D) denial

20. relevant
    (A) controversial
    (B) recent
    (C) applicable
    (D) impressive

21. sterile
    (A) antique
    (B) germ-free
    (C) unclean
    (D) perishable

22. imperative
    (A) impending
    (B) impossible
    (C) compulsory
    (D) logical

23. assist
    (A) malign
    (B) incur
    (C) advise
    (D) aid

24. maximum
    (A) greatest
    (B) limited
    (C) oldest
    (D) smallest

25. construe
    (A) violate
    (B) contradict
    (C) question
    (D) interpret

Test 4

Directions: Select the word or phrase closest in meaning to the given word.

1. customary
   (A) methodical
   (B) usual
   (C) curious
   (D) procedural

2. minute
   (A) quick
   (B) protracted
   (C) tiny
   (D) shrunken

3. preclude
   (A) arise from
   (B) account for
   (C) prevent
   (D) define

4. abundant
   (A) plentiful
   (B) accessible
   (C) concentrated
   (D) scattered

5. invoice
   (A) speech
   (B) bill
   (C) offense
   (D) liability

6. recreation
   (A) sport
   (B) recess
   (C) diversion
   (D) escapade

7. futile
   (A) medieval
   (B) unfortunate
   (C) wasteful
   (D) useless

8. expenditure
   (A) exhaustion
   (B) budgeting
   (C) conservation
   (D) spending

9. stamina
   (A) part of a flower
   (B) incentive
   (C) staying power
   (D) reservation

10. advantageous
    (A) profitable
    (B) winning
    (C) enterprising
    (D) shrewd

11. merchant
    (A) producer
    (B) executive
    (C) advertiser
    (D) storekeeper

12. observable
    (A) noticeable
    (B) understandable
    (C) keen
    (D) blatant

13. parole
    (A) sentence
    (B) conditional release
    (C) good behavior
    (D) granting of privileges

14. reveal
    (A) describe fully
    (B) make known
    (C) guess at
    (D) question seriously

15. fraud
    (A) guilt   (C) cheating
    (B) criminality   (D) disguise

16. asserted
    (A) decided   (C) contradicted
    (B) agreed   (D) declared

17. durable
    (A) thick   (C) lasting
    (B) waterproof   (D) costly

18. vindictive
    (A) revengeful   (C) aggressive
    (B) boastful   (D) impolite

19. bourgeois
    (A) middle-class   (C) decadent
    (B) affluent   (D) prevalent

20. absurd
    (A) careless   (C) impulsive
    (B) foolish   (D) regrettable

21. hospitable
    (A) careful   (C) relaxed
    (B) incurable   (D) welcoming

22. graft
    (A) undercover activity
    (B) political influence
    (C) illegal payment for political favor
    (D) giving jobs to relatives

23. emendations
    (A) illustrations
    (B) new problems
    (C) unexplained actions
    (D) corrections

24. punctuality
    (A) partiality   (C) precision
    (B) being on time (D) being delayed

25. fatigue
    (A) illness   (C) weariness
    (B) worry   (D) indolence

Test 5

Directions: Select the word or phrase closest in meaning to the given word.

1. affluence
    (A) persuasion   (C) inspiration
    (B) power   (D) wealth

2. related
    (A) subordinated   (C) detached
    (B) connected   (D) finished

3. designate
    (A) name   (C) accuse
    (B) illustrate   (D) change

4. chagrin
    (A) enjoyment   (C) smirk
    (B) disappointment (D) disgust

5. anomalous
    (A) out of place   (C) similar
    (B) vague   (D) unknown

6. altitude
    (A) outlook   (C) distance
    (B) height   (D) magnitude

7. precise
    (A) short   (C) exact
    (B) picky   (D) trivial

8. ignominy
    (A) fame   (C) bad luck
    (B) disgrace   (D) despair

9. normal
    (A) comfortable   (C) usual
    (B) right   (D) necessary

10. increase
    (A) decline   (C) quantity
    (B) plenty   (D) growth

11. collate
    (A) destroy   (C) assemble
    (B) separate   (D) copy

12. authorize
    (A) permit   (C) train
    (B) write   (D) constrain

13. platitude
    (A) data
    (B) length
    (C) theory
    (D) trite remark

14. entrenched
    (A) firmly established
    (B) at war
    (C) eternal
    (D) earthy

15. constant
    (A) absent
    (B) unchanging
    (C) perpetrated
    (D) tiring

16. misnomer
    (A) wrong address
    (B) mistaken identity
    (C) crime
    (D) wrong name

17. monitor
    (A) preserve
    (B) warn
    (C) keep ahead of
    (D) keep track of

18. sufficient
    (A) interesting
    (B) enough
    (C) excessive
    (D) accepting

19. prior
    (A) previous
    (B) official
    (C) conflicting
    (D) important

20. anticipated
    (A) required
    (B) revised
    (C) expected
    (D) extraordinary

21. substitute
    (A) excuse
    (B) replacement
    (C) arrangement
    (D) pretense

22. rapidity
    (A) idleness
    (B) delay
    (C) speed
    (D) efficiency

23. verify
    (A) control
    (B) line up
    (C) confirm
    (D) decide

24. attain
    (A) mar
    (B) exhaust
    (C) reach
    (D) attack

25. requisition
    (A) payment
    (B) written order
    (C) formality
    (D) cancellation

Test 6

Directions: Select the word or phrase closest in meaning to the given word.

1. fundamental
   (A) serious
   (B) emphasized
   (C) essential
   (D) difficult

2. experiment
   (A) refinement
   (B) test
   (C) patent
   (D) plan

3. sporadic
   (A) occasional
   (B) restless
   (C) unpredictable
   (D) seeded

4. forwarded
   (A) returned to sender
   (B) detained
   (C) sent on
   (D) cancelled

5. larceny
   (A) homicide
   (B) levity
   (C) theft
   (D) corruption

6. bibliography
   (A) list of books on a subject
   (B) spelling dictionary
   (C) geographical index
   (D) thesaurus

7. disclose
   (A) lock up
   (B) uncover
   (C) unhinge
   (D) set free

8. chronological
   (A) in time order
   (B) recent
   (C) schematic
   (D) in order of importance

9. participate
   (A) supervise
   (B) depend on
   (C) divide up
   (D) join in

10. vacant
    (A) empty
    (B) preoccupied
    (C) quiet
    (D) available

11. remit
    (A) confess
    (B) send a bill
    (C) pay
    (D) delete

12. sundry
   - (A) valuable
   - (B) specific
   - (C) miscellaneous
   - (D) general

13. irritating
   - (A) unnerving
   - (B) annoying
   - (C) unbearable
   - (D) nervous

14. secure
   - (A) convenient
   - (B) nearby
   - (C) safe
   - (D) secret

15. document
   - (A) outline
   - (B) agreement
   - (C) blueprint
   - (D) record

16. reprehensible
   - (A) censurable
   - (B) above reproach
   - (C) dim-witted
   - (D) without precedent

17. assemble
   - (A) mark
   - (B) put in order
   - (C) bring together
   - (D) locate

18. compelled
   - (A) tempted
   - (B) forced
   - (C) persuaded
   - (D) commanded

19. affiliation
   - (A) connection
   - (B) juncture
   - (C) parental guidance
   - (D) sincere affection

20. privileged
   - (A) covert
   - (B) indoctrinated
   - (C) autocratic
   - (D) honored

21. deliberate
   - (A) clever
   - (B) considerate
   - (C) intentional
   - (D) daring

22. slate
   - (A) schedule
   - (B) postpone
   - (C) diagram
   - (D) quench

23. disclaim
   - (A) cry out
   - (B) query
   - (C) argue
   - (D) deny

24. principal
   - (A) main
   - (B) only
   - (C) authoritarian
   - (D) current

25. vandalism
   - (A) boyish prank
   - (B) willful destruction
   - (C) petty thievery
   - (D) juvenile delinquency

Test 7

Directions: Select the word or phrase closest in meaning to the given word.

1. supervise
   - (A) acquire
   - (B) oppress
   - (C) oversee
   - (D) restrain

2. mitigate
   - (A) lessen
   - (B) incite
   - (C) measure
   - (D) prosecute

3. logical
   - (A) reasoned
   - (B) calm
   - (C) fixed
   - (D) cold

4. peculiar
   - (A) sensitive
   - (B) special
   - (C) arbitrary
   - (D) indefensible

5. effect
   - (A) raise
   - (B) put on
   - (C) bring about
   - (D) pass

6. utilize
   - (A) offer
   - (B) employ
   - (C) ponder
   - (D) enjoy

7. analogous
   - (A) similar
   - (B) hidden
   - (C) metallic
   - (D) unreasonable

8. uniformity
   - (A) costume
   - (B) sameness
   - (C) custom
   - (D) boredom

9. legible
   - (A) printed
   - (B) allowed
   - (C) typed
   - (D) readable

10. augment
    (A) adopt   (C) modify
    (B) increase   (D) predict

11. complex
    (A) group of buildings   (C) neighborhood
    (B) tower   (D) corporation

12. request
    (A) tell   (C) suspect
    (B) ask   (D) complain

13. courteous
    (A) fast   (C) impersonal
    (B) polite   (D) royal

14. tenacity
    (A) firmness   (C) sagacity
    (B) temerity   (D) discouragement

15. insignificant
    (A) useless   (C) low
    (B) unrewarding (D) unimportant

16. execute
    (A) resign   (C) carry out
    (B) affect   (D) harm

17. unique
    (A) sole   (C) certain
    (B) odd   (D) valuable

18. utensil
    (A) machine   (C) tool
    (B) fork   (D) object

19. outcome
    (A) result   (C) premise
    (B) aim   (D) statistic

20. dogmatic
    (A) manual   (C) canine
    (B) doctrinaire (D) unprincipled

21. basic
    (A) fundamental (C) simplistic
    (B) outstanding (D) strange

22. plenary
    (A) progressive (C) temporary
    (B) unusual   (D) full

23. insert
    (A) put in   (C) fold up
    (B) copy   (D) mail

24. accumulate
    (A) get used to (C) pile up
    (B) shut off   (D) care for

25. median
    (A) midpoint in a series
    (B) numerical score
    (C) bar graph
    (D) first item in a series

Test 8

Directions: Select the word or phrase closest in meaning to the given word.

1. discard
    (A) ignore   (C) refuse
    (B) throw away   (D) fire

2. capitulate
    (A) repeat   (C) finance
    (B) surrender   (D) retreat

3. extenuating
    (A) excusing   (C) incriminating
    (B) opposing   (D) distressing

4. degraded
    (A) assorted   (C) receded
    (B) declassified (D) debased

5. dynamic
    (A) noisy   (C) forceful
    (B) static   (D) magnetic

6. criterion
    (A) charge   (C) standard
    (B) theater   (D) requirement

7. intercept
    (A) cut off   (C) ask
    (B) speak   (D) break away

8. memorandum
    (A) formal letter
    (B) command
    (C) note
    (D) minutes of a meeting

9. instruct
    (A) teach
    (B) work
    (C) build
    (D) study

10. assure
    (A) normalize
    (B) insist
    (C) persist
    (D) make certain

11. affidavit
    (A) arraignment
    (B) written statement made under oath
    (C) enforceable promise
    (D) invoice

12. determine
    (A) convince
    (B) find fault with
    (C) find out
    (D) bring about

13. xenophobic
    (A) susceptible to disease
    (B) fearing strangers
    (C) opposed to gambling
    (D) fearing dogs

14. extensive
    (A) thorough
    (B) arbitrary
    (C) superficial
    (D) leisurely

15. unreliable
    (A) late
    (B) untrustworthy
    (C) independent
    (D) temporary

16. compensation
    (A) remuneration
    (B) fulfillment
    (C) appreciation
    (D) promotion

17. common
    (A) occasional
    (B) frequent
    (C) rare
    (D) debased

18. affected
    (A) influenced
    (B) arrogant
    (C) reduced
    (D) caused

19. employ
    (A) restore
    (B) use
    (C) use up
    (D) plan

20. exempt
    (A) excused
    (B) withdrawn
    (C) selected
    (D) honored

21. liaison
    (A) connection
    (B) lie
    (C) opportunity
    (D) officer

22. recur
    (A) get well
    (B) run away
    (C) happen again
    (D) give back

23. engender
    (A) make inanimate
    (B) imperil
    (C) manage skillfully
    (D) produce

24. pamphlet
    (A) novel
    (B) prospectus
    (C) advertisement
    (D) booklet

25. corroborate
    (A) connect
    (B) confirm
    (C) cooperate
    (D) rust

Test 9

Directions: Select the word or phrase closest in meaning to the given word.

1. sanctioned
    (A) standardized
    (B) carefully planned
    (C) officially approved
    (D) publicly announced

2. proper
    (A) many
    (B) appropriate
    (C) straight
    (D) obsolete

3. reprimand
    (A) engraved invitation
    (B) investigation
    (C) formal rebuke
    (D) revocation of privileges

4. demonstrable
    (A) describable
    (B) able to be shown
    (C) under control
    (D) able to be taught

5. insure
   (A) determine
   (B) pay for
   (C) retain
   (D) assure against loss

6. alarm
   (A) frighten   (C) endanger
   (B) confuse   (D) insult

7. obligatory
   (A) optional   (C) inconsequential
   (B) advisable   (D) compulsory

8. tangential
   (A) on target
   (B) merely approximate
   (C) off the main subject
   (D) geometrical

9. legacy
   (A) game of chance   (C) inheritance
   (B) tax-free gift   (D) benefit

10. diffidence
    (A) disgust   (C) shyness
    (B) confidence   (D) lack of harmony

11. mingle
    (A) visit   (C) mix
    (B) dance   (D) sing

12. explicitly
    (A) exclusively   (C) casually
    (B) specifically (D) intelligibly

13. infraction
    (A) violation   (C) minor fracture
    (B) uneven division (D) fraying

14. urgent
    (A) desirous   (C) startling
    (B) sudden   (D) pressing

15. residue
    (A) remainder   (C) refund
    (B) tenant   (D) delta

16. distinct
    (A) impressive   (C) regular
    (B) loud   (D) clear

17. procure
    (A) legalize   (C) obtain
    (B) serve   (D) possess

18. neglectful
    (A) unworthy   (C) unfit
    (B) inattentive   (D) abandoned

19. cull
    (A) separate out (C) search out
    (B) deceive   (D) think over

20. plausible
    (A) hesitant
    (B) seeming reasonable
    (C) evoking applause
    (D) tranquil

21. recapitulate
    (A) surrender again
    (B) seize again
    (C) be insubordinate
    (D) summarize

22. deceive
    (A) trick   (C) mitigate
    (B) undermine   (D) infuriate

23. exorbitant
    (A) alien   (C) excessive
    (B) eccentric   (D) modest

24. origin
    (A) direction   (C) beginning
    (B) model   (D) end

25. exhibit
    (A) suppress   (C) publicize
    (B) promote   (D) display

Test 10

Directions: Select the word or phrase closest in meaning to the given word.

1. culpable
   (A) dangerous   (C) blameworthy
   (B) soft   (D) easily perceived

2. status
   (A) departure   (C) stature
   (B) position   (D) interference

3. rehabilitate
   (A) restore
   (B) reiterate
   (C) realize
   (D) parole

4. eligible
   (A) incompetent
   (B) unreadable
   (C) suitable
   (D) lawless

5. annul
   (A) make a commitment
   (B) recall
   (C) make void
   (D) subscribe

6. contract
   (A) formal agreement
   (B) license
   (C) commission
   (D) influence

7. partially
   (A) seemingly
   (B) always
   (C) particularly
   (D) not entirely

8. abridged
   (A) linked
   (B) expanded
   (C) shortened
   (D) alphabetized

9. unethical
   (A) morose
   (B) dishonest
   (C) fine
   (D) bent

10. considerable
   (A) large
   (B) frequent
   (C) common
   (D) potential

11. assume
   (A) argue
   (B) display
   (C) suppose
   (D) hope

12. feasible
   (A) impossible
   (B) payable
   (C) guilty
   (D) practicable

13. deplete
   (A) omit
   (B) exhaust
   (C) deposit
   (D) replenish

14. delegate
   (A) fire an employee
   (B) give to a subordinate
   (C) notify
   (D) elect to an office

15. deny
   (A) injure
   (B) hinder
   (C) blockade
   (D) refuse

16. legislature
   (A) court of appeals
   (B) high office
   (C) lawmaking body
   (D) session

17. complete
   (A) mediocre
   (B) fancy
   (C) entire
   (D) retired

18. forum
   (A) majority rule
   (B) place for public discussion
   (C) means of escape
   (D) consensus

19. motivation
   (A) intervention
   (B) reservation
   (C) argument
   (D) reason

20. arbitrary
   (A) responsible
   (B) despotic
   (C) conciliatory
   (D) argumentative

21. auspicious
   (A) questionable
   (B) well-known
   (C) free
   (D) favorable

22. recipient
   (A) receiver
   (B) donor
   (C) carrier
   (D) borrower

23. premature
   (A) too easy
   (B) too small
   (C) too early
   (D) too late

24. probability
   (A) guess
   (B) theory
   (C) preference
   (D) likelihood

25. adequate
   (A) long
   (B) acceptable
   (C) required
   (D) equal

Test 11

Directions: Select the word or phrase closest in meaning to the *italicized* word in the sentence.

1. To say that the work is *tedious* means, most nearly, that it is
   (A) technical   (C) tiresome
   (B) interesting   (D) confidential

2. An *innocuous* statement is one which is
   (A) forceful   (C) offensive
   (B) harmless   (D) brief

3. To say that the order was *rescinded* means, most nearly, that the order was
   (A) revised   (C) misinterpreted
   (B) canceled   (D) confirmed

4. A *recurring* problem is one that
   (A) replaces a problem that existed previously
   (B) is unexpected
   (C) has long been overlooked
   (D) comes up from time to time

5. A *homogeneous* group of persons is characterized by its
   (A) similarity   (C) discontent
   (B) teamwork   (D) differences

6. Courage is a trait difficult to *instill*.
   (A) measure exactly
   (B) impart gradually
   (C) predict accurately
   (D) restrain effectively

7. A *conscientious* person is one who
   (A) feels obligated to do what he believes right
   (B) rarely makes errors
   (C) frequently makes suggestions for procedural improvements
   (D) has good personal relationships with others

8. There was much *diversity* in the suggestions submitted.
   (A) similarity   (C) triviality
   (B) value   (D) variety

9. The survey was concerned with the problem of *indigence*.
   (A) poverty   (C) intolerance
   (B) corruption   (D) morale

10. He was surprised at the *temerity* of the new employee.
    (A) shyness   (C) rashness
    (B) enthusiasm   (D) self-control

11. A *vindictive* person is one who is
    (A) prejudiced   (C) petty
    (B) unpopular   (D) revengeful

12. The vehicle was left *intact*.
    (A) a total loss
    (B) unattended
    (C) where it could be noticed
    (D) undamaged

13. The *remuneration* was unsatisfactory
    (A) payment   (C) explanation
    (B) summary   (D) estimate

14. The faculty of the Anthropology Department agreed that the departmental program was *deficient*.
    (A) excellent   (C) demanding
    (B) inadequate   (D) sufficient

15. His friends had a *detrimental* influence on him.
    (A) favorable   (C) harmful
    (B) lasting   (D) short-lived

16. When you study from an ARCO book, you get *accurate* information.
    (A) correct   (C) ample
    (B) good   (D) much

17. The speaker was urged to *amplify* his remarks.
    (A) soften   (C) enlarge upon
    (B) simplify   (D) repeat

18. An eighteen-year old is legally *competent* to enter into a contract.
    (A) expert   (C) rival
    (B) ineligible   (D) able

19. There is a specified punishment for each *infraction* of the rules.
    (A) violation   (C) interpretation
    (B) use   (D) part

20. The aims of the students and the aims of the faculty often *coincide*.
    (A) agree
    (B) are ignored
    (C) conflict
    (D) are misinterpreted

21. The secretary of the Sociology Department was responsible for setting up an index of *relevant* magazine articles.
    (A) applicable   (C) miscellaneous
    (B) controversial   (D) recent

22. One of the secretary's duties consisted of sorting and filing *facsimiles* of student term papers.
    (A) bibliographical listings
    (B) exact copies
    (C) summaries
    (D) supporting documentation

23. Your *numerical* rating is based upon your test score, education, experience, and veteran's status.
    (A) orderly   (C) employment
    (B) actual   (D) number

24. Please consult your office *manual* to learn the proper operation of our copying machine.
    (A) labor   (C) typewriter
    (B) handbook   (D) handle

25. No computational *device* may be used during the exam.
    (A) calculator   (C) mathematical
    (B) adding   (D) machine

Test 12

Directions: Select the word or phrase closest in meaning to the *italicized* word in the sentence.

1. The secretary *complied with* the boss's wishes.
   (A) objected to   (C) followed
   (B) agreed with   (D) disobeyed

2. A passing grade on the special exam may *exempt* the applicant from the experience requirements for that job.
   (A) excuse   (C) subject
   (B) prohibit   (D) specify

3. The civil service dictation test differs from the *conventional* dictation test.
   (A) agreeable   (C) large-scale
   (B) public   (D) usual

4. Improper office lighting may cause *fatigue*.
   (A) uniform   (C) tiredness
   (B) eyestrain   (D) overweight

5. The professor explained that the report was too *verbose* to be submitted.
   (A) brief   (C) general
   (B) specific   (D) wordy

6. The faculty meeting *preempted* the conference room in the dean's office.
   (A) appropriated   (C) filled
   (B) emptied   (D) reserved

7. The professor's credentials became a subject of *controversy*.
   (A) annoyance   (C) envy
   (B) debate   (D) review

8. The professor developed a different central theme during every *semester*.
   (A) bi-annual period of instruction
   (B) orientation period
   (C) slide demonstration
   (D) weekly lecture series

9. The College offered a variety of *seminars* to upperclassmen.
   (A) reading courses with no formal supervision
   (B) study courses for small groups of students engaged in research under a teacher
   (C) guidance conferences with grade advisors
   (D) work experiences in different occupational fields

10. The person who is *diplomatic* in his relations with others is
    (A) well dressed
    (B) very tactful
    (C) somewhat domineering
    (D) deceitful and tricky

11. Action at this time would be *inopportune*.
    (A) untimely
    (B) premeditated
    (C) sporadic
    (D) commendable

12. An incentive that is *potent* is
    (A) impossible
    (B) highly effective
    (C) not immediately practicable
    (D) a remote possibility

13. He presented a *controversial* plan.
    (A) subject to debate
    (B) unreasonable
    (C) complex
    (D) comparable

14. He sent the *irate* employee to the personnel manager.
    (A) irresponsible
    (B) untidy
    (C) insubordinate
    (D) angry

15. The secretary was asked to type a rough draft of a college course *syllabus*.
    (A) directory of departments and services
    (B) examination schedule
    (C) outline of a course of study
    (D) rules and regulations

16. Information may be *obtained* at your public library.
    (A) learned
    (B) read
    (C) gotten
    (D) distributed

17. Lateness will *bar* you from taking the test.
    (A) stick
    (B) prevent
    (C) harm
    (D) assist

18. An *ambiguous* statement is one which is
    (A) forceful and convincing
    (B) capable of being understood in more than one sense
    (C) based upon good judgment and sound reasoning processes
    (D) uninteresting and too lengthy

19. A solvent will assist the paint in the *penetration* of porous surfaces.
    (A) covering
    (B) protection
    (C) cleaning
    (D) entering

20. Iron oxide makes a very *durable* paint.
    (A) cheap
    (B) long-lasting
    (C) easily applied
    (D) quick-drying

21. For precision work, center punches are ground to a fine *tapered* point.
    (A) conical
    (B) straight
    (C) accurate
    (D) smooth

22. There are *limitations* to the drilling of metals by hand power.
    (A) advantages
    (B) restrictions
    (C) difficulties
    (D) benefits

23. A painter should make sure there is *sufficient* paint to do the job.
    (A) enough
    (B) the right kind of
    (C) the proper color of
    (D) mixed

24. The investigation unit began an *extensive* search for the information.
    (A) complicated
    (B) superficial
    (C) thorough
    (D) leisurely

25. They were not present at the *inception* of the program.
    (A) beginning
    (B) discussion
    (C) conclusion
    (D) rejection

Synonyms—Answer Key for Practice Tests

TEST 1

1.	C	6.	B	11.	B	16.	C	21.	D
2.	B	7.	C	12.	D	17.	C	22.	B
3.	D	8.	B	13.	A	18.	B	23.	D
4.	A	9.	C	14.	B	19.	A	24.	A
5.	C	10.	C	15.	D	20.	D	25.	A

TEST 2

1.	D	6.	C	11.	B	16.	A	21.	C
2.	A	7.	D	12.	C	17.	A	22.	D
3.	D	8.	B	13.	D	18.	D	23.	A
4.	A	9.	C	14.	B	19.	B	24.	C
5.	B	10.	D	15.	C	20.	A	25.	D

TEST 3

1.	A	6.	A	11.	A	16.	C	21.	B
2.	C	7.	C	12.	D	17.	C	22.	C
3.	B	8.	D	13.	A	18.	B	23.	D
4.	D	9.	D	14.	C	19.	A	24.	A
5.	A	10.	B	15.	D	20.	C	25.	D

TEST 4

1.	B	6.	C	11.	D	16.	D	21.	D
2.	C	7.	D	12.	A	17.	C	22.	C
3.	C	8.	D	13.	B	18.	A	23.	D
4.	A	9.	C	14.	B	19.	A	24.	B
5.	B	10.	A	15.	C	20.	B	25.	C

TEST 5

1.	D	6.	B	11.	C	16.	D	21.	B
2.	B	7.	C	12.	A	17.	D	22.	C
3.	A	8.	B	13.	D	18.	B	23.	C
4.	B	9.	C	14.	A	19.	A	24.	C
5.	A	10.	D	15.	B	20.	C	25.	B

TEST 6

1.	C	6.	A	11.	C	16.	A	21.	C
2.	B	7.	B	12.	C	17.	C	22.	A
3.	A	8.	A	13.	B	18.	B	23.	D
4.	C	9.	D	14.	C	19.	A	24.	A
5.	C	10.	A	15.	D	20.	D	25.	B

TEST 7

1. C	6. B	11. A	16. C	21. A
2. A	7. A	12. B	17. A	22. D
3. A	8. B	13. B	18. C	23. A
4. B	9. D	14. A	19. A	24. C
5. C	10. B	15. D	20. B	25. A

TEST 8

1. B	6. C	11. B	16. A	21. A
2. B	7. A	12. C	17. B	22. C
3. A	8. C	13. B	18. A	23. D
4. D	9. A	14. A	19. B	24. D
5. C	10. D	15. B	20. A	25. B

TEST 9

1. C	6. A	11. C	16. D	21. D
2. B	7. D	12. B	17. C	22. A
3. C	8. C	13. A	18. B	23. C
4. B	9. C	14. D	19. A	24. C
5. D	10. C	15. A	20. B	25. D

TEST 10

1. C	6. A	11. C	16. C	21. D
2. B	7. D	12. D	17. C	22. A
3. A	8. C	13. B	18. B	23. C
4. C	9. B	14. B	19. D	24. D
5. C	10. A	15. D	20. B	25. B

TEST 11

1. C	6. B	11. D	16. A	21. A
2. B	7. A	12. D	17. C	22. B
3. B	8. D	13. A	18. D	23. D
4. D	9. A	14. B	19. A	24. B
5. A	10. C	15. C	20. A	25. D

TEST 12

1. C	6. A	11. A	16. C	21. A
2. A	7. B	12. B	17. B	22. B
3. D	8. A	13. A	18. B	23. A
4. C	9. B	14. D	19. D	24. C
5. D	10. B	15. C	20. B	25. A

ANTONYMS

An antonym is a word with the opposite meaning to another word; in fact, antonym questions on civil service tests are sometimes called "opposites." Antonym questions are similar to synonym questions in that they test your understanding of words. However, antonym questions challenge you to demonstrate your mental flexibility as well as your verbal skills. At first, the task appears to be a simple one: Define the key word and pick its opposite. However, on closer examination of the questions, the task may be more complicated. When there is no true opposite, you must choose the word or phrase that is *most nearly opposite*. When there appears to be two or more opposites, you must choose the *best* opposite. You must guard against choosing an associated word or phrase that is different in meaning but is not a true opposite. After struggling to define a word, you must take care then not to choose the word or phrase that is most similar in meaning.

How to Answer Antonym Questions

There is a system to answering antonym questions. First read each word and its possible opposites very carefully. Then follow a step-by-step approach to find your answer. This approach may lead you through a number of different procedures depending on your initial reaction to the word.

Possibility #1. You know the meaning of the key word. You read all of the choices, and the *best* opposite is clear to you. Mark your answer quickly and go on. You will need more time for questions that are more difficult for you.

Possibility #2. You know the meaning of the key word, but no answer choice seems correct.

- Perhaps you misread the word. Does it have a close look-alike with a different meaning? For example, did you read *revelation* for *revaluation* or *compliment* for *complement*?

- Perhaps you read the word correctly but accented the wrong syllable. Some words have alternative pronunciations with vastly different meanings. Remember *de•sert'* and *des'•ert*.

- Perhaps you are dealing with a single word that can be used as two different parts of speech and has two entirely unrelated meanings in those two roles. A *moor* (noun) is a boggy wasteland. To *moor* (verb) is to tie up a vessel. The proper noun *Moor* refers to the Moslem conquerors of Spain.

- Perhaps the word can appear as many parts of speech with numerous meanings and shades of meaning within each of these. *Fancy* (noun) can mean "inclination, love, notion, whim, taste, judgment, imagination." *Fancy* (verb) can mean "to like, to imagine, to think." *Fancy* (adjective) can mean "whimsical, ornamental, extravagant.'' Your task is to choose from among the four choices one word or phrase that is the opposite of *one* of these meanings of the word *fancy*.

Possibility #3. You do not know the meaning of the word, but it appears to contain prefix, suffix, or root clues. Let us suppose that the question looks like this:

inaudible
(A) invisible
(B) bright
(C) loud
(D) clear

You do not know the meaning of the word *inaudible,* but you recognize a part of *audio* and you know that the audio of your TV is the sound. You also know that the prefix *in* generally means *not.* You might also see *able* in *ible* and thereby reconstruct *not soundable* or *not heard.*

BEWARE! This is the point at which your reasoning can easily lead you astray. If you associate the word with your TV, you may think, ''The opposite of *not heard* is *not seen* or *invisible,* (A).'' Wrong. These are not true opposites. Or you might associate *not heard* with *not seen* and choose the opposite of *not seen, bright* (B). Wrong again. Or you might think of *inaudible* as *hard to hear* and choose (D), *clear. Clear* would not be a bad answer, but (C), *loud,* is better and is indeed the *best* answer. The *best* opposite of *inaudible* is *loud.*

Possibility #4. You do not know the meaning of the word and can see no clues, but you have a feeling that the word has some specific connotation—sinister, gloomy, or positive. Play your hunch. Choose a word with the opposite connotation.

Possibility #5. You are stumped. Make an educated guess and move on. Do not waste time on a question for which you cannot figure out the answer, but do mark the question in the question book so that you can return and try again if you have time after your first run through the exam or section.

Sample Questions

Look now at some examples.

1. black
   (A) gray
   (B) bright
   (C) white
   (D) red

White, (C), is the accepted opposite of *black. Red* is only occasionally an opposite of *black,* as in checkers. *Gray* is a gradation between black and white. *Bright* is a quality of color rather than a color.

2. wet
   (A) humid
   (B) damp
   (C) acrid
   (D) dry

Dry, (D), is the opposite of *wet. Humid* and *damp* are degrees of wetness and are not true opposites. *Arid* is a synonym for *dry,* but word choice (C) is *acrid,* which means ''bitter'' (read carefully).

3. tepid
   - (A) milky
   - (B) clean
   - (C) exuberant
   - (D) firm

Exuberant, (C), means "enthusiastic" or "extreme in degree." It is the opposite of *tepid*, which means "unenthusiastic" or "lukewarm." You probably are familiar with the word *tepid* as it is applied to bathwater or milk. Can you see how "lukewarm" can be the opposite of *exuberant*?

4. hot
   - (A) warm
   - (B) cool
   - (C) cold
   - (D) temperate

The true and absolute opposite of *hot* is *cold*, (C). Remember that you must choose the *best* opposite.

5. brave
   - (A) sensible
   - (B) cowardly
   - (C) strong
   - (D) weak

A person may be both *brave* and *sensible*, both *brave* and *strong*, or both *brave* and *weak* (even though *weak* is the opposite of *strong*). A person cannot be both *brave* and *cowardly*, (B), because the two traits are opposites.

6. illicit
   - (A) inadvisable
   - (B) wise
   - (C) public
   - (D) permitted

That which is *illicit* is "illegal or forbidden." That which is *not illicit* is *permitted*, (D). Illicit deeds are often performed in private, but the word *public* is not the opposite of the word *illicit*.

7. expunge
   - (A) efface
   - (B) amplify
   - (C) soften
   - (D) jostle

To *expunge* is "to erase," so the best opposite given here is (B), *amplify*, which means "to enlarge."

8. inane
   - (A) serious
   - (B) energetic
   - (C) speechless
   - (D) capable

Inane means "empty" or "silly." If no better antonym were offered, one might make a case for *capable* as a possible near opposite. However, (A), *serious*, is a far better antonym for *inane*.

Antonyms—Practice Tests

Circle the letter before each answer you choose. Answer Key on page 241.

Test 1

Directions: Select the word that means the opposite or most nearly the opposite of the given word.

1. affluent
   (A) glamorous
   (B) scanty
   (C) stable
   (D) charitable

2. trepidation
   (A) fearlessness
   (B) anger
   (C) honesty
   (D) vigor

3. commodious
   (A) disengaged
   (B) rich
   (C) mourned
   (D) small

4. endearment
   (A) attachment
   (B) strangeness
   (C) hostility
   (D) thriftiness

5. affectation
   (A) hatred
   (B) vanity
   (C) security
   (D) modesty

6. credulity
   (A) doubt
   (B) understanding
   (C) muscularity
   (D) commendation

7. alienate
   (A) unfurl
   (B) befriend
   (C) banish
   (D) encourage

8. vulnerable
   (A) reverent
   (B) innocent
   (C) unassailable
   (D) inflated

9. abatement
   (A) addition
   (B) lessening
   (C) guarantee
   (D) denial

10. estrange
    (A) allow
    (B) release
    (C) recognize
    (D) reconcile

11. frivolity
    (A) distraction
    (B) seriousness
    (C) warmth
    (D) exactness

12. imperturbable
    (A) disrespectful
    (B) relaxed
    (C) rattled
    (D) penetrable

13. abhorrence
    (A) revelation
    (B) detachment
    (C) engagement
    (D) admiration

14. garish
    (A) dull
    (B) sweet
    (C) damp
    (D) closed

15. gregarious
    (A) sour
    (B) unsociable
    (C) free
    (D) shortened

16. diverse
    (A) happy
    (B) understandable
    (C) definite
    (D) similar

17. unkempt
    (A) tidy
    (B) tied
    (C) sloppy
    (D) enclosed

18. luminous
    (A) solar
    (B) unknown
    (C) unimaginative
    (D) dim

19. rescind
    (A) provide
    (B) reinstate
    (C) cancel
    (D) mutilate

20. affable
    (A) unbent
    (B) untruthful
    (C) unfriendly
    (D) unable

21. caustic
    (A) sleepy
    (B) sharp
    (C) soothing
    (D) unintelligent

22. adamant
    (A) effeminate
    (B) prayerful
    (C) yielding
    (D) courageous

23. unadulterated
    (A) diluted
    (B) childlike
    (C) destroyed
    (D) disgraced

24. mundane
    (A) livable
    (B) spiritual
    (C) unprotected
    (D) critical

25. rectitude
    (A) disillusionment
    (B) dishonesty
    (C) decisiveness
    (D) wholesomeness

Test 2

Directions: Select the word that means the opposite or most nearly the opposite of the given word.

1. timorous
    (A) timid
    (B) tardy
    (C) punctual
    (D) bold

2. debilitate
    (A) harass
    (B) annoy
    (C) strengthen
    (D) affix

3. mutable
    (A) changeable
    (B) lovable
    (C) constant
    (D) spiteful

4. deprecate
    (A) plead for
    (B) dishonor
    (C) contrive
    (D) aver

5. secular
    (A) musical
    (B) worldly
    (C) sacred
    (D) hospitable

6. vocation
    (A) purification
    (B) novitiate
    (C) silencing
    (D) hobby

7. recondite
    (A) confused
    (B) hidden
    (C) clear
    (D) unpaid

8. parsimonious
    (A) frugal
    (B) obdurate
    (C) officious
    (D) extravagant

9. nascent
    (A) terminating
    (B) commencing
    (C) erecting
    (D) halting

10. culpable
    (A) blameless
    (B) childless
    (C) courteous
    (D) trustworthy

11. nepotism
    (A) favoritism
    (B) indifference
    (C) impartiality
    (D) apathy

12. surfeit
    (A) plenty
    (B) insufficiency
    (C) distaste
    (D) affect

13. odious
    (A) hateful
    (B) sinful
    (C) inoffensive
    (D) spiteful

14. inveterate
    (A) habitual
    (B) experienced
    (C) dauntless
    (D) inexperienced

15. indolent
    (A) industrious
    (B) lazy
    (C) opulent
    (D) corpulent

16. venial
    (A) sacrificial
    (B) unrealistic
    (C) unpardonable
    (D) sanguine

17. rubicund
    (A) ruddy
    (B) pale
    (C) rotund
    (D) roseate

18. espouse
    (A) support
    (B) relate
    (C) sue
    (D) oppose

19. lugubrious
    (A) doleful   (C) happy
    (B) mournful   (D) malicious

20. emend
    (A) improve   (C) correct
    (B) worsen   (D) ignore

21. accelerate
    (A) stop   (C) quicken
    (B) slow   (D) hasten

22. docile
    (A) active   (C) probable
    (B) healthy   (D) teachable

23. candor
    (A) frankness   (C) deception
    (B) doubt   (D) enthusiasm

24. nomadic
    (A) secret   (C) stationary
    (B) anonymous   (D) famous

25. humble
    (A) simple   (C) hurt
    (B) just   (D) proud

Test 3

Directions: Select the word that means the opposite or most nearly the opposite of the given word.

1. succumb
   (A) arrive   (C) eat
   (B) yield   (D) conquer

2. divert
   (A) instruct   (C) bore
   (B) include   (D) amuse

3. assent
   (A) agree   (C) climb
   (B) disagree   (D) fall

4. diminish
   (A) lessen   (C) complete
   (B) begin   (D) expand

5. brazen
   (A) frozen   (C) rustproof
   (B) humble   (D) leaky

6. intent
   (A) alfresco   (C) disinterested
   (B) busy   (D) shy

7. smother
   (A) cuddle   (C) aerate
   (B) expel   (D) rescue

8. lavish
   (A) filthy   (C) squander
   (B) elegant   (D) conserve

9. aloof
   (A) sociable   (C) public
   (B) humble   (D) ignorant

10. elated
    (A) on time   (C) ideal
    (B) tardy   (D) depressed

11. furnish
    (A) dress   (C) remove
    (B) decorate   (D) polish

12. ostracize
    (A) include   (C) hide
    (B) shun   (D) delight

13. exorbitant
    (A) priceless   (C) fair
    (B) worthless   (D) straight

14. chastise
    (A) dirty   (C) praise
    (B) cleanse   (D) straighten

15. profit
    (A) gain   (C) suffer
    (B) money   (D) disgust

16. defy
    (A) desire   (C) fight
    (B) embrace   (D) abscond

17. gorge
    (A) duck   (C) stuff
    (B) diet   (D) valley

18. curtail
    (A) curry
    (B) open
    (C) shorten
    (D) extend

19. initiate
    (A) instruct
    (B) begin
    (C) terminate
    (D) invade

20. grant
    (A) confiscate
    (B) money
    (C) land
    (D) swamp

21. clamor
    (A) ugliness
    (B) beauty
    (C) silence
    (D) dishonor

22. rouse
    (A) lull
    (B) alarm
    (C) complain
    (D) weep

23. credible
    (A) believable
    (B) unbelievable
    (C) honorable
    (D) dishonorable

24. thorough
    (A) around
    (B) circumvented
    (C) sloppy
    (D) slovenly

25. wooden
    (A) iron
    (B) slippery
    (C) rubbery
    (D) green

Test 4

Directions: Select the word that means the opposite or most nearly the opposite of the given word.

1. sparkling
   (A) military
   (B) festive
   (C) lethal
   (D) grimy

2. loyal
   (A) lovely
   (B) unfaithful
   (C) unlucky
   (D) unusual

3. refuse
   (A) reheat
   (B) accept
   (C) reveal
   (D) tidy

4. acquire
   (A) solo
   (B) buy
   (C) release
   (D) collect

5. scant
   (A) sparse
   (B) scoundrel
   (C) abundant
   (D) straight

6. pinnacle
   (A) bridge
   (B) base
   (C) wall
   (D) rummy

7. corpulent
   (A) bulky
   (B) singular
   (C) company
   (D) slender

8. naive
   (A) rural
   (B) dull
   (C) sophisticated
   (D) funny

9. depression
   (A) incline
   (B) valley
   (C) hill
   (D) oppression

10. diminish
    (A) trim
    (B) augment
    (C) decorate
    (D) decrease

11. abandon
    (A) abdicate
    (B) keep
    (C) refer
    (D) encourage

12. abhor
    (A) pour
    (B) waft
    (C) desire
    (D) hate

13. finite
    (A) endless
    (B) final
    (C) done
    (D) galaxy

14. homogeneous
    (A) similar
    (B) foolish
    (C) mixed
    (D) pasteurized

15. detrimental
    (A) favorable
    (B) harmful
    (C) lasting
    (D) short

16. divergent
    (A) spontaneous
    (B) differing
    (C) apparent
    (D) alike

17. sporadic
    (A) perpetual   (C) sudden
    (B) irregular   (D) disturbing

18. opportune
    (A) convenient   (C) commendable
    (B) premeditated   (D) untimely

19. flourish
    (A) nourish   (C) flounder
    (B) blossom   (D) wave

20. candid
    (A) predictable   (C) frank
    (B) written   (D) confidential

21. meager
    (A) overdue   (C) abundant
    (B) valuable   (D) scanty

22. commencement
    (A) graduation   (C) beginning
    (B) end   (D) diploma

23. reluctance
    (A) eagerness   (C) consolation
    (B) consultation   (D) energy

24. potent
    (A) practical   (C) possible
    (B) weak   (D) impossible

25. cogent
    (A) convincing   (C) opposite
    (B) confusing   (D) unintentional

Test 5

Directions: Select the word that means the opposite or most nearly the opposite of the given word.

1. vindictive
   (A) forgiving   (C) beautiful
   (B) petty   (D) unattractive

2. tedious
   (A) technical   (C) interesting
   (B) boring   (D) confidential

3. various
   (A) dangerous   (C) uniform
   (B) spotted   (D) several

4. ambiguous
   (A) alike   (C) eager
   (B) definite   (D) dexterous

5. precisely
   (A) roughly   (C) poorly
   (B) costly   (D) doubly

6. sturdy
   (A) strong   (C) unintelligent
   (B) dark   (D) decrepit

7. response
   (A) activity   (C) question
   (B) breath   (D) duty

8. rescind
   (A) grow   (C) cancel
   (B) affirm   (D) burn

9. assistance
   (A) handout   (C) insufficiency
   (B) aid   (D) hindrance

10. consume
    (A) create   (C) want
    (B) sell   (D) begin

11. hazardous
    (A) annoying   (C) inflammable
    (B) flammable   (D) secure

12. decelerate
    (A) climb   (C) speed up
    (B) slide   (D) unwind

13. subsequently
    (A) before   (C) wisely
    (B) afterward   (D) stupidly

14. decompose
    (A) choose   (C) create
    (B) pick up   (D) break

15. restrict
    (A) hit again   (C) limit
    (B) sit down   (D) activate

16. augment
    (A) stop
    (B) disagree
    (C) decrease
    (D) decant

17. utilize
    (A) improve
    (B) waste
    (C) praise
    (D) criticize

18. toxic
    (A) abnormal
    (B) beneficial
    (C) improved
    (D) ashen

19. sufficient
    (A) hurt
    (B) choke
    (C) better
    (D) lacking

20. abatement
    (A) reduction
    (B) revival
    (C) tax
    (D) tide

21. analyze
    (A) study
    (B) submit
    (C) accept
    (D) dissect

22. adjacent
    (A) distant
    (B) neighboring
    (C) added
    (D) subtracted

23. dismantle
    (A) inspect
    (B) appraise
    (C) assemble
    (D) cover

24. extensive
    (A) prolonged
    (B) adulterated
    (C) exaggerated
    (D) narrow

25. ultimate
    (A) penultimate
    (B) original
    (C) final
    (D) most

Test 6

Directions: Select the word that means the opposite or most nearly the opposite of the given word.

1. dense
   (A) sparse
   (B) stupid
   (C) cloudy
   (D) shallow

2. negligible
   (A) careful
   (B) serious
   (C) loved
   (D) abandoned

3. dwindle
   (A) extinguish
   (B) twinkle
   (C) multiply
   (D) flirt

4. salient
   (A) prominent
   (B) salty
   (C) bland
   (D) unnoticeable

5. improvised
   (A) worn out
   (B) agreed to
   (C) planned
   (D) wealthy

6. retained
   (A) tied up
   (B) educated
   (C) debriefed
   (D) dismissed

7. contamination
   (A) radioactivity
   (B) pollution
   (C) purification
   (D) release

8. approaches
   (A) trespasses
   (B) disperses
   (C) encroaches
   (D) exits

9. extricate
   (A) intricate
   (B) bury
   (C) specialize
   (D) puzzle

10. facilitate
    (A) copy
    (B) use
    (C) destroy
    (D) deter

11. indiscriminate
    (A) selective
    (B) blind
    (C) haphazard
    (D) unclear

12. proximity
    (A) location
    (B) distance
    (C) certainty
    (D) neighborhood

13. supplement
    (A) vitamin
    (B) subtraction
    (C) addition
    (D) diet

14. indigent
    (A) healthy
    (B) wealthy
    (C) wise
    (D) active

15. procrastinate
    (A) eulogize
    (B) invest
    (C) expedite
    (D) mediate

16. intrepid
    (A) surreptitious
    (B) monotonous
    (C) paranoid
    (D) pedestrian

17. mend
    (A) give back
    (B) change
    (C) clean
    (D) destroy

18. abstract
    (A) art
    (B) absurd
    (C) concrete
    (D) asphalt

19. rambunctious
    (A) lazy
    (B) friendly
    (C) calm
    (D) fearful

20. invalid
    (A) alert
    (B) trusting
    (C) brave
    (D) true

21. discord
    (A) reward
    (B) music
    (C) punishment
    (D) harmony

22. articulate
    (A) decisive
    (B) direct
    (C) general
    (D) silent

23. verve
    (A) cowardice
    (B) ability
    (C) lethargy
    (D) agility

24. indolent
    (A) rich
    (B) happy
    (C) generous
    (D) ambitious

25. queasy
    (A) nautical
    (B) partial
    (C) broken
    (D) confident

Test 7

Directions: Select the word that means the opposite or most nearly the opposite of the given word.

1. consolidate
   (A) strengthen
   (B) diversify
   (C) liquefy
   (D) educate

2. arrogant
   (A) meek
   (B) exaggerated
   (C) clever
   (D) inept

3. eulogize
   (A) bury
   (B) engrave
   (C) defend
   (D) berate

4. tranquility
   (A) harm
   (B) concern
   (C) discord
   (D) knowledge

5. fortuitous
   (A) unfortunate
   (B) strong
   (C) designed
   (D) fearful

6. candid
   (A) painted
   (B) dishonest
   (C) unimportant
   (D) expected

7. loathe
   (A) reprieve
   (B) formalize
   (C) provoke
   (D) love

8. placid
   (A) rigid
   (B) agitated
   (C) pliable
   (D) demure

9. graphic
   (A) tabular
   (B) painted
   (C) obscure
   (D) incorrect

10. excise
    (A) forgive
    (B) insert
    (C) deny
    (D) imprint

11. sentimental
    (A) unresponsive
    (B) unwilling
    (C) unreliable
    (D) unpardonable

12. diffident
    (A) assertive
    (B) happy
    (C) companionable
    (D) easygoing

13. consequence
    (A) truth
    (B) plan
    (C) cause
    (D) retaliation

14. inflammable
    (A) soaked
    (B) fireproof
    (C) on fire
    (D) flammable

15. surcharge
    - (A) commence
    - (B) receipt
    - (C) bill
    - (D) discount

16. lethargic
    - (A) silky
    - (B) limpid
    - (C) vigorous
    - (D) metallic

17. pacify
    - (A) assault
    - (B) tremble
    - (C) conceal
    - (D) exhibit

18. incarcerate
    - (A) remit
    - (B) extinguish
    - (C) decline
    - (D) release

19. temerity
    - (A) verve
    - (B) beginning
    - (C) humility
    - (D) strength

20. tractable
    - (A) retarded
    - (B) brilliant
    - (C) airborne
    - (D) stubborn

21. benign
    - (A) relevant
    - (B) tumorous
    - (C) malevolent
    - (D) precarious

22. soporific
    - (A) painkilling
    - (B) narcotic
    - (C) awakening
    - (D) drying

23. mutiny
    - (A) exchange
    - (B) valor
    - (C) hesitation
    - (D) obedience

24. obscure
    - (A) disclose
    - (B) dishonor
    - (C) discover
    - (D) disavow

25. assail
    - (A) pretend
    - (B) conceal
    - (C) commend
    - (D) despise

Test 8

Directions: Select the word that means the opposite or most nearly the opposite of the given word.

1. tenacity
    - (A) vacillation
    - (B) inspiration
    - (C) retaliation
    - (D) equalization

2. slovenly
    - (A) youthful
    - (B) intelligent
    - (C) swift
    - (D) tidy

3. exorbitant
    - (A) axiomatic
    - (B) astral
    - (C) reasonable
    - (D) disobedient

4. antecedent
    - (A) opposition
    - (B) prerequisite
    - (C) preventative
    - (D) subsequent

5. captivate
    - (A) alienate
    - (B) stipulate
    - (C) stimulate
    - (D) indicate

6. compliance
    - (A) profanity
    - (B) strictness
    - (C) approval
    - (D) rebellion

7. intelligible
    - (A) dull
    - (B) unclear
    - (C) unteachable
    - (D) faulty

8. consensus
    - (A) poll
    - (B) disharmony
    - (C) converence
    - (D) miscount

9. obtuse
    - (A) sensitive
    - (B) slim
    - (C) ill-mannered
    - (D) angular

10. endorse
    - (A) allot
    - (B) invest
    - (C) elect
    - (D) denounce

11. fuse
    - (A) obey
    - (B) regulate
    - (C) sever
    - (D) negate

12. embellish
    - (A) simplify
    - (B) overeat
    - (C) abstain
    - (D) signify

13. prohibition
    - (A) amendment
    - (B) illegality
    - (C) pretense
    - (D) endorsement

14. extraneous
    - (A) immigrant
    - (B) emigrant
    - (C) irregular
    - (D) inherent

15. torsion
    (A) compressing
    (B) sliding
    (C) spinning
    (D) straightening

16. cognizant
    (A) afraid
    (B) ignorant
    (C) capable
    (D) optimistic

17. magnanimous
    (A) insolent
    (B) shrewd
    (C) selfish
    (D) threatening

18. judicious
    (A) foolish
    (B) biased
    (C) illegal
    (D) limited

19. zealous
    (A) awkward
    (B) enthusiastic
    (C) reluctant
    (D) skillful

20. dubious
    (A) cheerful
    (B) questionable
    (C) unacceptable
    (D) assured

21. morose
    (A) curious
    (B) morbid
    (C) impatient
    (D) optimistic

22. terse
    (A) detailed
    (B) harsh
    (C) vague
    (D) concise

23. spurt
    (A) spill
    (B) seep
    (C) stream
    (D) stalk

24. supple
    (A) soft
    (B) stale
    (C) lazy
    (D) rigid

25. perturbed
    (A) disrespectful
    (B) tractable
    (C) cheerful
    (D) relaxed

Antonyms—Answer Key for Practice Tests

TEST 1

1.	B	6.	A	11.	B	16.	D	21.	C
2.	A	7.	B	12.	C	17.	A	22.	C
3.	D	8.	C	13.	D	18.	D	23.	A
4.	C	9.	A	14.	A	19.	B	24.	B
5.	D	10.	D	15.	B	20.	C	25.	B

TEST 2

1.	D	6.	D	11.	C	16.	C	21.	B
2.	C	7.	C	12.	B	17.	B	22.	A
3.	C	8.	D	13.	C	18.	D	23.	C
4.	A	9.	A	14.	D	19.	C	24.	C
5.	C	10.	A	15.	A	20.	B	25.	D

TEST 3

1.	D	6.	C	11.	C	16.	B	21.	C
2.	C	7.	C	12.	A	17.	B	22.	A
3.	B	8.	D	13.	C	18.	D	23.	B
4.	D	9.	A	14.	C	19.	C	24.	C
5.	B	10.	D	15.	C	20.	A	25.	C

TEST 4

1.	D	6.	B	11.	B	16.	D	21.	C
2.	B	7.	D	12.	C	17.	A	22.	B
3.	B	8.	C	13.	A	18.	D	23.	A
4.	C	9.	C	14.	C	19.	C	24.	B
5.	C	10.	B	15.	A	20.	D	25.	B

TEST 5

1.	A	6.	D	11.	D	16.	C	21.	C
2.	C	7.	C	12.	C	17.	B	22.	A
3.	C	8.	B	13.	A	18.	B	23.	C
4.	B	9.	D	14.	C	19.	D	24.	D
5.	A	10.	A	15.	D	20.	B	25.	B

TEST 6

1.	A	6.	D	11.	A	16.	C	21.	D
2.	B	7.	C	12.	B	17.	D	22.	D
3.	C	8.	D	13.	B	18.	C	23.	C
4.	D	9.	B	14.	B	19.	C	24.	D
5.	C	10.	D	15.	C	20.	D	25.	D

TEST 7

1.	B	6.	B	11.	A	16.	C	21.	C
2.	A	7.	D	12.	A	17.	A	22.	C
3.	D	8.	B	13.	C	18.	D	23.	D
4.	C	9.	C	14.	B	19.	C	24.	A
5.	C	10.	B	15.	D	20.	D	25.	C

TEST 8

1.	A	6.	D	11.	C	16.	B	21.	D
2.	D	7.	B	12.	A	17.	C	22.	C
3.	C	8.	B	13.	D	18.	A	23.	B
4.	D	9.	A	14.	D	19.	C	24.	D
5.	A	10.	D	15.	D	20.	D	25.	D

SENTENCE COMPLETIONS

In sentence completion questions, you are given a sentence containing one or more blanks. A number of words or pairs of words are suggested to fill the blank spaces. You must select the word or pair of words that will *best* complete the meaning of the sentence as a whole.

Sentence completion questions are more complex than synonym questions and antonym questions. They test not only your knowledge of basic vocabulary but also your ability to understand what you read. While studying individual words may be helpful, the best way to prepare for this type of question is to read as much as possible. A dictionary will tell you what a word means; reading will teach you how it is actually used.

In a typical sentence completion question, *any* of the answer choices might be inserted into the blank spaces and the sentence would be technically correct, but it might not make sense. Usually, more than one choice would make sense, but only one choice completely carries out the full meaning of the sentence. There is one *best* answer.

How to Answer Sentence Completion Questions

1. Read the sentence. Try to figure out what it means.

2. Look at the blank or blanks with relation to the meaning of the sentence. Is a negative connotation called for or a positive one? If there are two blanks, should the pair be comparative, contrasting, or complementary? Are you looking for a term that best defines a phrase in the sentence?

3. Eliminate those answer choices that do not meet the criteria you established in step two.

4. Read the sentence to yourself, trying out each of the remaining choices, one by one. Which choice is the most exact, appropriate, or likely considering the information given in the sentence? Which of the choices does the *best* job of completing the message of the sentence?

5. First answer the questions you find easy. If you have trouble with a question, leave it and go back to it later. If a fresh look does not help you to come up with a sure answer, make an educated guess.

Before you go on to the practice tests, study the following examples:

1. Trespassing on private property is ———— by law.
   - (A) proscribed
   - (B) warranted
   - (C) prescribed
   - (D) eliminated

The most likely and therefore correct answer is (A), *proscribed,* which means "forbidden" or "outlawed." *Warranted,* (B), may remind you of a *warrant* for arrest, which might be a result of trespassing. *Warranted,* however, means "justified," which would make the given sentence obviously untrue. Choice (C), *prescribed,* looks similar to the correct answer but means "recommended." Like *warranted,* it makes nonsense out of the given sentence. Choice (D), *eliminated,* is also less likely than (A). The law may be intended to eliminate trespassing but it can never be completely successful in doing so.

2. Despite the harsh tone of her comments, she did not mean to _____ any criticism of you personally.
    (A) infer
    (B) aim
    (C) comply
    (D) imply

The correct answer is (D), *imply,* which means "suggest indirectly." Choice (A), *infer,* is a word often confused with *imply.* It means "conclude from reasoning or implication." A speaker implies; a listener infers. Choice (C), *comply,* meaning "obey," makes no sense in this context. Choice (B), *aim,* is more likely, but it doesn't work in the sentence as given. You might say, "she did not mean to *aim* any criticism *at* you," but you would not normally say, "she did not mean to *aim* any criticism *of* you."

3. The department's _____ does not allow for unlimited copying by all of the instructors in the program. Each instructor can be reimbursed for copying expenses only up to ten dollars.
    (A) paperwork
    (B) staff
    (C) organization
    (D) budget

Since the concern here is with money, the correct answer is (D), *budget.* It is a *budget* that puts limits on spending. Choices (A), *paperwork*, and (B), *staff*, are not appropriate to the meaning of the passage. Choice (C), *organization*, is barely possible but only because it is so vague. *Budget* both makes sense and is much more exact.

4. If the company offered a settlement commensurate with the damages sustained, the couple would _____ their right to a hearing.
    (A) cancel
    (B) ensue
    (C) waive
    (D) assert

The correct answer is (C), *waive*, which means "forgo" or "give up." One *waives* something to which one is entitled, such as a right. Choice (A), *cancel,* is similar in meaning but is not used in this way. One can cancel a hearing but not a right. Choice (B), *ensue*, may mislead you by its similarity to *sue*. The sentence does imply that the couple is suing or planning to sue the company for damages of some sort. However, *ensue* simply means "follow as a result" and so makes no sense in this context. Choice (D), *assert*, here means the opposite of *waive*. One can assert a right, but the meaning of the first part of the sentence makes this choice unlikely.

5. Those who feel that war is stupid and unnecessary think that to die on the battlefield is _____.
    (A) courageous
    (B) pretentious
    (C) useless
    (D) illegal

The correct answer is (C). The key to this answer is the attitude expressed—that war is stupid and unnecessary. Those who are antagonistic toward war would consider a battlefield death to be *useless*. While it is true that giving one's life on the field of battle is courageous, (A), that is not the answer in the context of this sentence. Choice (B), *pretentious*, meaning "affectedly grand or ostentatious" does not go along with the idea that war is stupid. Choice (D) does not make sense in relation to a battlefield death.

6. An unruly person may well become _____ if he is treated with _____ by those around him.
   (A) angry . . . kindness
   (B) calm . . . respect
   (C) peaceful . . . abuse
   (D) interested . . . medicine

The correct answer is (B). This sentence is about a person's behavior and how it is affected by the way he is treated. "Unruly" is a word with a negative connotation, but this unruly person may become _____. The implication is that the person will change for the better, so the first blank should be filled by a word with a positive connotation. The sentence further implies a cause-and-effect relationship. Positive treatment logically results in a positive change in behavior, so the second blank should also be filled by a word with a positive connotation. Choices (A) and (C) contain words that have a negative connotation. Of the remaining choices, (B) makes more sense than (D).

7. A _____ person cannot be expected to resist _____.
   (A) wealthy . . . money
   (B) religious . . . temptation
   (C) starving . . . food
   (D) peculiar . . . quarreling

The correct answer is (C). The sentence implies that there is a necessary connection between something that cannot be resisted and an implied need. Of the four pairs of words, only (A) and (C) show a real, logical connection between the two words. Since a wealthy person has no need for money, clearly (C) is the *best* answer. A *starving* person cannot be expected to resist *food*.

8. The _____ on the letter indicated that it had been mailed in Chicago three weeks _____.
   (A) address . . . ago
   (B) stamp . . . in advance
   (C) postmark . . . previously
   (D) water stains . . . late

The correct answer is (C). The sentence clearly asks you to fill the blank with the clue that tells the date that a letter was mailed. The only possible answer to fill the first blank is *postmark*. Since the second blank may quite reasonably be filled by *previously*, this question is an easy one to answer.

9. Since movies have become more _____, many people believe television to be _____.
   (A) helpful . . . utilitarian
   (B) expensive . . . necessary
   (C) common . . . inadequate
   (D) costly . . . useless

The correct answer is (B). Movies and television are both media of entertainment. The sentence compares the two media in terms of their cost, stating that many people believe television (which is free after the initial investment in the set) is *necessary* because movies have become so *expensive* (and therefore out of reach for many people.)

10. Unethical landowners used to _____ gold in old mines to _____ naive speculators who would pay high prices for nearly worthless land.
    (A) hide . . . escape
    (B) find . . . repay
    (C) discover . . . anger
    (D) imbed . . . entice

The correct answer is (D). That the landowners are described as unethical (immoral) indicates that they would attempt to deceive the naive speculators. This would best be accomplished by *imbedding* gold in worthless mines in order to *entice* (tempt or lure) the speculators to pay inflated prices.

Sentence Completions—Practice Tests

Use the answer sheet on page 154 to mark the letter of the answer you choose. Answer Key on page 255.

Test 1

Directions: Select the word or word pair that best completes each sentence.

1. Although her lips wore a smile, her eyes wore a _____.
   (A) veil
   (B) laugh
   (C) shadow
   (D) frown

2. Martha's _____ handling of the steaks caused us to amend our plans for dinner and eat out.
   (A) ingenious
   (B) disingenuous
   (C) inverted
   (D) inept

3. The stigma attached to this job makes it _____ even at a(n) _____ salary.
   (A) unattractive—attractive
   (B) attractive—attractive
   (C) sybaritic—meager
   (D) uninviting—nominal

4. One man's meat is another man's _____.
   (A) dairy
   (B) flesh
   (C) poison
   (D) prerogative

5. Joseph's _____ handling of the Thompson account made him the laughingstock of the industry.
   (A) proper
   (B) maudlin
   (C) humorous
   (D) incompetent

6. Do not undertake a daily program of _____ exercise such as jogging without first having a physical checkup.
   (A) light
   (B) spurious
   (C) hazardous
   (D) strenuous

7. The police received a(n) _____ call giving them valuable information that led to an arrest. The caller refused to give his name out of fear of reprisals.
   (A) anonymous
   (B) asinine
   (C) private
   (D) candid

8. He was the chief _____ of his uncle's will. After taxes, he was left with an inheritance of $20,000,000.
   (A) exemption
   (B) beneficiary
   (C) contestant
   (D) winner

9. Don't be _____; I don't have time to split hairs with you.
   (A) spurious
   (B) childish
   (C) picayune
   (D) erudite

10. When his temperature climbed above 104 degrees, he became _____.
    (A) tepid
    (B) discordant
    (C) deceased
    (D) delirious

11. To climb at another's expense is to _____ yourself morally.
    (A) upbraid
    (B) energize
    (C) enervate
    (D) abase

12. We waited patiently for the storm to slacken; it _____ refused to _____.
    (A) persistently—strengthen
    (B) stoutly—abate
    (C) wanly—sublimate
    (D) sternly—mitigate

13. "Berty" decided to _____ when he found that he couldn't have his love and his throne at the same time; it was 1937.
    (A) prevaricate
    (B) alter
    (C) abrogate
    (D) abdicate

14. Although he was not ever at the scene of the crime, his complicity was uncovered; he had _____ and _____ in the robbery by acting as a fence.
    (A) stolen—sold
    (B) assisted—testified
    (C) witnessed—participated
    (D) aided—abetted

15. In view of the extenuating circumstances and the defendant's youth, the judge recommended _____.
    (A) conviction
    (B) a mistrial
    (C) leniency
    (D) hanging

16. A person who will not take "no" for an answer may sometimes be classified as a _____.
    (A) salesman
    (B) persistent
    (C) zealot
    (D) heretic

17. The children were told that they should be _____ of strangers offering candy.
    - (A) weary
    - (B) wary
    - (C) envious
    - (D) considerate

18. Politicians are not coerced into taxing the public; they do it of their own _____.
    - (A) reputation
    - (B) graft
    - (C) expediency
    - (D) volition

19. Elder statesmen used to be _____ for their wisdom when respect for age was an integral part of the value structure.
    - (A) known
    - (B) venerated
    - (C) exiled
    - (D) abused

20. The 45-minute sermon is a potent _____; it is an absolute cure for _____.
    - (A) astringent—drowsiness
    - (B) aphrodisiac—celibacy
    - (C) soporific—insomnia
    - (D) therapeutic—malaise

21. His cynicism was _____; it was written all over him.
    - (A) affected
    - (B) covert
    - (C) infamous
    - (D) manifest

22. Suffering from _____, she was forced to spend most of her time indoors.
    - (A) claustrophobia
    - (B) anemia
    - (C) agoraphobia
    - (D) ambivalence

23. We were not allowed to _____ our appetite until we had tidied up our living quarters.
    - (A) fill
    - (B) whet
    - (C) sate
    - (D) flag

24. If you don't badger the child, he may do what you want him to do without _____.
    - (A) pleasure
    - (B) pain
    - (C) pressure
    - (D) volition

25. You must see the head of the agency; I am not _____ to give out that information.
    - (A) nervous
    - (B) authorized
    - (C) programmed
    - (D) happy

Test 2

Directions: Select the word or word pair that best completes each sentence.

1. The local tavern was one of the last _____ of male supremacy.
    - (A) coffins
    - (B) places
    - (C) potentials
    - (D) bastions

2. The ship was in a(n) _____ position; having lost its rudder it was subject to the _____ of the prevailing winds.
    - (A) unintended— riptides
    - (B) untenable—vagaries
    - (C) dangerous—breezes
    - (D) favored—weaknesses

3. _____ shadows played over her face as the branches above her danced in the sunlight.
    - (A) Transient
    - (B) Prolonged
    - (C) Clandestine
    - (D) Sedentary

4. Alchemists expended their energies in an attempt to _____ base elements into gold.
    - (A) transfer
    - (B) raise
    - (C) translate
    - (D) transmute

5. Publication of the article was timed to _____ with the professor's fiftieth birthday.
    - (A) coincide
    - (B) amalgamate
    - (C) terminate
    - (D) interfere

6. The chariot _____ around the curve completely out of control when Thessalius dropped the reins.
    - (A) competed
    - (B) careened
    - (C) fell
    - (D) caromed

7. Don't _____; stick to the _____ of the issue so that we can take it to a vote.
    - (A) prevaricate—gist
    - (B) procrastinate—promptness
    - (C) delay—urgency
    - (D) digress—crux

8. The more the search proved fruitless, the more _____ the parents of the missing child became.
    - (A) disconsolate
    - (B) dislocated
    - (C) disappointed
    - (D) disheveled

9. When the unpopular war began, only a few men enlisted; the rest had to be _____.
    - (A) shot
    - (B) processed
    - (C) reassured
    - (D) conscripted

10. The runner advanced one base without stealing, hitting, or getting a walk; the pitcher had committed a _____.
(A) spitball
(B) balk
(C) dropped ball
(D) syllogism

11. They prefer to hire someone fluent in Spanish, since the neighborhood where the clinic is located is _____ Hispanic.
(A) imponderably
(B) sparsely
(C) consistently
(D) predominantly

12. A dark, cloudy sky is a _____ of a storm.
(A) remnant
(B) precursor
(C) belier
(D) proof

13. The Freedom of Information Act gives private citizens _____ government files.
(A) access to
(B) excess to
(C) redress of
(D) release from

14. His remarks were so _____ we could not decide which of the possible meanings was correct.
(A) ambiguous
(B) facetious
(C) improper
(D) congruent

15. His performance was _____; it made a fool of him.
(A) auspicious
(B) ludicrous
(C) luscious
(D) interlocutory

16. The upset furniture and broken window silently _____ to the fact that the apartment had been robbed.
(A) witnessed
(B) confirmed
(C) attested
(D) admitted

17. Although the warrior could cope with blows from swords, he was _____ to gunshots; his armor was not _____ to them.
(A) reachable—proof
(B) vulnerable—susceptible
(C) vulnerable—impervious
(D) invulnerable—susceptible

18. When she addressed the reporters, her beauty, bearing, and elegant garb were belied by the _____ words she uttered.
(A) untrue
(B) uncouth
(C) unemotional
(D) unfettered

19. "A stitch in time saves nine" and other such _____ expressions made his speeches insufferable.
(A) tried
(B) cryptic
(C) redundant
(D) trite

20. The new regulations turned out to be _____, not permissive.
(A) impermissive
(B) liberal
(C) stringent
(D) uniform

21. A person who commits a wrong may be required to _____ his property as a penalty.
(A) confiscate
(B) destroy
(C) forfeit
(D) assess

22. When the desk was placed facing the window, he found himself _____ from his work by the activity in the street.
(A) distraught
(B) destroyed
(C) distracted
(D) decimated

23. He said he didn't get the job done because he was incapacitated; in truth, he was _____.
(A) indigent
(B) indolent
(C) indulgent
(D) insipid

24. The "police" turned out to be clowns; it was all a _____.
(A) mystery
(B) mixup
(C) fracas
(D) hoax

25. The authorities declared an _____ on incoming freight because of the trucking strike.
(A) impression
(B) immolation
(C) embargo
(D) alert

Test 3

Directions: Select the word or word pair that best completes each sentence.

1. We are indeed sorry to hear of your mother's passing; please accept our sincerest _____.
(A) adulations
(B) congratulations
(C) condolences
(D) concatenations

2. While on a diet I remained lean, but once off it I became _____.
(A) scrawny
(B) remiss
(C) corpulent
(D) corporeal

3. With his gutter language and vile manner he was positively _____.
   (A) urbane
   (B) rural
   (C) liberal
   (D) boorish

4. The voters show their _____ by staying away from the polls.
   (A) interest
   (B) usury
   (C) apathy
   (D) serendipity

5. Being less than perfectly prepared, I took my exams with _____.
   (A) aplomb
   (B) confidence
   (C) trepidation
   (D) indifference

6. The grade was steep and the load heavy; we had to _____ the oxen in order to arrive home on time.
   (A) rest
   (B) feed
   (C) goad
   (D) slaughter

7. He was proved guilty; his alibi had been a complete _____.
   (A) attestation
   (B) fabrication
   (C) intonation
   (D) litany

8. He claimed to be deathly ill, although he looked perfectly _____ and _____ to us.
   (A) fine—fettle
   (B) sane—sound
   (C) hale—hearty
   (D) hectic—healthy

9. Although he had _____ about the weather, he had no _____ about his ability to navigate through it.
   (A) doubts—confidence
   (B) confidence—qualms
   (C) qualms—confidence
   (D) misgivings—qualms

10. The police department will not accept for _____ a report of a person missing if his residence is outside the city.
   (A) foreclosure
   (B) convenience
   (C) investigation
   (D) control

11. Rabbits, elephants, deer, and sheep are _____; they eat only plants.
   (A) omnivorous
   (B) herbivorous
   (C) carnivorous
   (D) ruminants

12. The rivals were fighting tooth and nail when suddenly, in the thick of the _____, the bell rang.
   (A) night
   (B) day
   (C) fray
   (D) ring

13. The judge _____ the union from blocking the accesses.
   (A) enjoined
   (B) ordered
   (C) forbade
   (D) unfrocked

14. The _____ on the letter showed it had been mailed in North Dakota two weeks previously.
   (A) address
   (B) stamp
   (C) postmark
   (D) envelope

15. It is easy to see the difference between the two photographs when they are placed in _____.
   (A) disarray
   (B) juxtaposition
   (C) composition
   (D) jeopardy

16. During colonial winters in America there was a _____ in every _____.
   (A) fire—hearth
   (B) stokes—pot
   (C) flintlock—chimney
   (D) tepee—Indian

17. He was a _____ salesman; he could sell refrigerators to Eskimos.
   (A) perverse
   (B) low-keyed
   (C) glib
   (D) fruitless

18. When the bomb exploded in front of the building, it destroyed the whole _____.
   (A) cellar
   (B) pontoon
   (C) facade
   (D) facet

19. He is expected to testify that he saw the _____ thief fleeing the scene of the crime.
   (A) convicted
   (B) delinquent
   (C) alleged
   (D) innocent

20. A child who has not slept well will be anything but _____.
   (A) intractable
   (B) docile
   (C) equine
   (D) bovine

21. What we thought was a _____ volcano suddenly erupted.
   (A) deceased
   (B) dactylic
   (C) dormant
   (D) disruptive

22. Cigarette smoking is _____ to your health.
   (A) disengaging
   (B) deleterious
   (C) delectable
   (D) irrespective

23. My uncle hardly ever needed a telephone; his voice was _____ from a distance of half a mile.
    (A) inaudible   (C) suspicious
    (B) audible   (D) visible

24. His parents never had to _____ him for being _____.
    (A) chide—industrious
    (B) ride—superfluous
    (C) chide—indolent
    (D) punish—independent

25. The current use of "_____" in place of "fat" is a euphemism.
    (A) ale   (C) porter
    (B) portly   (D) beer

Test 4

Directions: Select the word or word pair that best completes each sentence.

1. An accident report should be written as soon as possible after the necessary _____ has been obtained.
    (A) bystander   (C) information
    (B) formulation   (D) charter

2. A change in environment is very likely to _____ a change in one's work habits.
    (A) affect   (C) effect
    (B) inflict (D) prosper

3. With typical diplomatic maneuvering, the State Department used every _____ known to man to avoid expressing the avowed policy in _____ language.
    (A) trick—diplomatic
    (B) page—gobbledygook
    (C) circumlocution—concise
    (D) summary—plain

4. The U.N., like the League of Nations before it, is an exercise in _____; it begs the issues and brings no _____ deterrent to the impending cataclysm.
    (A) debating—acknowledged
    (B) self-rule—realistic
    (C) parliamentarianism—neutral
    (D) futility—substantive

5. The main reason for the loss of the Alamo was the _____ of Santa Ana's forces.
    (A) decline   (C) preponderance
    (B) felicitation (D) isolation

6. A cloudy suspension may be described as _____.
    (A) turbid   (C) suspicious
    (B) precipitous (D) auspicious

7. The flamenco dancer stood still, ready to perform, his arms _____.
    (A) blazing   (C) flailing
    (B) akimbo   (D) deadlocked

8. The woman sued the magazine, claiming that the article _____ her character.
    (A) demoted (C) defamed
    (B) deplored (D) whitewashed

9. To be a "joiner" is to be _____.
    (A) gregarious (C) hilarious
    (B) popular   (D) woodworking

10. As a result of constant and unrelenting eating, her figure changed from "pleasingly plump" to _____.
    (A) overrun   (C) oblate
    (B) parsimonious (D) obese

11. When you have _____ your palate with pickles, you want no more.
    (A) scarred (C) imbibed
    (B) satiated (D) covered

12. To protect the respondents' privacy, names and Social Security numbers are _____ the questionnaires before the results are tabulated.
    (A) referred to   (C) retained in
    (B) deleted from (D) appended to

13. TASS was the _____ for Telegrafnoe Agentsvo Sovietskoyo Soyuza.
    (A) homonym (C) heteronym
    (B) acronym (D) pseudonym

14. Murdered by a crazed derelict who was running amuck, he was the victim of a _____ crime.
    (A) gratuitous (C) pathological
    (B) gratifying (D) demented

15. To put off until tomorrow what you should do today is to _____.
    (A) prorate   (C) face
    (B) procrastinate (D) proscribe

16. Being a geologist, he tended to _____ his head as he walked along the path; he didn't want to overlook a single pebble.
    (A) mind
    (B) cover
    (C) shake
    (D) incline

17. _____ means an injustice so _____ that it is wicked.
    (A) Iniquity—gross
    (B) Lobotomy—inane
    (C) Perjury—mendacious
    (D) Bias—slanted

18. A _____ has a strong _____ to steal.
    (A) pyromaniac—urge
    (B) megalomaniac—phobia
    (C) dipsomaniac—aversion
    (D) kleptomaniac—proclivity

19. The _____ assumed for the sake of discussion was that business would improve for the next five years.
    (A) labyrinth
    (B) hypothesis
    (C) outlay
    (D) itinerary

20. I wish you wouldn't be so _____; you make faces at everything I say.
    (A) supercilious
    (B) insubordinate
    (C) disconsolate
    (D) superficial

21. New York's climate is not very _____; its winters give you colds, and its summers can cause heat prostration.
    (A) sanitary
    (B) salutary
    (C) salubrious
    (D) healthy

22. One who _____ another is laughing *at* him, not *with* him.
    (A) derides
    (B) defiles
    (C) irks
    (D) buffoons

23. To give in to the terrorists' demands would be a betrayal of our responsibilities; such _____ would only encourage others to adopt similar ways to gain their ends.
    (A) defeats
    (B) appeasement
    (C) appeals
    (D) subterfuge

24. It is hard to believe that the Trojans could have been so easily deceived by the _____ of the wooden horse.
    (A) tragedy
    (B) stratagem
    (C) strategy
    (D) prolixity

25. At the 1986 Academy Awards there was a veritable _____ of stars.
    (A) bevy
    (B) collection
    (C) galaxy
    (D) pride

Test 5

Directions: Select the word or word pair that best completes each sentence.

1. I felt as _____ as a fifth wheel.
   (A) rolled
   (B) round
   (C) superfluous
   (D) axillary

2. If we were to _____ our democracy with a _____, there would be no way, short of civil war, to reverse the change.
   (A) contrast—parliament
   (B) substitute—constitutional monarchy
   (C) supplant—dictatorship
   (D) reinforce—three-party system

3. A(n) _____ look came into the poodle's eye as a dachshund wandered onto his territory.
   (A) feline
   (B) bellicose
   (C) onerous
   (D) canine

4. Always the _____, she spent hours preening herself in the presence of her escort.
   (A) prude
   (B) croquette
   (C) coquette
   (D) diplomat

5. The annual _____ in his school attendance always coincided with the first week of fishing season.
   (A) sequence
   (B) hiatus
   (C) accrual
   (D) increment

6. His sermon opened with a few _____ remarks about the Golden Rule and closed with a homily that was equally unclear.
   (A) nebulous
   (B) Old Testament
   (C) concise
   (D) sanctimonious

7. Giving preference to his brother's son for that office smacks of _____ to me!
   (A) chauvinism
   (B) sycophancy
   (C) nepotism
   (D) nihilism

8. A _____'s arms are usually very strong as a result of performing some of the tasks that would be otherwise done by his now _____ legs.
   (A) paraplegic—useless
   (B) hemiplegic—undependable
   (C) quadriplegic—diplegic
   (D) somnambulist—comatose

9. Although the wind was quite dependable in those waters, the schooner had an inboard engine as a _____ just in case.
   (A) relief   (C) generator
   (B) substitute   (D) subsidiary

10. Being perfectly prepared, I took my exams with _____.
    (A) aplomb   (C) trepidation
    (B) pugnacity   (D) indifference

11. His speech was too _____; its meaning escaped me completely.
    (A) protracted   (C) sordid
    (B) concise   (D) abstruse

12. The "life" of some subatomic particles is so _____ it has to be measured in nano-seconds.
    (A) contrived   (C) ephemeral
    (B) finite   (D) circumscribed

13. When income taxes are repealed, the _____ will have arrived.
    (A) apocalypse   (C) milestone
    (B) holocaust   (D) millennium

14. If you hadn't _____ we might have won the argument; I wish you'd keep your _____ to yourself henceforth.
    (A) confessed—hands
    (B) obtruded—opinions
    (C) obfuscated—deviations
    (D) truncated—pruning

15. You'll _____ the day you voted for Zilch; he'll break every promise he's made to you.
    (A) regard   (C) obliterate
    (B) eschew   (D) rue

16. During the Revolutionary War, Hessian troops fought on the British side not as allies, but as _____. They were paid in money, not glory.
    (A) orderlies   (C) hors de combat
    (B) valets   (D) mercenaries

17. On and on they came, countless as the blades of grass in a field, _____ of them.
    (A) myriads   (C) dozens
    (B) dryads   (D) multitudinous

18. If you find peeling potatoes to be _____, perhaps you'd prefer to scrub the floors?
    (A) preferable   (C) infectious
    (B) onerous   (D) relevant

19. The offenders then prostrated themselves and _____ for mercy.
    (A) applauded   (C) imprecated
    (B) supplicated   (D) deprecated

20. Her rebelliousness was _____; it was written all over her.
    (A) exterior   (C) implicit
    (B) covert   (D) manifest

21. If he hasn't yet learned the importance of speaking well of others, he must be quite _____.
    (A) loquacious   (C) arcane
    (B) oblique   (D) obtuse

22. Louis XIV was the _____ of _____ elegance; he wore a different outfit for practically every hour of the day.
    (A) paragon—peripatetic
    (B) epitome—sartorial
    (C) acme—epicurean
    (D) architect—gastronomic

23. Favoring one child over another will only intensify _____ rivalry.
    (A) fraternal   (C) parental
    (B) sororal   (D) sibling

24. The man _____ the speaker at the meeting by shouting false accusations.
    (A) corrected   (C) disconcerted
    (B) interfered   (D) collapsed

25. If he continues to _____ liquor at this rate, he will end up as an alcoholic.
    (A) buy   (C) secrete
    (B) imbibe   (D) accumulate

Test 6

Directions: Select the word or word pair that best completes each sentence.

1. I'm glad to see you have _____; patience is a virtue!
   (A) arrived   (C) time
   (B) decided   (D) forbearance

2. As the fog came _____, visibility dropped to five feet.
   (A) often
   (B) silently
   (C) nigh
   (D) damp

3. A(n) _____ jogger, she could do 15 miles a day.
   (A) reluctant
   (B) indefatigable
   (C) outfitted
   (D) aged

4. To tame wild horses was what her fierce nature _____.
   (A) exuded
   (B) precluded
   (C) intruded
   (D) required

5. Wars seem to be _____; the end of one always tends to precipitate the beginning of the next.
   (A) parallel
   (B) concatenate
   (C) irrelevant
   (D) hell

6. If you _____ your energy wisely you will never lack for it; if you _____ it, you'll remain poor.
   (A) burn—cauterize
   (B) use—dissipate
   (C) husband—economize
   (D) expend—spend

7. The only fair way to choose who will have to work over the holiday is to pick someone _____ by drawing lots.
   (A) covertly
   (B) conspicuously
   (C) randomly
   (D) painstakingly

8. Richelieu achieved eminence under Louis XIII; few cardinals since have been so politically _____.
   (A) retiring
   (B) unassuming
   (C) prominent
   (D) hesitant

9. People started calling him a _____; he had broken a law.
   (A) conspirator
   (B) transgressor
   (C) transient
   (D) bystander

10. "_____ and _____," he said with a smile as he met his class for the new term.
    (A) Warm—welcome
    (B) Friends—countrymen
    (C) Hail—fairwell
    (D) Greetings—salutations

11. Although he is reputed to be aloof, his manner that day was so _____ that everyone felt perfectly at ease.
    (A) reluctant
    (B) gracious
    (C) malign
    (D) plausible

12. Speeding may be a _____, but fleeing from the scene of a crime is a _____.
    (A) mistake—nuisance
    (B) faux pas—crime
    (C) misdemeanor—felony
    (D) felony—misdemeanor

13. Among his _____ was the skill of escaping from any type of handcuffs.
    (A) virtues
    (B) crafts
    (C) habits
    (D) repertories

14. Gold is one of the most _____ elements; it can be hammered into sheets thinner than a human hair.
    (A) brittle
    (B) adamantine
    (C) soft
    (D) malleable

15. The impact of the situation failed to touch him; he remained _____ as a stone.
    (A) oppressive
    (B) reticent
    (C) immaculate
    (D) impassive

16. A(n) _____ lawyer will help his client _____ the law.
    (A) efficient—abrogate
    (B) honest—bend
    (C) unscrupulous—evade
    (D) clever—elect

17. Your banker may look at you _____ if you admit to not wanting to save money.
    (A) respectfully
    (B) only
    (C) askance
    (D) directly

18. The gossip-hungry readers combed through the article for every _____ detail.
    (A) lurid
    (B) common
    (C) nagging
    (D) recurring

19. Worshipping her every move, he was her most _____ admirer.
    (A) beneficent
    (B) fatuous
    (C) ardent
    (D) sophisticated

20. He was stubbornly persistent; nothing or nobody could _____ him form his self-appointed mission.
    (A) slow
    (B) pervade
    (C) arrest
    (D) dissuade

21. To be _____ was her lot; she was destined never to earn enough money to support herself.
    (A) importune   (C) impecunious
    (B) impulsive   (D) innocuous

22. You would not be so _____ if you worked out at the gym; you have loose fat all over!
    (A) flaccid   (C) avoirdupois
    (B) placid   (D) gaunt

23. A police officer's _____ job is to prevent crime.
    (A) primary   (C) only
    (B) compendium   (D) infrequent

24. The general couldn't attend, but he sent his _____.
    (A) commandant   (C) adjutant
    (B) commander   (D) superior

25. You can depend on a malingerer to _____ his duty.
    (A) perform   (C) shirk
    (B) pursue   (D) lack

Sentence Completions—Answer Key for Practice Tests

TEST 1

1. D	6. D	11. D	16. C	21. D
2. D	7. A	12. B	17. B	22. C
3. A	8. B	13. D	18. D	23. C
4. C	9. C	14. D	19. B	24. C
5. D	10. D	15. C	20. C	25. B

TEST 2

1. D	6. B	11. D	16. C	21. C
2. B	7. D	12. B	17. C	22. C
3. A	8. A	13. A	18. B	23. B
4. D	9. D	14. A	19. D	24. D
5. A	10. B	15. B	20. C	25. C

TEST 3

1. C	6. C	11. B	16. A	21. C
2. C	7. B	12. C	17. C	22. B
3. D	8. C	13. A	18. C	23. B
4. C	9. D	14. C	19. C	24. C
5. C	10. C	15. B	20. B	25. B

TEST 4

1. C	6. A	11. B	16. D	21. C
2. C	7. B	12. B	17. A	22. A
3. C	8. C	13. B	18. D	23. B
4. D	9. A	14. A	19. B	24. B
5. C	10. D	15. B	20. A	25. C

TEST 5

1. C	6. A	11. D	16. D	21. D
2. C	7. C	12. C	17. A	22. B
3. B	8. A	13. D	18. B	23. D
4. C	9. D	14. B	19. B	24. C
5. B	10. A	15. D	20. D	25. B

TEST 6

1. D	6. B	11. B	16. C	21. C
2. C	7. C	12. C	17. C	22. A
3. B	8. C	13. B	18. A	23. A
4. D	9. B	14. D	19. C	24. C
5. B	10. D	15. D	20. D	25. C

ANSWER EXPLANATIONS

TEST 1

1. **(D)** *Although* means "regardless of the fact that." Hence the missing noun must be contrary to the key word *smile*, that is, *frown*.

2. **(D)** What kind of handling of food would make them decide to forego Martha's cooking? (A) is wrong because such handling would have the opposite effects; it would make people eager to try.

3. **(A)** The word *stigma* indicates that the job puts the jobholder in a disgraceful or unenviable position so that it is *unattractive* (A) or *uninviting* (D). But *even* means "in spite of," so the salary must be quite unlike the job. This eliminates (D) and makes (A), with its exact opposite adjectives, the best choice.

4. **(C)** The sentence structure, balancing "one" against "another," suggests contrast. Contrast is not provided by (D), meaning right or privilege, since the first man already has the meat as his prerogative. (A) and (B) make no sense. (C) is a clear contrast, providing not nourishment but harm.

5. **(D)** What kind of handling of the account would make Joseph an object of jokes and ridicule? (A) would earn him respect. (C) would make the industry laugh *with* him, not *at* him. (B), meaning "effusively sentimental," might well elicit ridicule, but in respect to a business situation, (D) is the more logical answer choice.

6. **(D)** Think of how you categorize jogging. It is more than *light* but less than *hazardous*. And it is certainly not *spurious* or counterfeit.

7. **(A)** Since the caller refused to give his name, the call was *anonymous*, literally "without a name."

8. **(B)** There is nothing to suggest that he has been omitted from the will or will fight it in court. Although he may be described as a *winner*, the legal term used for an heir is *beneficiary*.

9. **(C)** *Picayune* means "petty," "mean," or "small-minded." It is an appropriate adjective for a *hair-splitter*.

10. **(D)** High fever can produce a state of mental confusion.

11. **(D)** The context implies you would be taking unfair advantage of another and so, in a moral sense, would lower or *abase* yourself. There is no indication that you would scold or censure yourself, as (A) would mean, or that it would stimulate (B) or weaken (C) you.

12. **(B)** This is an instance in which you might first try out the second word in each pair. The main idea is that the storm *refused to slacken*, so you must find for the second blank a close synonym for *slacken*. Of the two offered, *mitigate*, meaning "alleviate," sounds too affected; *abate*, meaning "diminish" or "subside," is more idiomatic and is a better choice.

13. **(D)** Which of the four words offered has much to do with deciding between love and the throne? Only *abdicate*, meaning "to give up a high position." *Prevaricate*, meaning "to lie," would not be a decision; (C) doesn't tell us how or what he would alter or change; (D) wouldn't say which one he would nullify or abolish, which is what *abrogate* would involve.

14. **(D)** A fence, by disposing of and paying for fruits of the crime, serves as an encouraging force to robbers thereby aiding and abetting the crime itself.

15. **(C)** "Extenuating circumstances" are those that make an offense less serious by providing partial excuses. Coupled with "the defendant's youth," such circumstances make it unlikely that the judge would have recommended conviction or hanging. A mistrial results from errors in the course of a trial or from inability of the jury to come to a conclusion. It is the province of a judge to recommend *leniency*, that is, mercy or restraint.

16. **(C)** A fanatical and uncompromising person is by definition a *zealot*. (B) is correct in terms of meaning but is ruled out as a correct answer because the blank requires a noun, not an adjective.

17. **(B)** The word *strangers* and the fact that they would be offering candy to children they do not

know strongly suggest that the children be cautious and watchful, that is, *wary*.

18. **(D)** The position taken is that people do things in response either to outer force or to inner will. *Volition* means "free will."

19. **(B)** The phrase "respect for age" calls for a statement of strong positive regard or reverence for elder statesmen. That they were *venerated* fills the requirement. (A) is too mild, while (C) and (D) are negative.

20. **(C)** The length of the sermon indicates its effect on the congregation: boredom and resultant passivity. Hence the metaphor of the sermon as a *soporific*, a sleep-inducing drug, which would cure *insomnia*, or chronic wakefulness.

21. **(D)** Since his cynicism was written all over him, it was obvious or *manifest*.

22. **(C)** The result of her condition identifies it as *agoraphobia*, fear of open spaces. *Claustrophobia* is the opposite, a fear of closed spaces. *Anemia*, a blood deficiency, and *ambivalence*, a state of conflicting emotions, do not necessarily keep their sufferers indoors.

23. **(C)** To *sate* is to "satisfy" or to "indulge." One assumes that the appetite is already *whetted*, and *fill* is simply an unidiomatic usage.

24. **(C)** The missing word must be a synonym for *badgering*. (B) is too strong, (A) the opposite, and (D) wrong because the speaker *wants* the child to do it on his own volition.

25. **(B)** The speaker is clearly working in a formal relationship to a superior. We learn this from the first clause. *Authorized* is the correct word for describing what he or she is and is not allowed to say.

TEST 2

1. **(D)** The point that the sentence is trying to make is that "male supremacy" is now on the defensive. *Places* is too neutral a word to convey this meaning. *Bastions*, defensive strongholds, is a highly appropriate word choice.

2. **(B)** Choice (D) is ridiculous, and (A) cannot be correct because winds do not have riptides. The first blank could easily be filled with *dangerous*

of choice (C), but *breezes* would not contribute to the danger. The *vagaries* or unexpected actions of the winds would make the position of a rudderless boat *untenable*.

3. **(A)** If the branches are moving, so are the shadows they cast. *Transient*, which means "passing," describes the action of the shadows.

4. **(D)** *Transmute* means "to change from one form, nature, or substance into another."

5. **(A)** It would be a nice birthday present to have an article appear in print right on the professor's fiftieth birthday. The other choices all fit grammatically, but make no sense in context.

6. **(B)** The chariot *careened* out of control; it lurched and swerved rapidly. *Caromed* means "bounced off" or "rebounded." The sentence makes no mention of a collision, so (D) does not constitute an acceptable completion.

7. **(D)** The sentence indicates that the speaker is urging someone not to *digress*, not to "stray," from the *crux* or "heart" of the matter. To *prevaricate* is to "lie." The sentence deals with time pressure, not with truthfulness.

8. **(A)** Under the circumstances, *disappointed* seems rather tame. These parents would surely have been *disconsolate*, that is, "beyond consolation" or "hopelessly sad."

9. **(D)** *Conscripted* means "drafted." Both volunteers and draftees would have to be *processed*, so (B) is not a good choice.

10. **(B)** You do not have to know baseball to choose this answer by elimination. A *syllogism* is a tactic in abstract logic. If you try out the verb with choices (A) and (C), you determine that these choices just don't fit well. (B) is the only possible choice even if you do not know that a *balk* is a false move by the pitcher for which the penalty is the award of one base for the runner.

11. **(D)** If everyone in the neighborhood spoke Spanish, as implied by choice (C), hiring a Spanish-speaking person would be mandatory and not a matter of preference.

12. **(B)** Dark clouds tend to come before the storm; they constitute a *precursor* or a forerunner. Sometimes, however, the clouds blow over and there is no storm, so (D) is incorrect.

13. **(A)** The title of the act alone is a dead give-away to the meaning of the correct answer. Incorrect choice (B) is based upon common mispronunciation and misspelling.

14. **(A)** "Possible meanings" is your main clue. Any statement with more than one meaning is, by definition, *ambiguous*.

15. **(B)** A performance by which one makes a fool of one's self causes scornful laughter; it is *ludicrous*.

16. **(C)** The blank requires a verb that is used with *to*. (D) creates an impossible situation. The state of the house gave unspoken testimony, *attested* to the fact that it had been robbed.

17. **(C)** "Although" alerts you to expect a reversal of results: they will be different for "blows" and "gunshots." Since the warrior could deal with swords, look for problems with gunshots and with his armor. *Vulnerable* means "susceptible to injury" or "not sufficiently protected" from. *Impervious* means "incapable of being penetrated." His armor was NOT *impervious*; therefore he was *vulnerable*.

18. **(B)** The word "belied" tells you that her words contradicted the beauty and elegance of her appearance. *Uncouth* means "crude, rude, and ungraceful."

19. **(D)** Apparently it is the quality, not the content, of the expressions he uses that makes them "insufferable" or unbearable. *Trite* expressions are so overused, so overfamiliar, that they no longer command our interest.

20. **(C)** The opposite of "permissive" is *stringent*. *Impermissive*, while not incorrect, would create a redundant sentence. Regulations may be both permissive and *uniform*, so there is no contrast in (D).

21. **(C)** Destruction of property gives no benefit to either victim or state, so (B) is unlikely. The form of the sentence requires the wrongdoer to himself do something to the property. He cannot *confiscate* or "seize" his own property, but he can *forfeit* it or "give it up."

22. **(C)** The meaning alone should make this an easy completion. In addition, *distracted* is the only past participle here that is used with "from."

23. **(B)** He lied when he said he couldn't do the job because he was physically unable; actually he was *indolent* or just plain "lazy."

24. **(D)** "Police" in quotes is your clue that this was all a big joke or *hoax*.

25. **(C)** Think of what happens to incoming freight if there is a trucking strike and the cargo is piled up at the unloading dock. An *embargo* or "suspension or prohibition of trade" is the best solution until the strike is settled.

TEST 3

1. **(C)** *Condolences* are expressions of sympathy for another person's grief or pain.

2. **(C)** The "but" signals that an opposite must fill the blank. Logically, coming off an effective diet which kept one lean, one would then become *corpulent* or fat.

3. **(D)** A *boorish* person is rude, crude, and impolite. Being *liberal* has nothing to do with manners but rather with attitude.

4. **(C)** *Apathy* or lack of interest is shown by staying away from the polls; conversely, if the voters flock to the polls, they show interest.

5. **(C)** Being unprepared leaves the student taking exams filled with alarm, apprehension, maybe even some trembling—in short, with *trepidation*.

6. **(C)** The grade is steep and the load heavy; obviously we need those oxen to get home, so we cannot slaughter them. Time appears to be a factor. Resting and feeding the oxen would involve time that we do not have, but if we *goad* them, prod them with a stick, they may move faster.

7. **(B)** He was convicted because his "alibi," his explanation of where he was at the time of the crime, turned out to be false, a lie or *fabrication*.

8. **(C)** People do not tend to "look" perfectly *sane*, so eliminate choice (B). *Hale* means whole or free from defect, and *hearty* means vigorous and robust.

9. **(D)** Only choice (D) fulfills the requirement of the "Although . . ." construction. The others might make more sense with *and*; for example,

"He had doubts about the weather *and* he had no confidence."

10. **(C)** If they were to accept a missing person report, the only thing the police could do with it would be to *investigate*.

11. **(B)** By definition, an animal that eats only plants is *herbivorous*; it may also be called a "vegetarian."

12. **(C)** "The thick of the *ring*" does not ring true as an expression. "The thick of the *fray*," of the fight or the heated contest, is the idiomatic expression.

13. **(A)** The first three choices all convey the correct meaning, but structurally only *enjoined* works with "from blocking."

14. **(C)** Only the *postmark* gives date and location of mailing.

15. **(B)** The best way to compare two photographs is to place them in *juxtaposition*, that is, side-by-side.

16. **(A)** You can assume that given the technology of colonial America and the presence of virgin forests from coast to coast that wherever there was a hearth at all, there was a fire burning in it on cold winter days. Furthermore, none of the other choices makes any sense at all.

17. **(C)** The sentence depends upon your acceptance of the stereotype cliche that Eskimos, that is, persons who live in cold places, have no use for refrigerators. Then a salesperson who is a fast talker, who is very *glib*, could sell them what they do not need.

18. **(C)** The *facade* is the front of the building, logically the portion most likely to be destroyed by an explosion at the front of the building.

19. **(C)** If a witness is to testify against him, the case is still in the trial stage, and he has not yet been *convicted* nor adjudged *innocent*. An *alleged* thief is a person accused of thievery.

20. **(B)** The child who has not slept well will be cranky and unmanageable, in other words, *intractable*. Note the construction of the sentence; it calls for a completion that tells what the child will be "anything BUT." The child then will be the opposite of *intractable*, that is, anything but *docile*.

21. **(C)** If the eruption was unexpected, the volcano must have been considered inactive or *dormant*, literally "sleeping."

22. **(B)** Aside from the tobacco industry and its spokespersons, everyone agrees that cigarette smoking is harmful, that is, *deleterious*, to your health.

23. **(B)** *Audible* means "able to be heard." The sentence calls for a positive statement; (A) is the opposite of the meaning required.

24. **(C)** Any one of the choices might fill the first blank, but the second blank must be filled with a reason for which parents might *chide* or punish a child. Choice (B) makes no sense, and parents would take great pride in a child who was *industrious* or *independent*. However, parents might well *chide* or scold a child who was *indolent* or lazy.

25. **(B)** A "euphemism" is a word or phrase which is less direct and less expressive but considered less distasteful or offensive than the more blunt word. *Portly* means "stout in a dignified way." It is often used in place of "fat."

TEST 4

1. **(C)** Obviously you cannot write the report until you have gathered the *information*. Writing the report as soon as possible once you have the information makes sense; you can work with fresh facts, and those who receive the report can act upon it promptly.

2. **(C)** To *effect* is "to bring about or to produce as a result." The incorrect *affect* means "to have an influence on."

3. **(C)** A *circumlocution* is a "roundabout way of saying something." *Circumlocution* is certainly the way of avoiding *concise* expression. Choice (D) is half right. Diplomats do try to avoid *plain* language, but they don't use *summaries* to do it.

4. **(D)** Once you have picked up the sense of the second half of the sentence, even without filling the blank, completing the first half is easy. It is an exercise in *futility* that produces no results. Having filled the first blank, you need only confirm that the second part of choice (D)

also makes sense. It logically follows that an exercise in *futility* will bring no *substantive* deterrent to the impending cataclysm.

5. **(C)** If Santa Ana's forces had been in *decline* or *isolation*, the Alamo would not have fallen. Santa Ana won through sheer force of numbers. In fact, what makes the Alamo battle historic is that so small a force held out for so long against an army overwhelmingly superior in numbers, that is, a *preponderant* army.

6. **(A)** *Turbid* is a synonym for "cloudy."

7. **(B)** A process of elimination makes this an easy completion even with no knowledge of flamenco dancing. The dancer stood still, so his arms could not be *flailing*. *Akimbo*, hands on hips with elbows bent outward, is the only position for arms.

8. **(C)** The woman would not sue because her character had been *whitewashed*; she would sue because it had been "blackened" or *defamed*. To *defame* is to maliciously injure a reputation.

9. **(A)** A "joiner," in quotes, is one who joins. A person who joins groups or organizations does so in order to socialize with others of like interests. A *gregarious* person is sociable. Without the quotes, a joiner is a cabinetmaker.

10. **(D)** *Obese* means extremely fat. Here the sentence itself moves from a euphemism to a blunt statement of fact.

11. **(B)** If you want no more, you are *satiated* or satisfied. Your "palate" is the roof of your mouth.

12. **(B)** This is just plain common sense. To protect privacy the names and social security numbers must be removed or *deleted from* the questionnaires.

13. **(B)** A word made up of the initial letters of words in a phrase is an *acronym*. A *homonym* is a word that has the same sound as another word, but a different meaning. A *heteronym* is a word with the same spelling but different meanings with different pronunciations, cf. tēar, a drop from the eye; tĕar, a rip in a piece of clothing. A *pseudonym* is an assumed name.

14. **(A)** A criminal can be *demented*, as indeed this one may have been, or *pathological* but a crime cannot. *Gratuitous* in this context means unwarranted and unjustified.

15. **(B)** The test sentence itself provides a good definition of the verb to *procrastinate*.

16. **(D)** To accomplish his goal, he would want to lower his head toward the ground, perhaps as if to bow, that is, to *incline* it and to keep it steady. *Shaking* his head would make it harder to focus on individual stones.

17. **(A)** The words chosen must add up to a "wicked injustice." By definition, an *iniquity* is a *grossly* immoral act.

18. **(D)** The first blank requires someone with some feelings about stealing. A *kleptomaniac* has an uncontrollable urge to take things, without regard to need; in fact, a *kleptomaniac* has a strong *proclivity* to steal. A *pyromaniac* starts fires; a *megalomaniac* has delusions of omnipotence; a *dipsomaniac* is an alcoholic. None of these should have strong feelings about stealing stemming from their own peculiarities.

19. **(B)** A *hypothesis* is a proposition stated as the basis for argument or experiment.

20. **(A)** The blank must be filled with a word somehow related to making faces. Only *supercilious* is specifically related to facial expression. It suggests "being disdainful, raising the eyebrows, looking down the nose."

21. **(C)** The climate is allegedly not conducive to good health; it is not *salubrious*. *Sanitary* conditions have no effect on heat prostration. *Salutary*, meaning "conducive to improvement," goes beyond the scope of the second clause. As for *healthy*, climates can be "healthful," but only living things can be *healthy*.

22. **(A)** To *deride* is to "treat with contemptuous mirth"; it clearly refers to laughing *at*. A *buffoon* (noun not verb) is one who makes him or herself the object of laughter.

23. **(B)** Granting concessions to enemies in order to maintain peace is *appeasement*. In some sense, *appeasement* is giving in to blackmail.

24. **(B)** *Strategy* is the overall science of military planning. A *stratagem* is one small part of a strategy, a single maneuver or tactic. The use of the wooden horse to breach the Trojan gates was a clever *stratagem*.

25. **(C)** A large assembly of stars, film or otherwise, is a *galaxy*. A group of larks or bathing beauties is called a *bevy*; a group of lions is a *pride*.

TEST 5

1. **(C)** For purposes of stability, extra wheels are added to large vehicles in pairs. A fifth wheel added to a four-wheel vehicle would serve no useful purpose, would be *superfluous*. A person who feels as *superfluous* as a fifth wheel feels useless.

2. **(C)** *Parliaments, constitutional monarchies,* and *three-party systems* all contain internal mechanisms for legislated change. Only a *dictatorship* would destroy the mechanism for its own change, requiring civil war for reversal. Once the second blank is filled, completion for the first blank falls into place.

3. **(B)** Animals tend to be territorial, so when the dachshund encroached on its turf, the poodle was ready to fight. *Onerous*, which means "burdensome," would never be used to describe a look.

4. **(C)** Some sentence completions are easier than others. Provided you do not mistake *croquette*, a fried cake, for *coquette*, a vain flirt, you should have no trouble with this sentence.

5. **(B)** The start of the fishing season brings to mind immediately the phrase, "Gone fishin'!." A person who has "gone fishin'" is not there. A *hiatus* or gap in attendance is the result of the lure of the fishing season.

6. **(A)** The words "equally unclear" tip you off to the need for a word that means "unclear." *Nebulous*, which means "cloudy" or "vague," is the only choice to meet the test.

7. **(C)** *Nepotism* is showing favoritism to relatives for employment or for appointment to high places.

8. **(A)** Whose arms would be very strong? Not both arms of the *hemiplegic*, who suffers extreme weakness on one side of the body as a result of a stroke or other brain damage; not those of the *quadriplegic*, who suffers paralysis of all four limbs; and not those of the *somnam-* *bulist*, the sleepwalker whose legs are far from *comatose*. The *paraplegic*, who is paralyzed from the waist down, tends to develop extra strength in the arms to compensate for now *useless* legs.

9. **(D)** From "although" you infer the engine is there "just in case" the wind becomes less dependable and the ship needs an additional source of power. The implication is not so much that it needs a complete *substitute* as a *subsidiary*, a "supplement" or an "auxiliary."

10. **(A)** *Aplomb* is the poise that comes from self-confidence.

11. **(D)** The speech was difficult to understand, that is, *abstruse*. If the speech was also *protracted*, long and drawn-out, that might have contributed as well to its being *abstruse*.

12. **(C)** Since the lifetime is being measured in such tiny units, it must be extremely fleeting, short-lived, or *ephemeral*.

13. **(D)** The *millennium* is the eagerly awaited golden age of peace, prosperity, and happiness. To some, the repeal of income taxes will come with the *millennium*.

14. **(B)** Begin with the second blank. What should you keep to yourself? Only choices (A) and (B) are possible. However, keeping one's *hands* to oneself has nothing to do with *confessing*, so (A) must be eliminated. To *obtrude* is to force one's *opinions* on others.

15. **(D)** If you are disappointed, you will remember the day regretfully; you will *rue* it.

16. **(D)** Soldiers who rent themselves to foreign armies are called *mercenaries*.

17. **(A)** "Countless" calls for *myriads*, an indefinite number that surpasses *dozens*. *Multitudinous* conveys the correct meaning but is syntactically incorrect; "multitudes" would work here. *Dryads* are wood nymphs.

18. **(B)** Why would you prefer to scrub floors? You must find peeling potatoes to be very *onerous*, burdensome, or oppressive.

19. **(B)** The picture painted here is one of the offenders lying face down on the ground begging humbly, *supplicating*, for mercy.

20. **(D)** Her rebelliousness was written all over her; it was clearly and openly revealed; it was *manifest*. *Exterior* bears the correct meaning but is an inappropriate descriptor.

21. **(D)** A dull, insensitive or *obtuse* person would be slow to learn basic social niceties.

22. **(B)** The second clause indicates that we are here talking about Louis XIV with respect to elegance of dress. The adjective *sartorial* is used specifically to refer to men's clothing.

23. **(D)** The sentence is gender neutral, so the blank must be filled with a gender neutral word which includes brothers and sisters, that is, *siblings*.

24. **(C)** "False accusations" do not constitute *corrections*. The man did indeed *interfere*, but the sentence lacks the necessary preposition to utilize choice (B). To be correct, the completion would need to read "interfered with." What did happen was that the man *disconcerted* the speaker by upsetting, ruffling, or irking him or her.

25. **(B)** He will never become an alcoholic if he just *buys*, *secretes* (hides), or *accumulates* liquor. He must *imbibe* it; in other words, he must drink it too.

TEST 6

1. **(D)** Turn the sentence around. Patience is a virtue; I'm glad to see that you have (it), that is, *forbearance*. (C) would seem to be irrelevant. If the person has the *time*, then being patient does not represent great virtue. As for (A), since "have" can be either an independent verb taking an object or an auxiliary to a main verb, "have *arrived*" seems possible. But (A) makes for a self-congratulatory statement; (D) is the better choice.

2. **(C)** Only increasing nearness would reduce visibility. *Nigh* means "near" or "close."

3. **(B)** Of the choices offered, only the fact that she was an *indefatigable* or tireless jogger can explain that she covered 15 miles a day.

4. **(D)** The relationship between her nature and the activity described is best explained by saying the first *required* the second.

5. **(B)** *Concatenate* means "linked together as a chain." This is precisely the state of affairs described in the second clause. War is *hell*, but not according to the requirements of this sentence.

6. **(B)** The "if" construction of both clauses indicates that we are seeking two contrasting behaviors which produce contrasting results. Both (C) and (D) offer synonyms to fill the two blanks, while (A) makes no sense at all. To *dissipate* is to waste or to squander, quite the opposite of using wisely.

7. **(C)** "Lots" are "objects used to make a choice by chance," that is, *randomly*.

8. **(C)** "Eminence" is used to describe Richelieu. The only adjective that corresponds to "eminent" is *prominent*.

9. **(B)** A *transgressor* is someone who has simply crossed the boundary between legal and illegal activities.

10. **(D)** "*Greetings* and *salutations*" is a bit redundant, but it is a frequent form of address to a group, especially at an initial meeting. (C) would be ridiculous. The teacher would not begin by saying "Hello and goodbye." The "and" between the two words precludes choice (A). (B) would be acceptable if there were no other choice, but (D) is the more likely opener in this context.

11. **(B)** "Although" alerts you to look for the opposite of "aloof," which means "cool, distant, and uninvolved." The only opposite offered is *gracious*, which means "warm, courteous, and sympathetic."

12. **(C)** The "but" tells you that the two activities are being put into different categories. The word "crime" in juxtaposition with "speeding" suggests that the second blank is to be filled with a word carrying more serious import than the first. Choice (A) is clearly wrong, and (D) reverses the order of seriousness. As for (B), "speeding" is not a social blunder, which is what a *faux pas* is.

13. **(B)** *Craft* is a synonym for "skill"; it implies dexterity with the hands.

14. **(D)** *Malleable* means capable of being shaped, especially by hammering, pounding, or bend-

ing. The fact that gold is very *soft* contributes to its malleability, but "most soft element" is not good English.

15. **(D)** *Impassive* has more than one meaning. One meaning is motionless or still. By this meaning, *impassive* is the only choice that can describe a "stone." Another meaning of *impassive* is "devoid of feelings." Here "failed to touch him" means failed to affect his emotions, making him quite stonelike.

16. **(C)** Fill the second blank first. A client cannot *abrogate* (annul) or *elect* a law, so eliminate (A) and (D). Choice (B) fills the blanks so as to make an untrue statement. You are left with (C), which does happen now and then.

17. **(C)** To look *askance* is to look sidewise or with suspicion. Certainly a banker would wonder at anyone who claimed no interest in saving money.

18. **(A)** What the readers want is shocking, horrifying, or titillating, that is, *lurid* information.

19. **(C)** "Worshipping her every move" did not of itself make him her kindest, most *beneficent* admirer nor did it necessarily prove that he was more unconsciously stupid, *fatuous*, or more worldly and *sophisticated* than her other admirers. It qualified him only to be called her most devoted and most *ardent* admirer.

20. **(D)** Once you eliminate the totally irrelevant choice, *pervade*, meaning "to spread through," the other three words all express some degree of opposition. *Dissuade*, meaning "to discourage by persuasion," fits most gracefully into the syntax and conveys the most likely intent of the sentence.

21. **(C)** If she lacked money, she was *inpecunious*.

22. **(A)** The main clue is "loose fat," which indicates that the person being spoken to is flabby or *flaccid*. *Avoirdupois* does refer to weight, but it is a noun, and this blank calls for an adjective.

23. **(A)** Common sense is all you need to answer this question. Crime prevention is the officer's *primary*, first and most important, role, but the officer also instructs, assists, and reacts to crime in progress.

24. **(C)** The only person we are certain that the general had the authority to send in his place would have been his *adjutant*, his assistant.

25. **(C)** A "malingerer" is someone who pretends to be ill, hurt, or otherwise unavailable in order to avoid obligations—in other words, in order to *shirk* his duty.

BOOKS FOR JOB HUNTERS

CAREERS / STUDY GUIDES

Airline Pilot
Allied Health Professions
Automobile Technician Certification Tests
Federal Jobs for College Graduates
Federal Jobs in Law Enforcement
Getting Started in Film
How to Pass Clerical Employment Tests
How You Really Get Hired
Law Enforcement Exams Handbook
Make Your Job Interview a Success
Mechanical Aptitude and Spatial Relations Tests
Mid-Career Job Hunting
100 Best Careers for the Year 2000
Passport to Overseas Employment
Postal Exams Handbook
Real Estate License Examinations
Refrigeration License Examinations
Travel Agent

RESUME GUIDES

The Complete Resume Guide
Resumes for Better Jobs
Resumes That Get Jobs
Your Resume: Key to a Better Job

AVAILABLE AT BOOKSTORES EVERYWHERE

PRENTICE HALL